A History of Biology

Charles Darwin (1809-82)

A

History

O F

Biology

to about the year 1900

A General Introduction to the
Study of Living Things

Third and Revised Edition

BY Charles Singer

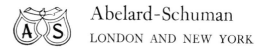

Abelard-Schuman
LONDON AND NEW YORK

'Εν πᾶσι τοῖς φυσικοῖς ἔνεστί τι θαυμαστόν.
'In all the things of Nature there is something marvellous.'

Aristotle, *Parts of animals* 645 a 16

ἀνδρῶν γὰρ ἐπιφανῶν πᾶσα γῆ τάφος
'The whole earth is the sepulchre of famous men.'

Thucydides II. 43, § 3.

PREFACE TO THE FIRST EDITION

THIS work attempts to give, in simple language, a critical survey of the historical development of biological problems. The bypaths and blind alleys of the subject are left unexplored. The immense extensions of detailed knowledge in many departments are discussed only in so far as they influence our general thinking about living things. Biographical detail is not included unless it is seen to have a direct influence upon the course of science. The mechanism of transmission of the biological tradition is but lightly touched upon. The exposition of the origin and development of the main problems of biology is the objective.

The author has striven to write in such a way as to demand only a minimum scientific training for the comprehension of his matter. Technical terms have been used as seldom as possible, and never without explanation. Overlapping has been largely avoided by cross references. The reader is advised to refer constantly to the table of contents. This will enable him to appreciate better his position at any point in the narrative as a whole. The 'Introduction' is perhaps more suitably read after and not before the book itself.

It has been possible to curtail considerably the length of the book since the author has already treated such topics as the influence of Galileo on science generally, Galenic anatomy and physiology, the circulation of the blood, the philosophical implications of scientific specialization and their relation to utilitarian thought, the discovery of the nature of the air, advances in knowledge of bacteriology, immunity and epidemics, in his *Short History of Medicine* which is, in some degree, a companion volume to this.

The book is written by one who finds mechanist interpretations of life unsatisfying. His attitude is prompted largely by the 'relativity of functions', that is by the conditioning of any one form of vital activity by innumerable concurrent forms, and this not only in the organism as a whole, but in each part susceptible of independent investigation. He recognizes, however, that the mechanist outlook has been responsible for countless far-reaching and important biological investigations, and he is aware that it remains indispensable for advances in many biological departments. He is, moreover, acutely conscious of the danger of confusing physical with metaphysical issues. He would, however, urge that there are many scientific propositions that are acceptable on one level of investigation but not on another. But the victories of the experimental method are too numerous, too complete, too general, for us to lose faith in its value because it fails to reveal a universe completely consistent with itself. We must accept with resignation the ineluctable fact that there are an increasing number of antitheses in the world of our experience which science exhibits no sign of resolving.

C. S.

January 1931

PREFACE TO THE SECOND EDITION

THIS book first appeared at Oxford in 1931. There has been no revision of the text hitherto but a few errors were corrected and it was reprinted in America under the title *The Story of Living Things*. Both the English and the American version have been out of print for some years. For this second edition I have had the advantage of the translation into French by Dr. Gidon, Paris (Payot), 1934, and of the translation into Spanish by Dr. Maximo Valentinuzzi, Buenos Aires (Espasa-Calpe Argentina, S.A.), 1947. Both these versions contain some additions.

During the past eighteen years many colleagues have sent me suggestions which are incorporated here. It is impossible to record all these friendly acts, but I owe much to E. S. Goodrich, C. A. Kofoid, E. A. Sharpey-Schafer, D'Arcy W. Thompson, W. H. Welch, and, among the living, to Dr. Agnes Arber, F.R.S., Professor F. S. Bodenheimer, Professor F. J. Cole, F.R.S., Dr. Clifford Dobell, F.R.S., Professor H. M. Evans, Professor J. P. Givler, Dr. Julian Huxley, F.R.S., Dr. Joseph Needham, F.R.S., and Professor Sir Charles Sherrington, F.R.S. In the revision of the last hundred pages I have had the kind help of Dr. W. Holmes of the Zoological Department, Oxford University. Mrs. Singer has helped me at every point in this, as in the previous edition.

CHARLES SINGER

'KILMARTH', PAR, CORNWALL, ENGLAND

March 1949

PREFACE TO THE THIRD EDITION

THE text of this edition has been revised throughout and many changes introduced. I have, however, avoided any attempt to present the views of the active generation of biologists. It thus remains a history but ends with the nineteenth century, though there are a few incursions into the twentieth where sense seems to demand it.

It had long been my intention to include in this work a general sketch of current biology. The immense labour needed caused me to delay in completing the task. I do not regret this because it has been done in *A Hundred Years of Biology*, 1952 (England, Duckworth; U.S.A., Macmillan) by Dr. Ben Dawes, Reader in Zoology at King's College, University of London. His book has made superfluous such an attempt as I had contemplated.

I have taken advantage of criticisms of previous editions made by the late Prof. F. J. Cole and by Dr. Dawes. I am grateful to them for having saved me from some errors, but for any that remain I alone am responsible. I have also to thank the Library of the Wellcome Historical Medical Museum, and especially Dr. F. N. L. Poynter, for several illustrations. These are acknowledged in the list of figures.

CHARLES SINGER

'KILMARTH', PAR, CORNWALL, ENGLAND

1st January 1959

TABLE OF CONTENTS

PART II. THE HISTORICAL FOUNDATIONS
OF MODERN BIOLOGY

IV. On the Inductive Philosophy and some of its Instruments

V. Rise of Classificatory Systems

PART III. EMERGENCE OF MAIN THEMES OF MODERN BIOLOGY

IX. Cell and Organism

X. Essentials of Vital Activity

Table of Contents

LIST OF ILLUSTRATIONS

List of Illustrations

List of Illustrations XXV

INTRODUCTION

§ 1. *Some implications of scientific specialization*

SCIENTIFIC specialists commonly insist that their current investigations are too abstruse, intricate, obscure, to be understood save by the specially trained. The reiteration of this statement, true in itself, has engendered a widespread and dangerous fallacy. Great scientific advances are not now, nor have they ever been, of their own nature specially difficult of comprehension. On the contrary, many highly significant scientific doctrines can be expressed as very simple statements or formulae. It is not the positive conquests of science that are peculiarly obscure, but rather the confused and active battle-front along which science is advancing at any given moment.

Nevertheless many scientific workers remain convinced that their results can never penetrate the mind of that singularly obtuse being, the 'general reader'. The very attempt to make their studies intelligible is still suspected by the scientifically initiated. It is, however, a corollary to the subdivision of the sciences that the exponent of one science becomes a 'general reader' for his colleague in another science. If those men of science be right who assume as inevitable their own unintelligibility to a public all too ready to accept this assumption, then is the outlook of our age gloomy indeed. As knowledge advances it must become more divided. As it becomes more divided, it will be appreciated by fewer and fewer. In the limit, the body of science will be so fragmented that each investigator will be intelligible only to himself. He can thus leave no scientific heir. The system must collapse. Science must perish from off the goodly land into which its specialist leaders have brought it from out the wilderness of medieval ratiocination. The experimental method must spontaneously disintegrate.

Let us glance at some of the elements in this cheerless prospect.

The progressing subdivision of the sciences is a product of the last hundred and fifty years, being made necessary by vast and rapid increase of knowledge. Only through the conviction of its value held by its exponents could this subdivision have become possible. It is doubtless psychologically necessary to most members of our fallible species that, for the doing of good scientific work, the worker should estimate the value of his method very highly. Every department of science is of the highest importance to those who work at it. And is not the historian of science even as other men?

Yet in fact the justification for splitting off a particular department from its parent stock is merely that certain specific problems may be solved or at least defined. Once those problems are solved—or exactly defined, which is nearly the same thing—the speciality has in part fulfilled its function, and that which its exponents have to purvey becomes incorporated into the general organon of knowledge.

Many illustrations might be given of this thesis. One, taken from the field of biology, will suffice. In the first half of the nineteenth century, with the advent of the cell theory and the improvement of the microscope, the study of microscopic organisms assumed a peculiar significance. Specialists of various kinds devoted themselves to the investigation of these organisms and at last succeeded in defining certain of them as beings, either plants or animals, consisting of but one cell. The knowledge of the structure and habits of microscopic animal forms, 'Protozoa' as they came to be called, now became part of the current body of biological knowledge. General zoologists included the group in their survey. Then in the later nineteenth century it was found that certain diseases that afflict man are of protozoal origin. Intensive investigation of the disease-producing Protozoa became urgent, and 'Proto-zoology' gained a name (1904) and became again a separate science. Much knowledge acquired by the new specialist, the protozoologist, has since been placed at the disposal of the general zoologist, while other products of his activity have joined the main stream of Pathology. Since the number of human diseases of protozoal origin must be limited, it is probable that a time will come when the protozoologist will cease to exist in fact, if not in name, and that we shall find him devoting himself to general problems of the cell or of pathogenic organisms.

Science as it exists to-day is made up of such more or less temporary special departments—'scientific outposts' as we may perhaps name them. No man can hope to grasp the nature of current progress in all of these. But that is no reason why the picture of nature that science reveals should be any less clear and sharp in our day than in previous ages. It would be a poor sort of science that made the world less and not more intelligible. The author of this work is convinced that it is possible to give a picture of the state of scientific knowledge as a whole. What is unfortunately lacking is a supply of writers with scientific and literary skill and time and experience who would apply themselves to the task of exposition.

How shall science be expounded? The mechanism has been fairly established for those who deal with any one of the numerous departments. The technique of the special-ized text-book has reached a high stage of development. But when an attempt is made to cover simultaneously a number of the sciences, we are still in the experimental stage. Attempts at a 'general survey' of the combined sciences are legion, but they create a very unsatisfactory impression on the trained and critical reader. They are often one-sided and always disjointed. This method of presentation has been repeatedly tried and found wanting, for continuity, the cardinal element of scientific thought, is lacking in these over-simplified schemes.

Must we then accept the dismal claim of the specialist as to his essential isolation? The author of this work believes that such a view is wrong, both in fact and in principle. He believes that it is due to a basic error as to the true nature of specialization. He believes that the error is philosophically baneful and scientifically injurious. He believes that the natural antidote to these evils is the proper use of the historical method in science. He believes that the way to cover any very wide scientific area is by the frank introduction of history. He believes that the proper crown of a scientific education is a general survey of the processes by which the most important current scientific ideas have reached their present state of development.

Moreover, even men of science who express little regard for history are not as neglectful of it as they think themselves. Oft they entertain an angel unawares. In the actual teaching of science, an historical method is almost invariably adopted, though its historical character may be unperceived by the teacher himself since the history with which he deals is very recent. Indeed, he cannot discuss scientific conclusions without discussing how they have been reached. And is not this history? Moreover, the wider he spreads his scientific net, the more varied the scientific conclusions that he discusses, the farther must he go back in history to obtain his viewpoint.

§ 2. *History in relation to scientific exposition*

The working man of science is accustomed to distinguish in his mind between the historical development

AGOOD THING

of his subject and the active process by which he is himself developing it. This distinction is among the fruits of specialization. We hear little of it from the less specialized but no less wise, scientific men of the seventeenth and eighteenth centuries. Reflection will bring conviction that the distinction cannot be maintained in the twentieth century, even on the 'common-sense' basis. The extension of knowledge must be based upon that already achieved. Now the limits of knowledge are by no means clearly demarcated. Between the unknown and the known is no well-surveyed frontier, but rather a ragged and ill-defined borderland. Before he reaches the firm ground on which alone foundations can be safely laid, the man of science must work backwards for a space until he is behind that disturbed and shifting area. As we widen the sphere of our scientific purview, so shall we need—if we would make our vision clear and free from distortion—to penetrate ever farther back in history. Unless we adopt a wide historical outlook, we shall find ourselves unable to grasp the nature of the great problems which we are discussing.

The trouble with the 'general surveys' is that, in abandoning the historical method, they have lost the connecting links between the departments of knowledge. It is as though one would seek to gain a complete picture of the form of a plant from cross sections of its numerous branches. The way in which the parts of the plant are related to each other can only be made apparent by tracing them from the roots from which they all spring. This is the task of history.

It is an essential element of the scientific method of investigation that the investigator separate a part of the universe and consider it in and for itself. In this sense specialism is certainly implicit in the scientific method. But the separation of part of the universe, *for purposes of investigation*, is a very different thing from the belief that

that part of the universe is in fact separate from other parts. Yet a survey of science that does not introduce the historical method must assume this absurd and untenable position. Surely the natural and proper way to survey the sciences is to treat them as arising seriatim in the course of the ages from that desire which is innate in every human being to know what Nature has to reveal. Thus expounded, the sciences become records of the process of human inquiry, and science itself coextensive with the history of science. *L'histoire de la science, c'est la science même.*

§ 3. *Methods of presenting the history of science*

Having discussed the error of him who neglects or seeks to neglect the historical method, we turn to consider the position of him who devotes himself to the history of the sciences. The first question that arises is as to what area of the sciences shall be covered in a single survey? Specialization has its dangers for the historian as for the investigator of natural phenomena.

Many writers have devoted themselves to the histories of one or other of the special sciences. But the sciences have developed and are developing not in watertight compartments but in intimate relation to each other. The limits of the separate sciences, as they are recognized in any age, are largely artificial and related to pressing practical needs. A more intellectually satisfying consideration than the history of a science is surely the history of a problem. The great problems of science demand for their solution many different sciences in turn. Problems are definite entities. Sciences—which at best are but special methods of research and at worst mere technical *expertises*—have no separate and permanent existence. To take a single example, the attempt to elucidate the mechanism of heredity, the problem which is foremost in the minds of contemporary biological thinkers, has drawn upon the whole gamut of the biological sciences. It would be impossible to restrict either the efforts or the conclusions of workers on heredity to the department of the zoologist or the botanist or the biochemist or the cytologist or the statistician or the experimental embryologist. It has thus seemed well, to construct this book on the basis of a consideration of biological problems.

At the outset of his effort, the historian of science finds himself in face of a dilemma which can be resolved only by an omniscience to which he seldom lays claim. There are two courses open to him. To both fundamental objections may be urged.

On the one hand, he can decide that his history shall terminate at some definite point of time in the past. He can then write a truncated account of the development of science which will avoid dangerous discussion of current problems, or at least partisanship in regard to them. The feat when accomplished—and it has often been performed —fails to arouse much consideration from the working man of science, or to deflect him from his view that history is the child's play of antiquarians, suitable, perhaps, to be dallied with in extreme old age when the grasshopper becomes a burden and laboratory work is perforce abandoned. Man goeth forth to his work and to his labour until the evening. In the plenitude of his power he must pass old-time records without regard, and must concentrate on the contemplation of the living issues with which he feels he is faced, for ancient history cannot solve modern problems.

On the other hand, the historian may decide to bring his story 'up to date'. To accomplish this he must perforce assume an unjustified position of authority. He must give much the same definiteness of tone to his interpretation of current and recent scientific advances as he has bestowed on those of the more remote past. Thus he must take sides in some matters, at least, in which he is ill-qualified to act as judge. His temerarious intervention in domestic differences will not fail to call down upon him the imprecations of those from whom he differs, while his ignorance may draw upon him the derision of those with whom he agrees. In any event he cannot cover, of his own knowledge, the whole area of the subject of which he treats. Thus his emphases and elisions will alike be subject to the criticism of working men of science.

With this choice of evils before him, the author, in the pride of his relative youth, more than thirty years ago, elected for the second and more dangeous task. Reflection of a third of a century has led him to withdraw to the first and safer alternative. He therefore now elects for the easier task, bringing his narrative to an end with the century in which he was born and grew to manhood.

§ 4. *Arrangement of the work*

The arrangement of the book is conditioned by the attempt to lead up to the consideration of major biological problems. Now in the history of problems—as in other forms of history—it soon appears that 'periods' are artificial though handy abstractions. It is difficult to establish their limits and impossible to make them exact. As abstractions they form convenient frameworks for setting forth our knowledge. But the framework must not be turned into a bed of Procrustes. Useful as *aide-mémoire*, it is dangerous as doctrine and wholly destructive as dogma. In what follows the periods will, in fact, be found constantly to overlap.

Bearing this qualification in mind, the subjects with which we have to deal arrange themselves, with comparative facility, under three main headings, corresponding to the three Parts of the book.

H.B.—2

We begin by treating of the 'Older Biology' (Part I). The knowledge of living things, elevated into a great system by Aristotle (Chapter I) and developed by Theophrastus, Galen, and others among the Ancients, dwindled during the Middle Ages (Chapter II), to be revived again in the sixteenth and early seventeenth centuries by such great investigators as Leonardo da Vinci, Vesalius, and Harvey (Chapter III).

At the dawn of the seventeenth century the inductive method makes a formal entry into the biological field (Part II). Francis Bacon and Descartes are its first philosophers. The assumption that the phenomena presented by living things can be brought under general laws, even as can the phenomena of the inorganic world, leads us to a position in which a new incentive is provided for the accumulation of biological data. The investigation of nature becomes organized. Scientific academics and journals arise. The microscope appears; it is the special instrument of the biological sciences around which a complex technique progressively gathers (Chapter IV). The recognition of the amazing multiplicity of living things calls for special classificatory treatment of which Linnaeus is the leading exponent (Chapter V). The details of organic structure, revealed by Cuvier and his predecessors and successors, prove themselves no less susceptible of orderly arrangement (Chapter VI). The distribution of plants and animals in space becomes gradually correlated with their distribution in time. This is the work of a host of explorers, travellers, geologists, and systematists among whom Alexander von Humboldt, Charles Lyell, A. R. Wallace, and J. D. Hooker deserve especially to be commemorated (Chapter VII). And then classificatory schemes, the data derived from comparative anatomy and the facts of distribution, are brought together under a grand generalization frequently adumbrated in earlier times but more significantly formulated by Charles Darwin. With Evolution as its keystone, the basal arch of classical biology is completed (Chapter VIII).

The foundations of a scientific biology being now truly laid, there emerge into clearer light those problems that have constituted the main themes of biological thought for the nineteenth century. We now enter on the third and longest section of our narrative. Here we must entirely neglect what are called the 'separate biological sciences' since the discussion of their individual history and development would obscure the issue that we seek. It is *problems* that must now claim our undivided attention. In attempting their solution, men of science have frequently sought aid in the researches of a remoter past. For the most part, however, the attempted solution and, what is no less important, the definite formulation of these problems is the product of the last century.

Among the numerous biological problems we distinguish seven for detailed discussion. To each we devote a chapter (IX–XV). Historically the earliest to emerge into modernity was that marvellous relation between the organism as a whole and the cells of which it is composed. What is the mysterious power that makes these living units subserve a common end? (Chapter IX). As we attack the question we find ourselves involved in a discussion of the essential nature of the process which we call Life, a subject of debate from the time when philosophy came first into being (Chapter X). How are the countless internal activities of the organism related to each other? (Chapter XI). And when and how was the beginning of that relationship that varies infinitely in detail, but is ever directed to the same ends—the preservation of the internal environment of the organism concurrently with the multiplication and extension of the organism with reference to its external environment? (Chapter XII). How may these activities be traced in the development of the individual organism? (Chapter XIII). As time has gone on, such problems have been referred to two main themes which have come to occupy the centre of the biological scene. These are the essential nature of sex with the marvel of generation (Chapter XIV) and the apparatus by which, in the sexual process, the nature of the parent organism is conveyed to or altered in the offspring (Chapter XV). 'Heredity' is the note on which we close our story. It is the outstanding topic of current biology.

Competing with and interwoven with the problems of heredity and reproduction are the problems of the chemistry of living matter and the origin of life itself. These questions will doubtless transcend all others in the second half of the twentieth century. When that century began, and our story ends, these were mere matters of speculation and had hardly entered the realm of scientific discourse.

If we thus close our own story, it must be accepted that the story itself is not and cannot be closed. History ends now but unfortunately the historian's difficulties increase as he advances towards his own time, since it becomes progressively more difficult for him to get his knowledge into perspective. A period is necessarily reached at which the history of science can be taken up only by one actively engaged in advancing it. For this purpose, a younger writer has an evident advantage. Happily the general trends of biological thought in the twentieth century have been outlined by Dr. Ben Dawes in his *A Hundred Years of Biology*. The lack of overlap in thought, as distinct from time between the two works emphasises the immense change of the direction of biological thinking which happened to coincide fairly exactly with the change of centuries.

PART I. THE OLDER BIOLOGY

I
THE RISE OF ANCIENT SCIENCE

§ 1. *Hippocrates* (*c.* 460–*c.* 370 B.C.)

SCIENTIFIC discussion begins with the people who came to call themselves the Hellenes, but whom we know as the Greeks. These were a collection of tribes of mixed origin that invaded the coasts and lands of the eastern part of the Mediterranean. They had settled there about the beginning of the second millennium before Christ.

Before the coming of the metal-using Greeks—or more accurately, of the Greek speaking peoples—the shores of the eastern Mediterranean were inhabited by peoples in a neolithic state of culture who were displaced or absorbed by them. The first of Greek tribes to attain a civilized state were the Myceneans and notably their colonial off-shoot in Crete known as Minoans (Fig. 2). With inter-necine warfare between the continental Myceneans and the Cretan Minoans the civilisation of both fell into decay from about 1400 B.C.

From the ruin left by this warfare there gradually emerged the distinctively Greek civilisation, the historical records of which have come down to us.

The first Greek tribes who have left any literary remains were the Ionians and the Dorians. The Ionians were especially fond of philosophy and were interested too in mathematics and astronomy. Less civilized but perhaps more practical were the Dorians. Just on the frontier between the two tribes, but within Dorian territory, was the island of Cos. There a medical school arose about 600 B.C. It is the earliest scientific institution the works of which have come down to us (Fig. 5).

We have no clear idea of the way in which this medical school was organized. A very considerable number of books written by members of the school has, however, survived. The earliest of them was composed about 500 B.C. The greatest member of the school was one Hippocrates,

FIG. 2.   Faience models from Crete of Minoan origin. Flying fish, paper nautilus, and other Marine objects are shown.

often spoken of as 'the father of medicine'. We must remember that all the biological sciences were first studied because of their bearing on medicine. Hippocrates, therefore, might well be called also 'the father of biology'.

Hippocrates was born on the island of Cos about the year 460 B.C., and he lived to be about a hundred years old. He learnt medicine on his native island, and he practised and taught it in various other islands and on the mainland of Greece. Such accounts as we have of him describe him as a man of very noble character, dignified bearing, and humane feeling.

A great many books bear the name of Hippocrates, but few if any of these can be really his work. Some of the best and most interesting of them are, however, of about his date. By studying them we can get a good idea of what he knew and of how he worked.

If we were to examine these early medical works, we should find that whole departments of knowledge which are now considered necessary for a doctor are entirely absent from them. Thus, for instance, they betray little or no anatomical, physiological, or chemical knowledge. The doctor of those times had no instruments for examining patients, such as listening tubes, thermometers, or magnifying glasses. He had only his own senses to guide him and he had very little record of what those who had gone before had seen of disease. On the other hand, his senses were well trained and he observed carefully and well, and put down what he saw with a wonderful eye for what was essential.

The real scientific value of these Hippocratic works lies in the careful record of what was seen. There were, of course, observations of scientific matters by peoples earlier than the Greeks. Thus we have, for example, notes on the positions and movements of stars by Assyrian scribes, written on clay tablets in the curious arrow-shaped writing known as *cuneiform*. Many of these notes are centuries earlier than Hippocrates. Moreover, from the writings of the Greeks themselves, we have information concerning men of science who were earlier than Hippocrates. Among them was the Ionian Thales of Miletus who predicted and observed the eclipse of the 28th of May 585 B.C. But the great interest of the Hippocratic writings is that they are complete works, while those of Thales and of the other predecessors of Hippocrates are either lost or survive only in fragments. The Hippocratic works not only record things seen, but they sometimes even tell us how these observations were made. Thus we can follow the physician into the sick-room and watch him at work.

Among the most striking of the Hippocratic works is a collection of short sayings or *Aphorisms* as they are called. They show that many well-known proverbs in use to-day, especially such as concern food and diet, come to us from remote antiquity. Among the Hippocratic Aphorisms which have passed into common speech are:

> Art is long and life is short.
> Desperate diseases need desperate remedies.
> Sleeping too much is as bad as waking too much.
> One man's meat is another man's poison.

Phrases such as these, however, are not science in the ordinary sense of the word. Their importance for us lies in the spirit in which they were made rather than in the observations themselves. To understand that spirit better we must compare it to that of the less scientific doctor of the time. Now in the days of Hippocrates a great many diseases were generally ascribed to the action of gods or demons or supernatural beings. One quite common disease was in fact actually called the *sacred* or *divine disease* on this account, and the popular opinion about it was shared by many physicians. One of the Hippocratic physicians studied the *sacred disease* and came to the conclusion that he had found out its actual cause. He says:

'It seems to me that the disease called sacred is no more divine than any other. It has a natural cause, just as other diseases have. Men think it divine merely because they do not understand it. But if they called everything divine which they did not understand, why there would be no end of divine things!

'Those who make a to-do about such things as being due to the gods appear to me, therefore, like certain magicians who pretend to be very religious and to know what is hidden from others. If you watch these fellows treating the disease, you will see them use all kinds of incantations and magic, but they are also very careful in regulating diet. Now if food makes the disease better or worse, how can they say it is the gods who do this? Nay, even in saying such a thing, they show impiety and suggest that there are no gods.

'The fact is that this invoking of the gods to explain diseases and other natural events is all nonsense. It doesn't really matter whether you call such things divine or not. In Nature all things are alike in this, that they can all be traced to preceding causes. Shall we say then that they are divine or not divine? Since they are all alike in this respect, it is really only a matter of words.' [*Somewhat paraphrased.*]

The fine book *On the sacred disease* from which this passage is taken was written about 400 B.C. It might justly be called the 'Charter of Science', for it sets forth the scientific method of assuming natural explanations for all observable events. While it is still ringing in our ears let us see how far men had got in those days in biological knowledge.

First of all, the artists had long been at work representing animal forms. Often they had made very exact studies, some of which disclose observation that might well be called scientific. The Greeks were essentially a maritime people. They nearly all lived within sight of the sea and a great many of them were employed as fishermen or sailors. It was natural therefore that they should take an especial interest in marine creatures, and many remarkably good representations of fishes have come down to us from the fourth and fifth and even from the sixth century B.C. (Fig. 3).

But the Greeks were not content to examine only the external forms of animals. A beginning had actually been made with dissection of the internal parts. Thus about 500 B.C. Alcmaeon, a Greek of Croton in Southern Italy,

H.B.—2*

had described the nerves of the eye and the tube that
connects the mouth with the ear. If you pinch your nose,
and, keeping your mouth closed, distend your cheeks by
blowing gently, you feel something move in your ear.
This is the membrane of the ear-drum (Fig. 102). The
cavity of this drum is connected with the mouth by a pipe,
the so-called *Eustachian tube*. In this action you have

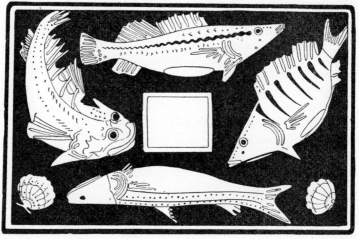

FIG. 3.   Rectangular plate of baked clay from Louvre decorated with the
figures of four marine fish and two shells. The work of a Greek artist of South
Italy of the fourth century B.C.

driven air from the mouth through the Eustachian tube
into the ear. Eustachi (p. 107), after whom this structure
is named, was a writer of the sixteenth century. Yet these
tubes were described by Alcmaeon two thousand years
earlier. This is one of many instances in which modern
discoveries were anticipated by the Greeks.

Alcmaeon made many other researches but only a most
fragmentary record of them has survived. He began the
study of the development of the young animal, *Embryology*

as it is now called, and he worked at it in much the same way that we do nowadays. He examined incubated eggs and watched the body of the chick developing.

Other Greek scientific men in the earlier part of the fourth century were following this same line of research. The work of one of these gives an intelligible account of the developing chick.

There is another department of biology which is very ancient indeed and was more widely though less satisfactorily pursued. This department is botany. Plants were studied by Hippocrates and his contemporaries and predecessors, though solely for their medical application. We have several works of the time of Hippocrates which tell us a good deal about how the physicians of the day used plants as drugs. These works show that a great number of different kinds of herbs were distinguished. Such writings are, however, of very little scientific value because they tell us hardly anything about the plants themselves.

§ 2. *Doctrine of the Four Humours*

Before we part with the Hippocratic writings we must refer to one topic which is of great importance for the effect that it had on the beliefs of the following ages. This was a very peculiar view of the constitution of the body.

The ancients supposed that all matter was made up of the four essential elements—*earth*, *air*, *fire*, and *water*. These, it was thought, were either in opposition or in alliance with each other. Thus, water was opposed to fire but allied to earth. These oppositions and affinities were associated with the view that each element was compounded of a pair of 'primary qualities', *heat* and *cold*, *moisture* and *dryness* (Fig. 4).

This strange conception was further developed by the

Greek medical writers. The Hippocratic works suppose that all living bodies are made up of four humours, *Blood* (sanguis), *Yellow Bile* (cholera), *Black Bile* (melancholia), and *Phlegm* (pituita). These four Humours had a special relationship with the four Elements (Fig. 4). Health depended upon the Humours being mixed, or to use the old word *tempered* or *complexioned*, in the right proportion. If one or other was in excess the patient suffered accord-

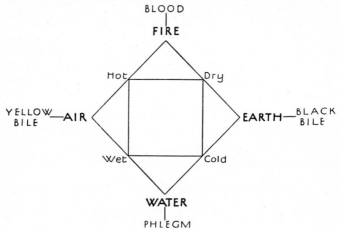

FIG. 4. The four *Elements* in association with the four *Humours* and the four *Qualities*.

ingly. This theory was thought to explain the nature of all disease. Diseases were classified according to the humour in excess as sanguine, choleric, melancholic, or phlegmatic. Moreover every individual, it was supposed, had a nature tending to one of the four types of disease.

Men of science have, of course, long ago abandoned this belief, yet it still finds a place in many expressions in current use. We still speak of a man's *temperament* or a lady's *complexion*, and we are all of us at times more or less *sanguine*, *choleric*, *melancholy*, or *phlegmatic*. The use of these words dates back to the time when they had a definite physiological meaning. Each humour became associated with a special organ, the blood with the liver, melancholy or black bile with the spleen, phlegm with the lungs, and choler with the gall bladder. To show how firmly impressed and lasting this theory was we may quote from an English peom by John Gower (1340–1408), a contemporary of Chaucer (1340–1400). The poem was dedicated to King Richard II (reigned 1377–99):

> The *spleen* is to Melancholy
> Assigned for herbergery [lodging].
>   The moist Phlegm with the cold
>   Hath in the *lungs* for his hold
>   Ordained him a proper stede [place]
>   To dwell there as he is bede [bid].
> To the Sanguine complexion
> Nature of his inspection
> A proper house hath in the *liver*
> For his dwelling made deliver [ready].
>   The drier Choler with his heat
>   By way of kinde [nature] his proper seat
>   Hath in the *gall* where he dwelleth
>   So as the philosopher telleth.'[1]

The philosopher mentioned in the last line is the greatest of all Greek men of science, and perhaps the greatest of all men of science, the philosopher *par excellence*. He was often spoken of in the Middle Ages simply as 'the philosopher'. That man is Aristotle, and to him we now turn.

§ 3. *Aristotle* (384–322 B.C.)

Aristotle was born in 384 B.C. at Stagira, a small town of the Chalcidice. Stagira was a Greek settlement, but it lay far from such centres as Athens or the Ionian cities

---

[1] The spelling is partially modernized.

where there had dwelt for centuries a population interested in intellectual matters. Moreover, Stagira lay on
the frontiers of the state of Macedon into which it was
soon to be absorbed. Athenians or Ionians looked upon
the inhabitants of Macedon much as well-bred and well-

FIG. 5. The States of ancient Greece.

educated people were accustomed to regard the inhabitants
of the 'back woods' a hundred years ago.

In the fourth century, however, this attitude was beginning to change. On the one hand, the older Greek
world was split up into very small states (Fig. 5) which
were constantly quarrelling. These minute states were
the poorer because individual liberty in some of them was
carried to an excess that prevented the adequate development of trade. The population in many of them had
become very dense but consisted largely of slaves. Most

people dwelt in or near the towns. The denseness of the population necessitated an intensive form of agriculture so that there was little wild or uncultivated land left.

On the other hand, the Macedonians were on the rise in most of the matters in which the older Greek states were on the decline. Macedon was, for one thing, by far the largest of the Greek states (Fig. 6). It was a kingdom governed by an absolute monarch to whom the 'liberty of the subject' seemed less important than the unity and power of his kingdom. The Macedonian lands were but half developed, large tracts of wild and forest country still remaining. But most important of all, Macedon came under a series of very able rulers. One of these was Philip, father of Alexander.

Philip was a statesman of the first rank, a man of high intellectual attainments, an admirer of the culture and civilization of the more settled parts of the Greek world. He tried to attract learned and cultured Greeks to Macedon, and he prided himself on being a Greek and not a semi-civilized barbarian.

The father of Aristotle was physician to Amyntas II, king of Macedon, who was himself the father of Philip. It was probably from his father, the physician, that Aristotle early acquired his taste for biological investigation.

Aristotle at seventeen was sent to what was the equivalent of a university. He went to Athens where he became a pupil of the great philosopher Plato. The institution where Plato taught was known as the *Academy*. The word was simply the name of a grove of trees where Plato and his pupils were wont to assemble. From that grove the school of Plato took its name, and from the school of Plato we have the modern meaning of the word 'Academy' as a place of learning.

There can be no doubt that Plato influenced his pupil Aristotle a great deal, but in certain philosophical matters there were profound differences between the two men. So far as science was concerned, we note that Plato had a strong bias towards mathematics while Aristotle was essentially a biologist. Aristotle, however, remained closely attached to his master and continued to be a member of his school till Plato's death in 347 B.C. Aristotle perhaps expected to become head of the Academy. In this, however, he was disappointed, and he left Athens to reside at the court of the sovereign of a small Greek state on the coast of Asia Minor opposite to the island of Lesbos.

Aristotle's bias towards natural history now had full play. He had ample leisure and he was in a good position for investigating marine animals. The results of his investigations at Lesbos come out in his biological treatises. These are, at least in outline, perhaps his earliest works.

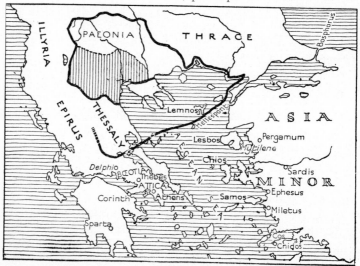

FIG. 6. Expansion of Macedon during the reign of Philip, father of Alexander the Great. The vertically shaded area shows the extent of Macedon at Philip's accession, the black outline its extent at his death.

In the year 342 B.C., when Aristotle was 42 years of age, he received an invitation from King Philip of Macedon to become tutor to his son Alexander. He accepted and settled in Macedon, where he remained for seven years.

Alexander must have owed much to his training by the greatest of biologists. It seems probable, however, that Alexander early passed out of Aristotle's hands. Thus, in the year 339 B.C., when Alexander was only 16 or 17, we hear of him taking charge of the kingdom during his father's absence. In the year 336 Philip was assassinated, Alexander succeeded to the throne, and Aristotle left Macedon to return again to Athens. There is evidence that tutor and pupil were not on good terms, though Alexander retained through life an interest in science. Aristotle and Alexander never met again.

Aristotle on returning to Athens in 336 was forty-eight years old. He began again to teach, but he did not return to the Academy. He set up his own school or university, which became known as the *Lyceum*, and he very soon gathered around him a large band of able pupils. It was during this second Athenian period that most of his philosophical work was done. His life remained uneventful, though marked by great mental activity, until the year 323 B.C.

It is pleasant to remember that, just as Plato taught in a grove, the *Academy*, so Aristotle taught in a garden. The word *Lyceum* is derived from an epithet of Apollo and was a familiar name for a temple to that god which stood hard by Aristotle's garden. The epithet *Lykeios* happens to be peculiarly appropriate to Aristotle, for its meaning is 'luminous' or 'light-bearing'. In the Lyceum Aristotle had his own favourite walk or *peripatos*, where he used to teach while walking. Hence his followers of later generations are often called *Peripatetics*.

During the thirteen years, from 336 to 323 B.C., while Aristotle was teaching in his garden, the power of Alexander was rapidly growing. By overthrowing the vast and immeasurably wealthy Persian Empire, which for two centuries had ruled nearly all the civilized Near East, including Egypt, he succeeded to the position of the Persian kings. He even crossed the Indus and fought successful battles in the Punjab (Fig. 23, p. 55). In spite of this the small Greek states still held jealously aloof from the upstart Macedonian power. Feeling ran high as these Greek states saw that they were inevitably to be engulfed into the huge Alexandrian realm.

Suddenly in 323 B.C. came the unexpected news of the death of the conqueror Alexander at Babylon at the age of thirty-three. The excitement was intense. The party most opposed to Macedon seized power in Athens. Aristotle, who had spent a long time in Macedon as tutor to Alexander, was under suspicion of having Macedonian sympathies. He thought it prudent to withdraw from Athens. He died in the following year, aged sixty-two. In his will he named his favourite pupil Theophrastus as head of the Lyceum, the school that he had founded (see pp. 44–53).

The surviving works of Aristotle place him as among the very greatest biologists of all time. He set himself to cover all human knowledge, and succeeded in this vast task in a way in which no one has succeeded before or since. He was a deeply original thinker, and he had an unrivalled capacity for arranging his own and other people's material. To these qualities he added first-class powers of observation and great shrewdness of judgment. No succeeding thinker has exercised so great an influence.

§ 4. *Aristotle's Biological Works*

Of Aristotle's biological works a number have survived, in a fairly complete, though disordered, state. Four are major treatises, and there are a number of less importance. The major works are most frequently referred to by their titles translated into Latin. They are:

(a) *On psyche* (De anima).
(b) *Histories about animals* (Historia animalium).
(c) *On the generation of animals* (De generatione animalium).
(d) *On the parts of animals* (De partibus animalium).

The very names of some of these are of interest and have themselves a history. Thus the first *On psyche* deals with what we may call the *living principle*, the quality or nature or essence, or whatever we may call it, that distinguishes living substance. This quality or nature or essence departs or ceases to exist or to act when living substance dies. Now various interpretations have been put on Aristotle's views in this matter. Gradually the word *psyche* that he uses for *living principle* has been transferred to other subjects. It has nowadays come to be used for a particular property of some living things, namely mind, or even the power to think. Thus we have the science of *Psychology* which deals with the properties of mind. Of this science, Aristotle may fairly be said to be the founder, and he devoted a large number of books to it.

But the word *psyche* had a long and adventurous history before Aristotle came to use it. The oldest Greek work we have is the poem known as the *Iliad* of Homer, which deals with the siege of the city of Troy by the Greeks. Parts of this poem date from about 1050 B.C. and therefore nearly 700 years before Aristotle. These parts were thus written about as long before him as the signing of Magna Charta is before us! This gives an idea of the time that Greek civilization had been growing. In the

Iliad of Homer we find the word *psyche* sometimes used in the sense of *breath*, which is its primary meaning.

Breathing is the most obvious sign of life, and when a man ceases to breathe we know that he is dead. So from *breath* the word *psyche* came to mean *life*, then the *principle of life*, and then the *soul* or again the *mind*. It is interesting to observe that in other ancient languages, as for instance Hebrew and Latin, the word for *soul* or *life* has gone through exactly the same history, being gradually changed from its original meaning of *breath*. A part of the story of this word is told for us in the Bible where we read in the book of Genesis 'And the Lord God formed man of the dust of the ground, and *breathed into his nostrils the breath of life, and man became a living soul*'.

The name of the second of the great biological works of Aristotle has also had a very peculiar and interesting history. We have translated the title *Histories about animals*, and the Greek title reads *Historiai peri ta zōa*. Now our English word *history* is derived from the Greek word *historia*, but its meaning has altered somewhat in the passage. At first the Greek word really meant the process of learning by inquiry, and this is the sense in which Aristotle uses it. It next came to mean the knowledge that was obtained by inquiry, and finally it was taken to mean the setting forth of this knowledge. The title of the work of Aristotle should, therefore, perhaps be translated *Inquiries about animals*. To give it a modern ring we might well render it *Zoological researches*, for that is what the book really contains. And what are researches but inquiries about things? When the word 'history' first entered the English language, it had the original Greek meaning of either inquiry or the results of inquiry. The word History in the ordinary sense, unqualified, has now come to mean the story of the past. It still retains its old meaning, however, in our term *Natural*

History. The word *story* is only a shortened form of the word *history.*

The actual contents of Aristotle's biological works may be divided into two parts (a) Observations, and (b) Deductions or Theories. These two elements are closely interwoven in all important biological works. There can be no important series of observations unless the observer is prompted by some theory. Conversely, there can be no valuable theory that is not based on a careful accumulation of observations. The two elements cannot be separated in the mind of the man of science. There are those who pride themselves on collecting facts without any reference to theory. In making this claim, the collector is exhibiting his ignorance both of the history of science and of the object of science. It can easily be shown that some theory is, and must be, at the back of all effective collecting. If no theory is there, then the collecting is aimless and unscientific. Science has neither need nor place for human jackdaws.

It is interesting to recall that Darwin, the greatest biological successor of Aristotle, laid great stress on the value of the formation of a theory to justify and stimulate the collection of observations. The condition of success is, of course, that the theory must be abandoned or changed freely to fit the observations.

Darwin in his *Autobiography* has some thoughts on this point which suggest the value of hypothesis in scientific discovery. 'From my youth', he says, 'I have had the strongest desire to understand or explain whatever I observed, that is, to group all facts under some general laws.' But he adds, 'I have steadily endeavoured to keep my mind free so as to give up any hypothesis, however much beloved (and I cannot resist forming one on every subject), as soon as facts are shown to be opposed to it.' Although observations and theory are thus inseparable, we must nevertheless distinguish them for the purpose of our narrative. In what follows, we shall first discuss some of the more interesting or important of Aristotle's actual observations and then deal with his theories.

§ 5. *Aristotle on the Habits of Fish*

Aristotle was fond of watching animals and their ways. It is evident too, that he was careful in keeping notes of what he saw. We shall first glance at his remarkable account of the breeding habits of the cat-fish (Figs. 7 and 8):

'The cat-fish deposits its eggs in shallow water, generally close to roots or close to reeds. The eggs are sticky and adhere to the roots. 'The female cat-fish having laid her eggs, goes away. The male stays on and watches over the eggs, keeping off all other little fishes that might steal the eggs or fry. He thus continues for forty or fifty days, till the young are sufficiently grown to escape from the other fishes for themselves. Fishermen can tell where he is on guard for, in warding off the little fishes, he sometimes makes a rush in the water and gives utterance to a kind of muttering noise. Knowing his earnestness in parental duty the fishermen drag into a shallow place the roots of water plants to which the eggs are attached, and there the male fish, still keeping by the young, is caught by the hook when snapping at the little fish that come by. Even if he perceive the hook, he will still keep by his charge, and will even bite the hook in pieces with his teeth.' [*Abbreviated.*]

Aristotle draws attention also to another peculiarity of this fish, namely its power of giving forth sound. 'Fishes can produce no voice for they have no lungs or windpipe, but certain of them, as the cat-fish in the river Achelous, emit inarticulate sounds and squeaks by a rubbing motion of their gill covers.'

The Achelous is a river of continental Greece which runs into the sea opposite the island of Ithaca and just at the mouth of the gulf of Corinth (Fig. 5). For many centuries the only notice taken of this account of the cat-fish of the Achelous river was to laugh at it. The passages in which there are references to it were thought to be spurious or to be simply erroneous. The cat-fish that were known in Europe do not look after their young in this fashion, though some can make a noise with their gill covers.

This was the state of knowledge of the habits of cat-fish until the middle of the nineteenth century. About that time the matter was taken up by the Swiss-American

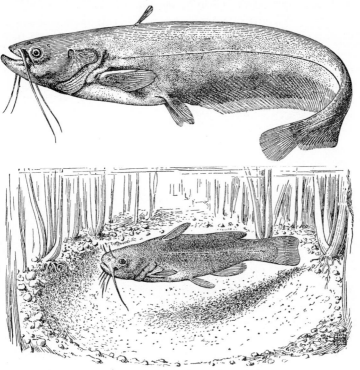

Fig. 7. Aristotle's cat-fish, *Parasilurus Aristotelis*.
Fig. 8. Male American cat-fish (*Ameiurus*) on guard over nest.

naturalist, Louis Agassiz (1807–73, pp. 309, 489) of Harvard. Now in America there are cat-fish, though of different species to those known in Europe. Agassiz observed that the male American cat-fish look after their young just as described by Aristotle (Fig. 8). This made him suspect that the story of Aristotle might be true also for Greek cat-fish. It happened that he had cat-fish sent to him from the river Achelous (1856). These, he found, were a peculiar species, different from the other European cat-fish and from the American form in which he had observed the male guarding the young. Working twenty-two centuries after Aristotle, and in a continent unknown to that great naturalist, Agassiz therefore called his newly discovered cat-fish after Aristotle. Unfortunately his description was overlooked by naturalists. It was not until fifty years ago (1906) that *Parasilurus Aristotelis* (Fig. 7) became properly known to men of science.

Probably most people have heard of the electric fish known as the *Torpedo* from which the instrument of destruction has been given its name. A great deal of research has been devoted to this animal, and we now know how it produces its electric discharge. It is characteristic of all muscle substance that, at the moment of contraction, it gives rise to an electric disturbance. In ordinary muscle this discharge can only be detected by means of delicate instruments. In the Torpedo fish the anterior part of the body is expanded and flattened (Figs. 9 and 10). This flat expanded body contains two special and peculiar kidney-shaped tracts of muscular substance (Fig. 9) in which the contractile power is reduced to a minimum but the power of producing an electrical disturbance is greatly increased. The electric shock from one of these creatures can easily numb and kill a smaller animal. The commonest Mediterranean species of this fish is *Torpedo ocellata* in which there are five large eye-like pigment spots on the dorsal surface (Fig. 10). This form was well known to the early Greeks and excellent figures of it are encountered on pottery from the sixth century B.C. onward.

Another peculiar form of fish common in the Mediterranean is the so-called *Angler fish*. This animal has, like the Torpedo, a somewhat flattened body. Some of the rays of the dorsal fins of the Angler are long and movable and terminate in a fleshy flap. These flaps are used by the fish as baits to attract smaller creatures. The animal is 'protectively' coloured, that is to say it is of the colour of the ground on which it lives, and is therefore almost completely invisible. It has an enormous head and huge mouth. The Angler lies still at the bottom. When small fishes are attracted by the baits, the great mouth opens and the little victims are engulfed. Owing to the enormous extension of the head of the Angler fish, the anterior, or so-called 'pectoral' fins, which are attached to the bony skeleton of the head, are carried backward right behind the posterior, or so-called 'pelvic' fins.

Here is Aristotle's description of the habits of these two animals, the Torpedo and the Angler (Figs. 9–11):

'In marine creatures one may observe many ingenious devices adapted to the circumstances of their lives. The accounts commonly given of the *Torpedo* and *Angler* are perfectly true.

'In both these fishes the breadth of the anterior part of the body is much increased. In the *Torpedo* the two lower fins (i.e. the so-called "pelvic" fins) are placed in the tail. This fish uses the broad expansion of its body to serve the office of a fin. In the *Angler* the upper fins (i.e. the "pectoral" fins) are placed behind the under fins (i.e. the pelvic fins). The latter are placed in this fish close to the head.

'The *Torpedo* stuns the creatures that it wants to catch, overpowering them by the force of shock in its body and feeding upon them.

It hides in the sand and mud, and catches all the creatures that swim within reach of its stunning power. The *Torpedo* is known to cause a numbness even to human beings.

'The *Angler* stirs himself up a place where there is plenty of mud and sand and hides himself there. He has a filament projecting in front of his eyes. This filament is long, thin and hair-like and rounded at the tip. It is used as a bait. The little creatures on which this fish feeds swim up to the filament taking it for a bit of the seaweed that they eat. Then the *Angler* raises the filament, and when the little fishes strike against it he sucks them down into his mouth.

'That these creatures get their living thus is evident from the fact that while sluggish themselves they are yet often found with mullets in their stomachs and mullets are very swift fish. Moreover the *Angler* is usually thin when taken after having lost the tips of his filaments.' [*Abbreviated from several passages.*]

In these and in many other passages in his biological works we see Aristotle, the first and in many ways the greatest of all naturalists, actually watching the creatures he loves. He is leaning out of a boat in the great gulf that indents the island of Lesbos, intent on what is going on at the bottom of the shallow water. In the bright sun and in the still, clear water of the Mediterranean every detail, every movement, can be discerned. Hour after hour he lies there, motionless, watching, absorbed, and he has left for us his imperishable account of some of the things that he has seen with his own eyes.

§ 6. *Handicaps of Early Naturalists*

Still clear water and bright sun are advantages that

Description of Figs. 9, 10, and 11.

Fig. 9.   Diagram of the electric organs of the Torpedo-fish.

Fig. 10.   *Torpedo ocellata* from Rondelet (1554, see p. 93).

Fig. 11.   Angler-fish from Rondelet (1544). Note the wide gaping mouth and the fishing filaments above it. Only the large pectoral fins are seen. The pelvic fins are in front of the pectorals, concealed by the anterior expansion of the body.

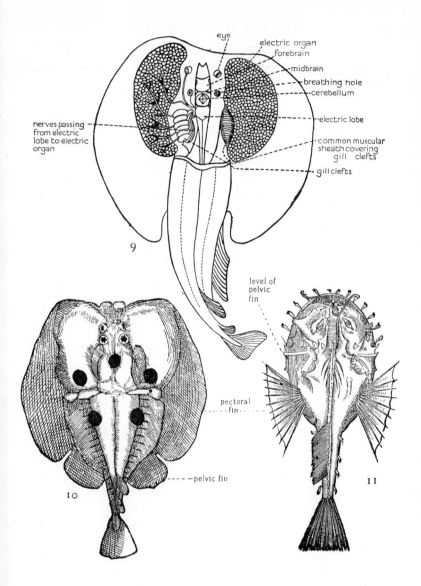

eye

electric organ
forebrain

midbrain

breathing hole

cerebellum

electric lobe

common muscular
sheath covering
gill clefts

gill clefts

nerves passing
from electric
lobe to electric
organ

9

level of
pelvic
fin

pectoral
fin

pelvic fin

10

11

FIGS. 9, 10, and 11. See note opposite.

Aristotle enjoyed over the naturalists of our own time and country. But if he had these advantages he had also many handicaps from which we have been freed, and freed partly by him. Only by keeping in mind these disadvantages under which Aristotle worked can we judge such mistakes as he made. Remember that his study was new. He had no books to consult, no training for such work, no instruments to help him, no learned societies, and only such learned companions as he had himself trained, to encourage and support him. He was as yet without a library—indeed the very idea of a great library in our sense, comes to us from him.

Moreover, Aristotle suffered from a drawback from which the Greeks never wholly freed themselves but from which their work has delivered us. He was without a special *scientific nomenclature*. The special words that we use in the sciences, ugly and difficult as they are, save an immense amount of labour and, in the end, make the subject easier to understand. Thus, if we had been writing the description of the fishes of which we have read Aristotle's account, we should have spoken of their *pectoral* and *pelvic* fins. Any naturalist accustomed to deal with fish would know at once exactly what was meant. The words *pectoral* and *pelvic* applied to a fish's fins have a definite anatomical meaning that would be quite unmistakable to a modern naturalist because he has traced these limbs through a whole series of fishes and has seen how they vary in form and position, and has compared them to the limbs of amphibians, reptiles, birds and mammals.

But the Greeks were beginning their subject. They had had no time to trace these fins through whole series of fishes. They could only describe them by their position, 'upper' and 'lower'. We who come after the Greeks, when in need of a new scientific term, turn to their language or to that of their heirs, the Latins. The Greeks had no literature that was classical to them upon which to draw. Nevertheless we shall see them making a beginning of a scientific nomenclature (p. 47). It is a process, however, which has been carried much further in our time, and through it Greek is still a *living* language.

There is yet another matter in which we have the advantage of Aristotle, but in which our advantage is based and founded on his work. In reading or listening to any account of biological matters we are greatly helped by drawings or diagrams. In fact, without such drawings or diagrams biological investigation would be impossible. Aristotle appears to have been the first to illustrate a biological treatise. In his works diagrams are sometimes referred to. Unfortunately the figures have long since disappeared, but his descriptions are occasionally of such a character as enables us to reconstruct them.

There are several places in Aristotle's great *Histories about animals* in which the lost diagram can be restored with confidence. An admirable example is his description of the general structure of the urinary and reproductive organs in the mammalian group, the animals, that is, that suckle their young. The ducts and vessels in connexion with these organs have a very complicated relationship. Aristotle describes them well and in a way that shows that he had dissected them. He actually refers to 'the accompanying diagram'. It is therefore a great satisfaction, in reading his text, to be able to reconstruct this diagram from the objects themselves and to find that the facts fit his description (Fig. 12). We shall come across other cases where we can supply the figures from his text (Figs. 13, 14).

§ 7. *Aristotle on Octopuses and their Allies*

Aristotle was extremely interested in the problem of how creatures come into existence. If he had lived now-

adays we should, perhaps, have called him an *embryologist.*
Embryo is a Greek word and means, in that language,
the young animal or plant before it has left the body
of the mother or before it has come out of its egg or

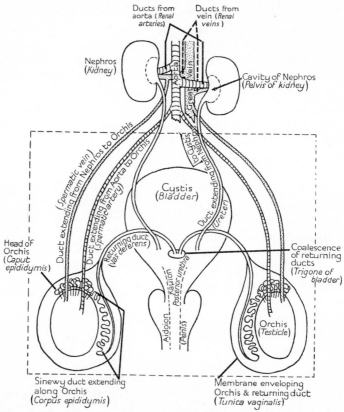

FIG. 12. The generative and urinary systems of a mammal as described by
Aristotle. The part framed in a dotted rectangle is a restoration of a lost diagram
prepared by Aristotle and described in his *Historia animalium.* The legends in
brackets are the modern scientific terms. The others are either transliterations of
the terms used by Aristotle or literal translations of such terms. The figure is
drawn as though the parts were viewed from behind.

seed, and while still incapable of independent life. In connexion with embryology Aristotle refers to his diagrams several times. There is in his works a delightful description of the development of the *Octopus* and of the *Sepia* or cuttlefish, forms of marine molluscs that are provided respectively with eight and ten 'arms'. These creatures, now described by men of science as belonging to the group of *Cephalopoda* (see pp. 28–31, Figs. 48–50, p. 95), specially interested Aristotle. We shall quote his description and associate with it some suitable drawings:

'The *octopus* breeds in spring, lying hid for about two months. The female, after laying her eggs, broods over them. She thus gets out of condition since she does not go in quest of food during this time. The eggs are discharged into a hole and are so numerous that they would fill a vessel much larger than the animal's body. After about fifty days, the eggs burst. The little creatures creep out, and are like little spiders, in great numbers. The characteristic form of their limbs is not yet visible in detail, but their general outline is clear. They are so small and helpless that the greater number perish. They have been seen so extremely minute as to be absolutely without organization [remember Aristotle had no lens or microscope] but nevertheless when touched they moved. The eggs of the *sepia* look like a bunch of grapes (Fig. 48, p. 95), and are not easily separated from one another. They are stuck together by some moist sticky substance exuded by the female. They are white at first but larger and black after this sticky substance has been exuded.

The sixteenth-century naturalist, Guillaume Rondelet (p. 93), based his great work on Mediterranean marine creatures upon Aristotle. Rondelet's drawings of the octopus and of the sepia are excellent, and his figure of the grape-like egg-clusters of the sepia is especially appropriate here since he used it to illustrate this identical passage of Aristotle above. To this day Mediterranean fishermen call the eggs 'grapes of the sea'. Aristotle says that

'The young *sepia* is first distinctly formed inside the egg out of the white substance. When the egg bursts it comes out. The inner part is formed as soon as the egg is laid and is something like a hail-stone. Out of this substance the young sepia grows, being, however, attached by the head in the same way as a developing bird is attached by the belly. [This refers to observations which Aristotle had made on the chick developing within the egg, see Figs. 13–14, and cp. Figs. 35 and 63.] As the young *sepia* grows the white substance dwindles

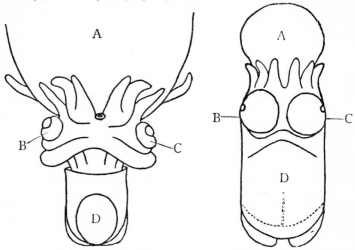

FIGS. 13 and 14. Diagrams of a younger and an older stage in the development of the *Sepia* or cuttlefish as described by Aristotle.

and, moreover, as with the yolk in birds' eggs, it finally disappears. In the young *sepia*, as in most animals, the eyes are at first relatively very large. To illustrate all this by means of a figure let *A* represent the egg, *B* and *C* the eyes, and *D* the body of the little *sepia*.

If, when the young ones are fully formed, you cut the outer covering a moment too soon, the young creatures eject pigment, and their colour changes from white to red in their alarm.'

The habit of the Cephalopoda of ejecting their 'ink' to cover their retreat is well known. A certain type of painter's pigment is called 'sepia' and is still, in fact, often derived from that animal. Moreover, the colour-changes which flit across the skin of the creatures appear to be expressive of their emotions. These changes are, however, in large part protective and enable the animals to resemble their surroundings, and thus deceive their enemies or their prey.

The study of the Cephalopods affords us also a good example of how the Aristotelian works sometimes err.

60384

FIG. 15. The Paper Nautilus, *Argonauta Argo*, from Belon (1551, see pp. 90–93). The animal is drawn using its arms as oars and its membrane as a sail, an erroneous idea derived from Aristotle.

One of the most beautiful and interesting of the Cephalopods is the co-called *Paper Nautilus* (to be clearly distinguished from the *Pearly Nautilus*). This exquisite creature has long been studied by naturalists, but we have acquired an accurate idea of its habits only in modern times. The Aristotelian work tells us that

'it is an octopus, but one peculiar both in its nature and habits. It often lives near the shore, and is apt to be thrown up on the beach where it is found with its shell detached.

'The Paper Nautilus rises from deep water and swims on the surface. Between its feelers is a web, like that between the toes of web-footed birds, but much thinner. When there is a breeze it uses this for a sail and lets down some of its feelers alongside as oars (Fig. 15). If startled it can fill its shell with water and sink. As regards the formation of the shell, knowledge is not yet satisfactory. The shell does not appear to be there from the beginning but to grow as in

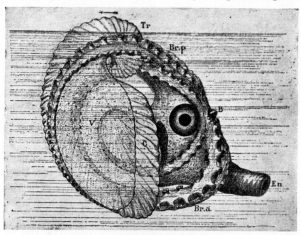

FIG. 16. Female of the Paper Nautilus, *Argonauta Argo*, swimming just below the surface of the water. The arrow shows the direction of movement. *B* is the parrot-like beak. *Tr* is the shell. *Br. p* is the base of one of the two special shell-carrying posterior arms. The other end of the arm is expanded into a wide membrane *V*. This membrane embraces the shell which can be seen extending beyond it both at *Tr* and at *C*. The membrane was supposed to form the 'sail' (see Fig. 15). *En* is the funnel which projects between the two anterior arms *Br. a*. The animal moves by the force of a jet of water driven forth from the funnel. From Lacaze-Duthiers, 1892.

other shell-fish. Nor is it ascertained whether the creature can live without its shell.'

The use by the Paper Nautilus of the membrane as a sail and of the arms as oars is now known to be a myth, though many excellent modern naturalists have believed it. The legend has earned for the creature its scientific name of 'Argonaut' (*Argonauta argo*), the word meaning 'sailor of the Argo', the ship on which Jason embarked in quest of the golden fleece. Belon (pp. 90–93), contemporary of Rondelet, gives a fine figure of the *Argonauta* with its membrane set as a sail and using its arms as oars! (Fig. 15.)

Some of the questions asked by Aristotle about the *Argonaut* can now, after the fullness of time, be answered. It progresses neither by sailing nor rowing, but by ejecting a jet of water from its 'funnel' (Fig. 16). The shell is in no organic connexion with the body of the creature but is used as a sort of perambulator to support and aerate the developing young. Moreover, the animal does not willingly sink below the surface, and if forced to do so will rise again: it is indeed doubtful if it is able to sink at all by its own efforts.

§ 8. *Aristotle on Whales, Porpoises, and Dolphins*

Aristotle had a clear idea of the nature of the great classes of animals. He knew the important anatomical, physiological, and embryological differences between, for example, the mammals which form the class to which man belongs, and the fishes. Thus he knew that mammals have lungs, breathe air, and have warm blood. He knew that they bring forth their young alive—that is to say, that they are *viviparous*—that they suckle their young, and that their young, while within their mother's body, are attached to the womb by means of a navel-string and a structure now known as the *placenta*. He knew that none of these features are encountered in fishes.

Knowing these things he separated the *Cetaceans*, that is to say the group of animals that contains the whales, dolphins, and porpoises, from the fishes. He placed the Cetaceans with the other mammals, but he was almost the only writer that did this for nearly two thousand years. In discussing the general nature of mammals Aristotle says

'Among viviparous animals are man, the horse, the seal, and other animals that are hair-coated and also, of marine animals, the cetaceans. These latter creatures have a blow-hole and are provided with lungs and breathe. Thus the dolphin has been seen asleep with his nose above water, and snoring. The dolphin takes in water and discharges it through his blow-hole but he also inhales air into his lungs, so that if caught in a net he is quickly suffocated for lack of air. He can, however, live for some time out of water, but keeps up, all the while, a dull moaning sound. The cetaceans take in water only as incidental to their mode of feeding for, as they get their food from the water, they cannot but take it in.

'The dolphin and porpoise bear one at a time but sometimes two. The whale two at a time but sometimes one. These animals are provided with milk and suckle their young (Fig. 46, p. 92). The young of the dolphin grow rapidly, being full-grown at ten years of age. They bring forth their young in summer and never at any other season. The young accompany them for a considerable time and these creatures are in fact remarkable for the strength of their parental affection.' [*Abbreviated.*]

§ 9. *On the Placental Dog-fish* ('*Galeos*')

Now in view of the difference in the normal mode of development of mammals (including Cetaceans) on the one hand and of fishes on the other, there is a certain observation by Aristotle that is most significant and worthy of further attention. Aristotle knew quite well that fishes usually differ from the mammals in being *oviparous*, that is to say in bringing forth their young as eggs and not as independent moving creatures. But he knew that there is a group of fishes in a few of which the young are sometimes brought forth actively alive. This

Description of Fig. 17.

A. Young free-swimming *Mustelus vulgaris*. *B*. Embryo of *Mustelus laevis* with placental yolk-sac. *C*. Embryo of *Mustelus laevis* with yolk-sac attached to uterus. *D*. Egg of *Mustelus vulgaris* viewed as a transparent object. *E*. Dissection of the lower end of uterus of *Mustelus laevis* showing relations of two placentae to uterine wall.

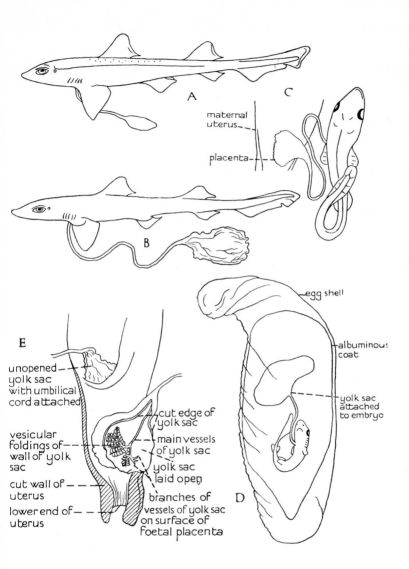

maternal
uterus

placenta

egg shell

albuminous
coat

yolk sac
attached
to embryo

E

unopened
yolk sac
with umbilical
cord attached

cut edge of
yolk sac

main vessels
of yolk sac

vesicular
foldings of
wall of yolk
sac

cut wall of
uterus

yolk sac
laid open

lower end of
uterus

branches of
vessels of yolk sac
on surface of
foetal placenta

D

FIG. 17. Development of Aristotle's placental dogfish ('Galeos') compared to
that of an allied species. From Johannes Müller, 1842. See note opposite.

group he called the *Selachē*, and he tells us that it includes the sharks and rays.   But among his *Selachē* he knew of one form in which the resemblance to the mammals goes even farther than the mere production of lively young.

In mammals the embryo before birth is connected with the mother's womb by a cord.   The point of attachment to the embryo can still be seen in the grown creature and is the *navel*.   At the other end the cord is attached to the womb by a large flat organ, which is detached after the young is born and is therefore called the *afterbirth*, otherwise known as the 'placenta'.   There are a few 'Selachia' —as scientific men now call Aristotle's *Selachē*—in which something similar takes place.

Aristotle tells us that there is a shark-like fish called 'Galeos' which, 'as it were, lays eggs within itself'.   These eggs being deposited in the womb acquire an attachment to it much like the placenta of a mammal.   In the *Galeos* he tells us that

'the eggs shift into the womb and descend, and the young develop with the navel-string attached to the womb. As the egg substance gets used up the young is sustained, to all appearances, just as in mammals. The navel-string is long and adheres to the under part of the womb, each navel-string being attached to the womb by a sucker [i.e. placenta] and to the young in the place where the liver is.   Each embryo, as in the case of mammals, is provided with separate embryonic membranes.'

Now this is a very extraordinary statement and one which seems at first sight most unlikely to be true. Naturalists for centuries passed it over or regarded it as a forgery.   It attracted, indeed, very little attention until, in the middle of the nineteenth century, the great German naturalist Johannes Müller (p. 393) proved that Aristotle had been perfectly correct in his description.   Müller showed that there are, in fact, a very few Selachia which in this respect resemble mammals. This was an amazing confirmation of an observation that had been made by Aristotle twenty-two hundred years before (Fig. 17).[1]

To prevent misunderstanding it should, however, be added that, though the placenta of *Galeos* bears a considerable resemblance to that of a mammal in both form and function, it is yet, in fact, derived from a different structure. The placenta of the mammal is derived from an embryonic structure known as the 'allantois', that of the *Galeos* from the yolk sac itself. The distinction will be made clearer if the reader will glance at Figs. 17, 35, 63 and 64.

§ 10. *The Aristotelian Bee-master*

Other Greek naturalists besides Aristotle were occupied in observing and recording the habits of animals. Such a one was the author of a fine treatise on bees which is usually printed as part of Aristotle's *Historia animalium*. It was evidently written by a practical bee-master and not by Aristotle himself.

The author of this account does not understand the nature of the queen bee. He is uncertain whether that creature should be regarded as a male or as a female. We know now that the animals he calls the 'ruler bees' are females or queens, that the drones are males, and that the worker bees, though unable under ordinary circumstances to produce any young are, in fact, females whose development has been arrested. This bee-master writes:—

'Very remarkable diversity is observed in the methods of working and general habits of bees. When the hive has been delivered to them clean and empty, they build their waxen cells, constructing the combs downwards from the top of the hive, and go down and down, linking the cells together until they reach the bottom.

'The cells, both those for honey and for the grubs, are double-doored,

---

[1] Rondelet (Fig. 47, p. 93) in the sixteenth and Stensen in the seventeenth century made imperfect observations on similar lines.

two cells being upon a single base, one pointing one way and one the other, after the manner of a double goblet. The two or three concentric rows of cells at the commencement of the comb and attached to the hives, are small and empty of honey. The cells that are well filled with honey are most thoroughly luted with wax. The ordinary bee is generated in the cells of the comb, but the ruler bees in cells apart from the rest. There are six or seven ruler bees which grow in a way quite different from the ordinary brood.

'The drones, as a rule, keep inside the hive. When they go out, they soar up in the air in a stream, whirling round and round in a kind of gymnastic exercise. This over, they come inside the hive and feed ravenously. The ruler bees never go outside the hive except with a "swarm" in flight, and during such a "swarming" all the other bees cluster around them. These rulers have the abdomen or part below the waist half as large again. They are called by some "mothers" from an idea that they generate the [worker] bees. As proof, they declare that drones appear even when there is no ruler bee in the hive, but that the [worker] bees do not appear in his absence. Others, again, assert that the drones are male and the [worker] bees female.

'Bees gather the beeswax with their front legs, wipe it off on to the middle legs, and pass it on to the hollow curves of the hind legs. Thus laden, they fly home, and by their manner of flight one may see plainly that their load is heavy. On a trip the bee does not fly from a flower of one kind to a flower of another, but from one violet, say, to another violet, and never meddles with another flower until it has got back to the hive.

'Whenever the working-bees kill an enemy they try to do so out of doors. Bees that die are removed from the hive, and in every way the creature is remarkable for its cleanly habits. Away from the hive they attack neither their own species nor any other creature, but in the close proximity of the hive they kill whatever they get hold of. Bees that sting die from their inability to extract the sting without at the same time extracting their intestines. The bee may recover, if the person stung presses the sting out; but once the bee loses its sting it must die.

'When the flight of a swarm is imminent, a monotonous and quite peculiar sound made by all the bees is heard for several days. For two or three days in advance a few bees are seen flying round the hive. When they have swarmed, they fly away and separate off to each of the rulers. If a small swarm happens to settle near a large one, it will shift to join this large one, and if the ruler whom they have abandoned follows them, they put him to death.' [*Abbreviated.*]

But for the vagueness as to the sex of the rulers and other kinds of bees, this account is almost unexceptionable.

§ 11. *Aristotle on the Nature of Life*

We have so far considered Aristotle as a practical working naturalist. When, however, we come to consider the inferences and theories that he deduces from his observations, we must feel that we are in the presence not only of an eminent naturalist but of a supremely great intellect.

The most fundamental division among biologists has been between the schools known as *Vitalists* and *Mechanists*. The difference between the thinkers of these two groups turns on the nature of life (Latin, *vita*) itself. As Biology is the science or rather the group of sciences concerned with living forms, this question seems fundamental. Nevertheless, it had been so long debated that it had lost immediate interest by the dawn of the twentieth century. In our time it looks as though the distinction may even be losing meaning. But historically the division between vitalist and mechanist has been of the highest importance.

What is a vitalist? The answer, so far as it can be given at this stage, is that a vitalist thinks that living things contain some principle of a quite peculiar nature and quite different from anything that is found in matter that is not living. A vitalist must further believe that this principle not only exists but has some manner in which it shows itself to us. A mechanist denies the truth of both these beliefs of the vitalist, and claims that all the actions and movements and activities of living things can be expressed as the outcome of forces which can be investigated in matter that is not living. In this sense of the word there can be no doubt that Aristotle was a vitalist.

Looking round for some word for the principle that differentiated living from not living substance, Aristotle hit, as we have seen, upon the word *psyche* (p. 15). Now when he began to examine different living things he reached the conclusion that there were different kinds or orders of psyche or soul. He thus came to differentiate between a vegetative, an animal, and a rational soul.

The first or lowest was the vegetative soul. Aristotle regarded plants as the lowest living forms, and the qualities of life that he distinguished in plants he regarded as alone essential for this lowest form of soul. These qualities seemed to him to be nutrition, growth, and the power of reproduction.

He considered that while animals had these qualities, they had also the power of movement, and the movements that they made seemed to correspond to what they felt. Animals then possessed, he thought, a second order of soul, to which he ascribed the sensitive and motive powers.

Lastly, man had all these qualities exhibited by the lower creation, both plants and animals, but he had certain others. He could reason and his movements and actions were dictated by his thoughts. He was therefore equipped, in Aristotle's view, not only with a vegetative and an animal soul but also with a rational or intellectual soul.

This distinction between the vegetable or nutritive, the animal or sensitive, and the human or rational soul is to be found over and over again in the literature of all subsequent ages. It still lingers in popular speech, much in the same way as does the doctrine of the humours

(p. 8).   Thus, to give an instance, perhaps the most widely read scientific work of the Middle Ages was the treatise *On the Properties of Things* by a Franciscan monk, one Bartholomew of England, who wrote it about 1230. This work contains many references to the three kinds of soul of Aristotle.   It was translated into English by a Cornishman, John of Trevisa (1326–1412), whose works are particularly interesting as specimens of early English prose.   This translation was among the first books printed in London.   It was produced by Wynkyn de Worde in 1495 and is certainly one of the finest volumes published in England in the fifteenth century.   In it we read:

'In divers bodies be three manner souls: *Vegetabilis*, that giveth life and no feeling as in plants and roots; *Sensibilis*, that giveth life and feeling and not reason in unskilfull beasts; *Rationalis*, that giveth life, feeling and reason in men.'[1]

Similarly, to this day, we speak of *vegetative* existence, of *sensual* feelings, of *rational* behaviour, all reminiscences of Aristotle's point of view.

But to return to Aristotle himself.   Early in the course of his thinking he completely separated man from lower creatures.   As his knowledge increased he seems to have become less inclined to make this distinction absolute. Indeed he came to admit that animals share to some extent with man in the possession of a rational soul. His final position demanded no fundamental distinction between life or soul, and mind.   This is the position of an important school of modern biologists.

It is probable that in ascribing to animals certain human qualities, Aristotle was influenced by his advance towards what would nowadays be called a belief in *Evolution*, a topic that we shall have to discuss more fully later (chap. viii).   It cannot be said that he ever definitely

---

[1] Spelling partially modernized.

attained to the 'evolutionary' point of view. But it is evident that he was moving in that direction, and perhaps if he had lived another ten years he might have reached it. But without claiming him as an Evolutionist, it is quite easy to read an evolutionary meaning into some of his biological writing. To do so is to develop but not to force his meaning.

In Aristotle's biological works nothing is more remarkable than his effort to reach some sort of way of exhibiting the relationships of living things. The method that he actually adopts for doing this is to arrange them in a serial order, *a scala naturae* or 'ladder of nature', as the naturalists of a hundred or so years ago would have called it. This *scala naturae* of Aristotle, as the earliest of its kind, is a subject of great interest and is worthy of all possible respect (Fig. 18). This is how he describes it:

'Nature', he says, 'proceeds little by little from things lifeless to animal life in such a way that it is impossible to determine the exact

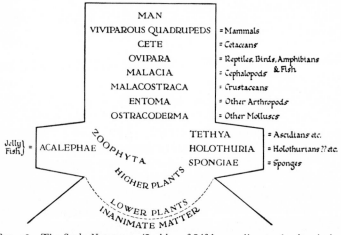

FIG. 18. The *Scala Naturae* or 'Ladder of Life' according to the descriptions of Aristotle.

line of demarcation, nor on which side thereof an intermediate form should lie. Thus, next after lifeless things in the upward scale comes the plant. Of plants one will differ from another as to its amount of apparent vitality. In fact the whole of plant-kind, whilst devoid of life as compared with animal-kind, is endowed with life as compared with other forms of matter. Indeed there is in plants a continuous scale of ascent towards the animal. Thus one is at a loss to say of certain beings in the sea whether they be animal or vegetable.'

'A sponge completely resembles a plant, in that it is attached to a rock, and that when separated from this it dies. Slightly different from the sponges are certain other sea creatures, the Holothurians which, through free and unattached, yet have no feeling, so that their life is but that of a plant separated from the ground. For even among land-plants there are some that are independent of the soil— or even entirely free. Such, for example, is the plant which is found on Parnassus, and which some call the *Epipetrum* [i.e., the common house-leek]. This you may hang up and it will yet live for a considerable time. It is therefore sometimes a matter of doubt whether a given organism should be classed with plants or with animals. Thus there are the *Tethya*, marine organisms that so far resemble plants as that they never live free and unattached, but, on the other hand, inasmuch as they have a certain fleshlike substance, they must be supposed to possess some degree of sensibility.'

'The *Acalephae* or "Sea-nettles" lie outside the recognized groups. Their constitution, like that of the *Tethya*, approximates them on the one side to plants, on the other to animals. Some can detach themselves and can fasten on their food, and they are sensible of objects which come in contact with them, and in this sense they have an animal nature. Yet they are closely allied to plants, firstly by the imperfection of their structures, secondly by being able to attach themselves to the rocks, and thirdly by having no visible residuum (after drying) notwithstanding that they possess a mouth.'

'Thus Nature passes from lifeless objects to animals in such un-broken sequence, interposing between them beings which live and yet are not animals, so that scarcely any difference seems to exist between two neighbouring groups owing to their close proximity.' [*Abbreviated from several passages.*]

§ 12. *Classification of Animals derived from Aristotle*

Aristotle does not make any formal classification of animals. But scattered through his works are many terms employed in a way which suggests that they might be developed for classifacatory purposes. By examining his definitions of these terms we are enabled to draw up the arrangement of animal forms which we may reasonably regard as the Aristotelian classificatory scheme. With his great interest in reproduction, this must be based on the process of generation.

ENAIMA=*VERTEBRATES*
(Having red blood and either viviparous or oviparous)

Viviparous in the internal sense.

1. Man.
2. Cetaceans.
3. Viviparous quadrupeds.
    (a) Ruminants with cutting teeth in lower jaw only, and with cloven hoofs=Sheep, Oxen, &c.
    (b) Solid-hoofed animals=Horses, Asses, &c.
    (c) Other viviparous quadrupeds.

Oviparous though sometimes externally viviparous.

With perfect eggs

4. Birds.
    (a) Birds of prey with talons.
    (b) Swimmers with webbed feet.
    (c) Pigeons, doves, &c.
    (d) Swifts, martins, &c.
    (e) Other birds.
5. Oviparous quadrupeds=Amphibians and most Reptiles.
6. Serpents.

With imperfect eggs

7. Fishes—
    (a) Selachians=Cartilaginous fishes ('Galeos' an exception.)
    (b) Other fishes.

ANAIMA=*INVERTEBRATES*

(*Without red blood and viviparous, vermiparous, budding or spontaneous*)

With perfect eggs	{ 8. Cephalopods. { 9. Crustaceans.
With eggs of special type.	10. Insects. spiders, scorpions, &c.
With generative slime, buds, or spontaneous generation.	11. Molluscs (except Cephalopods), Echinoderms, &c.
With spontaneous generation.	12. Sponges, Coelenterates, &c.

Some of the elements in this classification are fundamentally unsatisfactory in that they are based on negative characters. Such is the group of *Anaima* which is paralleled by our own equally convenient yet negative and biologically meaningless equivalent *Invertebrates*. It is but fair to remember that even this term Invertebrata held a meaning biological up to about a hundred years ago yet other Aristotelian groupings, such as the subdivisions of the viviparous quadrupeds, can only be somewhat forcibly extracted out of Aristotle's text. But there are still others, such as the separation of the cartilaginous from the bony fishes, that exhibit true genius and betray a knowledge that can only have been reached by careful investigation. Remarkably brilliant, too, is his treatment of Molluscs.

In our modern systems of classification we employ certain technical terms which are of Greek or Latin origin. The most important of these are the words *genus* and *species*, two words which are simply Latin translations of Greek words used by Aristotle. We owe our idea of the *species* in modern biology directly to Aristotle. Our use of the word *genus* differs considerably from his, but there can be no doubt that it also is rooted in his works. We shall consider the conceptions of *genus* and *species* later (pp. 177 ff.).

Aristotle was not content with investigations of the structure and habits of animals in the field which we nowadays call 'Zoology'. He wrote also works on the functions of the organs and parts of the body, that is on Physiology, and also, perhaps, on Botany. In the former subject, Physiology, it must be confessed that his efforts were less successful than in Zoology. He was not, in fact, an experimenter but rather an observer, and he had no exact body of knowledge of experimental physics and chemistry on which to build. Some of his successors among the Greeks, e.g. Galen (p. 60), far excelled him experimentally. We shall therefore pass over Aristotle's physiological views. His theory of the nature of generation, however, is of great importance and is considered elsewhere (p. 505).

No botanical treatise by Aristotle has come down to us. We have, however, very full botanical works by his pupil and successor as head of the Lyceum, Theophrastus, to whose work we now turn.

§ 13. *Theophrastus (c. 380–287 B.C.) and his Botanical Works*

Theophrastus was born about 380 on the island of Lesbos, where perhaps Aristotle first met him. Leaving his native land he went, like his master, to Athens and there, like him, he attended first the school of Plato. Later he transferred to the Lyceum. He became devoted to Aristotle and carried on his teacher's work after Aristotle had left Athens for the last time. This affection for Aristotle expresses itself in his will, in which he inserted special clauses for the preservation of the busts of his master. It is possible, though hardly probable, that a certain ancient bas-relief portrait is really that of the master.

The whole active life of Theophrastus may be said to be a commentary on Aristotle. He had not his master's vast attainments, but his abilities were good, his industry very great, and his interests wide. He lived to a very advanced age. His writings covered a wide range of subjects, but we are concerned only with the biological works. These are devoted to Botany, which is the more fortunate in that no botanical work of Aristotle has survived.

Theophrastus has left what are usually called *two* treatises on Botany. One of them, however, which is by far the more interesting of the two, the so-called *History of Plants*, is in fact not one work, but a collection of works. One part of this collection is really a scrap-book made up of such folk-lore about plants as was going about in the time of Theophrastus. It is a very entertaining book and it is important too, for our purpose, as showing the inferior kind of material on which scientific men of the day had to build. We have seen how, in the Hippocratic works, plants are only considered for their uses in medicine. In this Theophrastian scrap-book we are told a little more about them but in a most superstitious and gossiping vein. A few anecdotes from this section are worth relation.

We learn a good deal about the gathering of drugs from the botanical collections of Theophrastus, for they are essentially practical works. Thus we hear how frankincense and myrrh are collected. The story tells how certain travellers landed to look for water while on a coasting voyage from the 'Bay of Heroes' on the coast of Arabia. They saw the gums being collected.

'They reported that incisions were made on the stems and branches to extract gum. Palm leaf mats were sometimes put underneath, to collect the resin. The whole mountain range in that region, they said, is under the sway of the Sabaeans. These people are honest in dealing with one another. Myrrh and frankincense are collected from all parts into the Temple of the Sun which is very sacred and guarded by armed Sabaeans. Here each man piles up his store. On his pile he puts a tablet stating the measure and the price. When the merchants come they look at the tablets and, whichever pile pleases them, they put down the price on the spot. Then the priest takes a third part of the price for the god and leaves the rest. This remains safe for the owners till they claim it. The sailors, however, greedily took some of the frankincense and myrrh when no one was about and sailed away.' [*Abbreviated.*]

There are many such travellers' tales. A number of the accounts of plants, morever, are the merest superstition. These yarns Theophrastus had evidently collected not because he believed in them but because he thought that these idle notions might themselves have some interest —as indeed they have to the student of folk-lore. Here is such a story:

'Druggists and herb diggers enjoin that in cutting certain roots one should stand to windward. Thus if you cut *thapsia*, facing other than windward, your body will swell up. Again, some roots should be gathered at night, others by day, and yet others, as for instance the honeysuckle, before the sun strikes them. Hellebore makes the head heavy, and men therefore cannot dig it up for long unless they first eat garlic and take a draught of wine. The peony should be gathered by night, for, if a man is seen by a woodpecker while collecting the fruit, he is in danger of going blind. While cutting fever-wort beware of the buzzard-hawk.' [*Abbreviated.*]

There is much nonsense of this sort. Students of folk-lore have found superstitions of a like kind among English and continental peasants. It is a debated question to what extent such beliefs are really native and to what extent they are merely further degraded forms of classi-cal material of this type.

But we must turn from such gossip to the serious work of Theophrastus. He had not the inspiration of Aristotle. Being mainly an observer and collector of facts, rather than a great theorizer, he suffered especially from the lack of technical scientific terms. To describe accurately a plant or even a leaf in the language in ordinary use would often take pages and tax the powers of a first-class descriptive writer. Modern botanists have invented an elaborate terminology that took its origin with Jung (p. 181), the use of which saves a vast amount of mental and clerical labour. This question of scientific nomenclature among the Greeks has already been touched upon (p. 24).

There are cases in which Theophrastus seeks to give a special technical meaning to words in more or less current use. Among such words are *carpos* = fruit, *pericarpion* = seed vessel = *pericarp*, and *metra*, the word used by him for the central core of any stem whether formed of wood, pith, or other substance. It is from the usage of Theophrastus that the botanical definitions of *fruit* and *pericarp* have come down to us.

We may easily discern also the purpose for which he introduced into botany the term *metra*—a word meaning primarily the *womb*—and the vacancy in the Greek language which it was made to fill.

'Metra', he says, 'is that which is in the middle of the wood, being third in order from the bark and thus like to the marrow in bones. Some call it the *heart*, others the *inside*, yet others call only the innermost part of the metra itself the *heart*, while others again call this *marrow*.'

He is thus inventing a word to cover all the different kinds of core and importing it from another study. This is the method of modern scientific nomenclature which hardly existed for botanists even as late as the sixteenth century of our era.

Theophrastus understood the value of developmental study, a conception derived from his master.

A plant', he says, 'has power of germination in all its parts, for it has life in them all, wherefore we should regard them *not for what they are but for what they are becoming.*'

The various modes of plant reproduction are distinguished by Theophrastus thus:

'The manner of generation of trees and plants are these: (1) spontaneous, (2) from a seed, (3) from a root, (4) from a piece torn off, (5) from a branch or twig, (6) from the trunk itself, or (7) from pieces of the wood cut up small.'

The marvel of generation, of how plants and animals come into being, must have awakened admiration from an early date. We have seen it occupying the attention of Aristotle. He, like Theophrastus and like all naturalists until the seventeenth century, believed in the existence of *spontaneous generation* in plants and in lower animal forms. That is to say, they believed that plants and animals could arise without the intervention of parents, anew, out of matter that could not be regarded as living. How could they believe otherwise? These men were without microscopes and without the means of minute investigation to which the advances of modern times have accustomed us. All antiquity believed in spontaneous generation, and the destruction of that belief is one of the most important advances made by modern biology. It is a subject to which we shall presently return (chap. xii).

In his description of the process of germination, Theophrastus has left us his views on the formation of the plant in the seed. It is the first account of the subject on record and the best that was made until the seventeenth century of the Christian era.

'In some plants [Dicotyledons] root and leaves germinate from the same point and in some [Monocotyledons] from opposite ends of the seeds. Thus in the bean and other leguminous plants [Dicotyledons] the root and stem arise from the same point, namely their place of attachment to the pod. Wheat and barley and such cereals [Monocotyledons], however, germinate from opposite ends corresponding to the position of the seed in the ear, the root from the stout lower part, the shoor from the upper.

'In all leguminous plants the seeds have plainly two lobes and are double. As the bud germinates within the seed and increases in size, the seeds split. All such seeds are in two halves [Dicotyledons]. In cereals, however, the seeds being in one piece, this does not happen, but the root grows a little before the shoot [Monocotyledons].

'Barley and wheat [Monocotyledons] come up with one leaf but peas, beans, and chick peas with more leaves [Dicotyledons]. All leguminous plants have a single woody root, from which grow slender side roots [Dicotyledons]. Wheat, barley, and other cereals, however, have numerous slender roots which are matted together [Monocotyledons]. There is thus a contrast between these two kinds, for the leguminous plants have a single root and many side growths above from the single stem [Dicotyledons], while the cereals have many roots and send up many shoots, but these have no side shoots [Monocotyledons].'

Can there be any doubt that we have here an excellent piece of first-hand observation on the behaviour of germinating seeds? The distinction between the dicotyledons and the monocotyledons (which we have indicated by the words in brackets) is quite clearly set forth, though the emphasis on various points is not quite what we should make to-day. The observations are, however, accurate and clear and might be used in company with the plates of a modern text-book. In fact the figures of the seventeenth-century naturalist Marcello Malpighi, whom we shall presently discuss (p. 151), illustrate this passage from Theophrastus admirably. The reader is advised to keep his eye on the beautiful figures of Malpighi (Figs. 19 and 20) while re-reading this account by

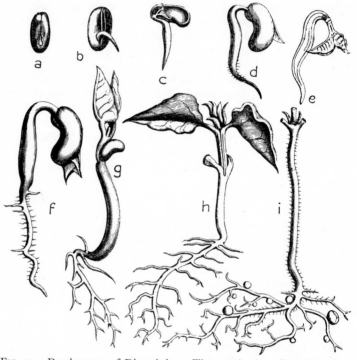

FIG. 19. Development of Dicotyledon. The germination of the bean from Malpighi (1679). In *b* the root is just appearing. *c* shows the seedling in section. *e* is *d* in section with the cotyledons removed. On the rootlets of *i* are seen the *root nodules*, and on its stem the first axillary buds. Root hairs are shown in *d, f,* and *i.* For the nature of the root nodules see p. 384.

Theophrastus of the manner of germination of Monocotyledons and Dicotyledons.

There is one very peculiar aspect of the question of generation of plants to which we shall now call attention. In ancient writings, plants are frequently described as male and female, and one writer, Pliny (A.D. 23–79, pp. 59–60), went so far as to say that some considered that all plants had sex. This sounds very modern, but exam-

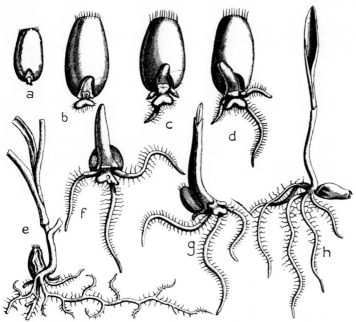

FIG. 20. Development of Monocotyledon. The germination of a grain of wheat from Malpighi (1679). *e* is the last of the series. In *f–h* the monocotyle-donous character is apparent. In *e* branches are appearing. 'The multiple character of the rootlets', on which in cereals Theophrastus lays stress, is apparent throughout from *b* to *f*. Root hairs are seen throughout.

ination of passages in ancient writings that refer to sex in plants shows that the so-called 'males' and 'females' are usually different species. In a few cases a sterile variety is described as the male and a fertile as the female. The real distinction between male and female parts of plants was not understood till modern times (pp. 508–519).

In the case of date palms, however, in which the male
and female organs are found on different trees, instead
of being on the same tree or combined in the same flower,
the nature of the difference between the two types of

FIG. 21.   Assyrian bas-relief from the walls of the palace of Assur-nasir-pal
(885–860 B.C.), discovered at Calah (Nimrûd) by Sir Henry Layard and now
in the British Museum.
Supernatural figures hold in the right hand the male inflorescence of the
date palm.   They shake these inflorescences over a conventionalized representation of the female palm, the form of which can be realized if the ornamentation around it be disregarded.   This ornamentation perhaps itself represents
a series of palm trees.   In that case we may interpret the central structure as
a male tree standing in a grove of female trees.

flower, male and female, was really grasped by Theophrastus and by other ancient writers.

'With dates,' says Theophrastus, 'the males should be brought to the
females, for the male makes them ripen and persist.   When the male
is in flower it is the custom to cut off the spathe with the flower and
shake the dust over the fruit of the female.   If treated thus the fruit
is not shed.'

This is a description of the artificial fertilization of the palm tree which was known to the ancient Babylonians and Egyptians as well as to the Greeks, and is still employed in the East. It was in fact practised centuries before Theophrastus (Fig. 21).

FIG. 22. 'Aristotle's Lantern', the five-rayed mouth-parts of the sea-urchin which, he says, 'looks like a lantern with the horn-panes left out all round'.

II

DECLINE AND FALL OF ANCIENT SCIENCE

§ 1. *Foundation of the Alexandrian School (about* 300 B.C.)

THE Athenian scientific school virtually came to an end in the generation after the death of Alexander in 323 B.C. His Empire broke up (Fig. 23), and Athens ceased to be an important centre. The different parts of the Empire of Alexander were seized by his generals. Egypt fell to Ptolemy. He founded a dynasty which reigned for nearly three hundred years. The last of the Ptolemaic sovereigns was Queen Cleopatra, who died in the year 30 B.C. These Ptolemaic sovereigns took much interest in science. The first established the learned tradition. The second instituted a library and museum at Alexandria, a city that had been founded and named after the great pupil of Aristotle. This city now became the centre of the scientific world. Learned men flocked to it and were supported by funds specially provided by the Ptolemaic rulers.

It was at Alexandria that Anatomy first became a recognized discipline. We may distinguish a definite 'Alexandrian period' in the history of science. It covers the last 300 years before the Christian era. Unfortunately the biological works of the Alexandrian school during the Ptolemaic period have perished. We have, however, fragments of the writings of two biologists of the time, Herophilus and Erasistratus.

The first important biological teacher of the school of Alexandria was a Greek, Herophilus (*c.* 280 B.C.). He was the first to dissect the human body openly. He recognized the brain as the centre of the nervous system, and he regarded it as the seat of the intelligence. There

are several parts of the brain which still bear the names
which Herophilus gave them. One part, which he called
the *wine-press*, in Latin *torcular*, is spoken of by anatomists
to this day by that name. It is the meeting-place of
four great veins or sinuses at the back of the head, which
gave rise, as was thought, to a circular or whirling
movement of the blood.

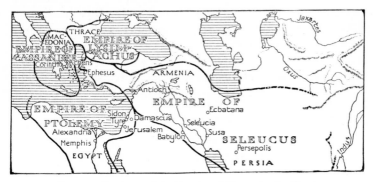

FIG. 23. Break-up of the Empire of Alexander the Great.

Herophilus made the first clear distinction between
arteries and veins. He observed that the arteries pulsate
vigorously while the veins do so little if at all. Herophilus
did not, however, ascribe this movement of the arteries
to the action of the heart, but considered it to be a
natural movement of the vessels themselves.

About contemporary with the anatomist Herophilus
was the physiologist Erasistratus, who was also a Greek
and also taught at Alexandria. The physiology of Erasistratus was based on the idea that every organ is made up
of a three-fold system of vessels—veins, arteries, and
nerves. In those days, and for long after, the nerves were
wrongly regarded as hollow. It was thought that their
cavities convey something material, the hypothetical

nervous fluid, in much the same way as the arteries and veins convey blood.

Erasistratus, like Herophilus, paid particular attention to the brain. He distinguished between the main brain, the so-called *cerebrum*, and the lesser brain, the so-called *cerebellum*. He observed the convolutions in the brain of both men and animals, and associated their greater complexity in man with his higher intelligence. He made a number of experiments on animals. By their means he distinguished between the anterior nerve-roots of the spinal cord, which convey the impulse of motion to the muscles, and the posterior roots, which convey the impressions from the surface of the body, some of which are felt as sensations (Fig. 155 and p. 417).

§ 2. *Beginnings of Scientific Plant Illustration* (50 B.C.)

After the death of Cleopatra, in 30 B.C., Egypt passed completely under the power of Rome, and became a Roman province. The school of Medicine in Alexandria continued for some hundreds of years but steadily lost its vitality. Indeed, with the advent of Roman rule almost all departments of science languished.

Apart from anatomy and physiology, there was only one aspect of biology in which any advance was made under the Roman Empire. This was Botany. With the cessation of scientific curiosity, medical men remained the only class that took any interest in Nature. For the purposes of their work they needed many drugs which they obtained from plants. Now one of the difficulties in dealing with these plants was that of identification of the different kinds. Since there was as yet hardly any special scientific terminology, plants were most easily identified by means of pictures.

The art of botanical drawing began to be practised towards the end of the first century B.C. Later copies of

figures of that period have come down with the name of
the artist, Crateuas, attached to them. This Crateuas
was a herb-gatherer as well as an artist, and his drawings

FIG. 24. 'Round Aristolochia',   FIG. 25. 'Argemone',
    Aristolochia pallida.   *Adonis aestivalis.*
Plants drawn by Crateuas about 50 B.C. and copied about A.D. 500.
Only the copies survive.

are peculiarly interesting as introducing a new art that was
combined with real study of Nature (Figs. 24–5). The
effect of this combination is to be discerned also in the
decorative art of the early Roman Empire, and especially
in those works which were executed under the first and
greatest of the Roman Emperors. In this so-called

'Augustan' art we often see peculiarly close and accurate
studies of animal and plant forms of a kind quite different
from that encountered in the earlier and more esteemed
Greek art.

§ 3.   *Dioscorides and Pliny* (1st century A.D.)

Along with the development of the Roman Empire
(Fig. 28) came greatly improved means of communication
and an increase in the complexity of life. Medical services
became organized, especially in connexion with the army,
and the traffic in drugs became regulated.   The old
difficulty of identification of plants repeatedly cropped up.
Special works were prepared to aid in identification. The
best-known of these was by Dioscorides, an Asiatic Greek
military surgeon in the army of the Emperor Nero (54–
68 A.D.)

The work of Dioscorides exercised great influence on
the ages which followed, and is to be traced into modern
botany.   Many names which he used are still employed
by botanists. His descriptions though always short are
occasionally good and often include the habit and locality
of the plant.   Dioscorides is recognized as one of the
founders of botany.

Despite the descriptions of Dioscorides, there was still
difficulty in identifying some of the plants in his drug
list. At a very early date, therefore, and probably during
his lifetime, copies of the herbal of Dioscorides were
prepared with pictures of the plants described. Thus the
work of Crateuas, begun in the previous century, was
extended (Figs. 26–7).

In the same century as Dioscorides, the Greek naturalist, there lived and worked the Roman naturalist, Pliny,
a man of an entirely different character and outlook.
Pliny was a well-born gentleman, and an able and efficient
civil servant.   He was a man of immense industry with

an enthusiasm for collection. He did not, however, collect natural history objects, but only information or rather misinformation about them. He put together a vast number of extracts from works concerning every aspect of

FIG. 26. *Erodium malachoides.*   FIG. 27. *Geranium molle.*

Figures from the so-called 'Juliana Anicia Manuscript' of Dioscorides written in Byzantium A.D. 512. These figures are from originals probably contemporary with or not much later than Dioscorides himself.

Nature. These he embodied in his famous book on *Natural History.* Unfortunately, Pliny's judgment was in no way comparable to his industry. He was excessively credulous. Thus his work became a repository of tales of wonder, of travellers' and sailors' yarns, and of superstitions of farmers and labourers. As such it is a very important source of information for the customs of antiquity, though as science, judged by the standards of his great predecessors, such as Aristotle or Theophrastus or Erasistratus, it is often laughable.

Despite the low quality of his material, Pliny's work was widely read during the ages which followed. He was the main source of such little natural history as was studied for a thousand years after his time. Many common superstitions have thus passed into current belief from Pliny.

One idea which comes down to us direct from Pliny is very commonly encountered among ignorant people. It is the belief that every animal, plant and mineral has some *use*; that is to say, was formed for the benefit of man. Not infrequently one hears ignorant people ask 'What is the use of flies?', 'What were stinging-nettles made for?' and so on. Such questions are based upon an obsolete and untenable view of the world. They ignore altogether the fact that flies and stinging-nettles have their own lives, which they live without regard to human beings; that they or their like existed before man appeared upon the earth and that they may continue to exist after he has ceased to be. They ignore the biological point of view, but express admirably the point of view of the gullible and industrious Pliny of nineteen hundred years ago.

§ 4. *Galen* (A.D. 130–200)

After Pliny there was only one important biological investigator in antiquity whose works have come to our time. This was the physician, Galen. He was born at Pergamum in Asia Minor. Pergamum and Alexandria were the two rival seats of learning in later antiquity. At fifteen Galen was attending philosophical lectures. At sixteen he began to study medicine in his native town. Even before he was twenty he appeared as an author, and he remained a very industrious writer throughout his life. The bulk of his works is enormous, and they cover every department of medicine.

When Galen was twenty he went travelling for study, as was the custom in his day. Among the many places that he visited for this purpose was Alexandria, where he learned anatomy at the medical school. When he was twenty-eight he returned to his native city, where for four years he acted as surgeon to the gladiators. In A.D. 161, when the great philosopher-emperor, Marcus Aurelius, had recently come to the throne, Galen migrated to Rome to seek a career, as many of his countrymen were doing. He was soon called to be physician to the emperor himself. At the same time his eminence as an anatomist and physiologist became recognized. Crowds came to his lectures on these subjects, on which he composed several important works.

Galen probably did not dissect the human body— at least not publicly, but frequently dissected animals. Thus he examined the structure of sheep, oxen, pigs, dogs, bears, and many other creatures. He recognized the resemblance between the anatomy of the human body and that of monkeys. Many of his anatomical descriptions which are intended to refer to the human subject are, in fact, taken from an examination of the body of the so-called Barbary ape. This animal is the only ape still existing in a wild state in Europe. It is found now only on the rock of Gibraltar, where a very few still roam at large. In Galen's time it was, however, commoner, and he had little difficulty in getting specimens. The animal is not unlike the little monkeys that are often carried by organ-grinders. Galen's descriptions of the muscular system are almost entirely from these or similar apes. In addition to the muscles, Galen gave good descriptions of the bones and joints, and many of the terms that he employed are still in use by anatomists.

Besides describing the structure of animals, Galen did a good deal of work on the functions of their organs. In

other words, he was a *physiologist.* Thus he investigated
the functions of the spinal cord, and he sought to find out
what were the purposes served by the act of breathing,
and by the movements of the heart. Among the curious
errors that he made in the course of these inquiries, we
may mention his belief that air enters the heart direct

FIG. 28. The Roman Empire at its greatest extent, about A.D. 100. The towns
marked were the main centres of learning at the time.

from the lungs, and that blood passes from one side of
the heart to the other through the septum between the
ventricles (Fig. 58, p. 107).

Galen so impressed the men of his time and of succeeding ages that for centuries his works were regarded
as almost infallible. No detail which he mentioned was
allowed to be altered, and all the errors that he made
were passed on to the men of the ages which followed him.
On the other hand, his works were too long and too
difficult for general study. Thus, whilst his errors were
remembered, much of the excellent work which he did
was forgotten. The anatomy and physiology of the ages

which followed represent a progressive deterioration from the standard of Galen, whose best works were very soon allowed to fall in oblivion.

Some explanation is necessary of the extraordinary respect in which the works of Galen were held during the Middle Ages. The reason is Galen's religion. At the end of the second century of the Christian era, when Galen was at work, the influence of Christianity was making itself generally felt in the Empire. Now Galen, though he remained a pagan, developed on his own account a religion which was not unlike early Christianity or Judaism. He believed in one God and had some distant knowledge of the Bible. He developed the idea that every organ in the human body was created by God in the most perfect possible form, with a view to the end that God had in mind for the use of that organ. This attitude fitted in well with that of the Christianity of his time and of the ages which immediately followed. It explains the immense respect in which the name and works of Galen were held for fifteen hundred years. In the Middle Ages it was even believed that Galen had been converted to Christianity.

§ *5. The Dark Ages* (A.D. 200–1200)

After Galen we encounter no biological activity for many centuries. There are some who expand the views of Galen, but they add nothing to him—except misunderstanding. The credulous Pliny, the most read of all the ancient writers, is constantly copied and his text as constantly corrupted. Ecclesiastical authors, with edification always in view, produce moralized and sometimes illustrated animal stories which exhibit no intelligent observation and are often childish to the verge of imbecility. The least fantastic of these productions are perhaps the so-called 'manuscript herbals'.

Dioscorides was now the favourite herbal author. He wrote in the Greek language understood by every educated man in the earlier centuries of the Roman Empire. As time went on, Greek became neglected in the west of Europe. For learned purposes it was replaced by Latin. Greek herbals were therefore translated. The work was done slavishly and with no improvement on the original. The figures too were copied from hand to hand, getting worse and worse and more and more formal at each stage. Finally they became unrecognizable. Nevertheless this unintelligent process continued until the sixteenth century.

The remarkable thing about all this copying was that neither the artist nor the scribe who wrote the text ever thought of looking for himself at the organisms he was describing. Thus, even very common plants, such as the blackberry (Figs. 29–30), are described and drawn in the most conventional way. This period of entire absence of observation we may rightly describe, so far as science is concerned, as the Dark Ages. The Dark Ages in Science last for at least a thousand years—from the death of Galen till the thirteenth century.

The causes of the decline of ancient science have often been discussed. The beginning of the process goes back very far. Many consider that there was a change for the worse between the writers of remoter antiquity such as Hippocrates, Aristotle, and Theophrastus, and those of the Alexandrian and Roman schools, such as Herophilus, Pliny, and Galen.

What caused this deterioration? It was, perhaps, connected with the fact that, at about the time of the death of Aristotle, the Greek states lost their independence. The habit of seeking knowledge for its own sake—'science' as we now call it—had been the peculiar creation of the inhabitants of certain small Greek states. Of these Attica, with its capital, Athens, was the chief. As these ceased to be independent, their citizens also lost their independence of thought. The scientific habit was transported to foreign soil and deteriorated in the process.

Nevertheless science lived on at Alexandria and at

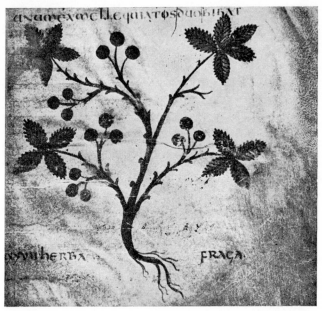

Fig. 29. Strawberry from a South French herbal, *c*. A.D. 550. The 'runners' are represented in excessive numbers. The characteristic arrangement of leaves in threes is exaggerated to fours and fives as in the nearly allied cinquefoil.

Rome. Not until the death of Galen did it cease to exhibit signs of activity. What was the cause of this cessation? It is a question to which various answers have been given.

It has often been said that the rise of Christianity was destructive of science. This view is untenable for two main reasons. On the one hand, science had begun its downward course long before the birth of Christianity. On the other hand, at the end of the second century of the

Christian era, when ancient science was effectively dead, the Christians formed still only a small and obscure sect, that had little or no influence in that class which might be expected to study science. It is true that the intense spiritual interest of Christianity in the early centuries did

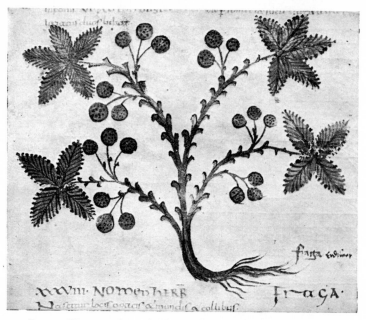

FIG. 30. The same plant as in Fig. 29 after many copyings as it appears in a Rhenish herbal *c.* 1050. The 'runners' have become thorns and the plant confused with a blackberry.

nothing to promote science, and later discouraged any revival of science. It is, however, untrue to say that the rise of Christianity had any definite relation to the fall of science.

A much more efficient factor in the destruction of ancient science was the habitual mental attitude of the ruling class in the Roman Empire. The Romans were essentially a practical people. Their best energies were devoted to ordering and governing their Empire. They produced great military leaders, great lawyers, great engineers. Such practical men seldom appreciate theoretical investigations. Now, whereas Science can be applied to the practical matters of life, yet this application is not in itself Science. Science cannot flourish without purely theoretical investigations undertaken for their own sake.

We may take an example in illustration of this point. The Roman military organization naturally required a medical service. The need for this service was clearly perceived by the military leaders. A service was instituted and was well and carefully organized. Thus far the Roman rulers showed their practical good sense. Yet these leaders never perceived the need of having their medical officers trained on scientific lines. They established no effective medical schools. No effective training in anatomy and physiology was to be found anywhere in the Roman Empire.

Had a great Roman military leader been questioned on this point, he would probably have replied, 'Of course doctors need anatomy, but isn't Galen's anatomy good enough?' But he would be wrong. It is not by reading that science is sustained; it is by contact with the object— by systematic observation and experiment. From these the Roman army doctor was shut off. One of the few Roman leaders who appreciated science was Julius Caesar. It was an ill day for science when he fell.

When we consider this Roman 'practical' attitude to abstract knowledge, it can cause no surprise that other departments of science were in like case to medicine. Even such a study as geography was allowed to languish. The brilliant and steadfast defences of the Empire would have been far more effectively and economically conducted had the Roman leaders had a grip of the geography of Europe. But their geography was of a purely military kind. It took no account of areas thought to be devoid of military importance. If a fraction of the cost of defence had been spent on exploration, the geographical results would have caused a fundamental alteration in the distribution of the Roman forces. But the frontiers remained unscientifically and extravagantly drawn (Fig. 27). They collapsed. The barbarian tribes shattered the barriers and wandered through Europe, wasting as they went and destroying, like children, all the beauty and order that had been stored from the past. The Empire is submerged, the civilization of antiquity is at an end, and the voice of science is silent for a thousand years.

The contempt for and neglect of science by the Roman leaders is significant, because it teaches a lesson which is of value to-day, and of value in every department of science and not least in Biology. The importance to the practical affairs of life of abstract science, studied for its own sake, is often forgotten. The whole mechanism by which our civilization is held together is based on science. Our trains and our motor-cars, our medical service and our food-supplies, depend ultimately on discoveries made without thought for their practical application. Unfortunately, the general direction of scientific research is largely determined by men who are mainly interested in its application. If a time should ever come when the importance of theoretical knowledge and investigation are neglected, our science will go the way of that of Greece and Rome and we should plunge again into an age of darkness.

§ 6. *Thirteenth-century Revival of Learning and Art*

While the knowledge and intelligence of men in Western Europe steadily deteriorated, there was one part of the world where the old traditions lingered. After the fall of ancient science, the most civilized part of the world was the Near East. In Syria, in Asia Minor, in Constantinople, men still read Greek scientific works. Then in the seventh century came the great movement of Islam (Fig. 31). It established Arabic as a literary language and the face of the world was changed.

Intellectual leadership thus passed, by the ninth century, to people of Arabic speech. It remained with them till the thirteenth century. The science of these Arabic-speaking peoples was based on translations of the Greek works into Arabic. These were prepared in large numbers

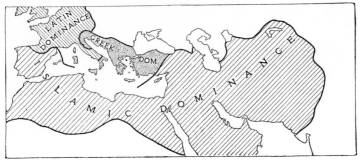

FIG. 31. Distribution of power at the end of the eighth century.

and ardently studied throughout Islamic countries. Islam was a rapidly conquering religion and soon extended not only through the Near East but also over the whole of North Africa, as well as Spain, Portugal, South Italy, Sicily, and many Mediterranean islands. The European peoples came to recognize their intellectual inferiority to these Eastern peoples. They made attempts to secure translations of their scientific works. These were made from the Arabic into Latin and were achieved largely with the help of Jews.

The movement for the translation of Arabic works into Latin began in the eleventh century. Such versions did not, however, become common until the beginning of the thirteenth century.

So far as biology is concerned, the most important Latin translations from the Arabic were the biological works of Aristotle. They were turned into Latin early in the thirteenth century by Michael Scot (d. *c.* 1235), the wizard of whom we read in Scott's *Lay of the Last Minstrel*. Michael had journeyed to Sicily where Arabic was spoken, and had obtained a Jewish assistant. His learning in all

FIG. 32. '*Basilisca*', perhaps the Basil, representing the extreme formal degradation of plant painting. From an Anglo-Normal herbal written about A.D. 1200. Contrast with Fig. 33.

FIG. 33. Naturalistic stone carving of Columbine of about A.D. 1260. From a panel in the south porch of Chartres Cathedral.

kinds of mysterious books, which few but he could read, earned him his reputation as magician. The superstitious people of the time attached a similar magical glamour to many others who had a smattering of Arabic or Hebrew. Men who had thus acquired some knowledge of Arabian science were commonly thought to have had commerce with the Devil.

Besides the biological works of Aristotle, some of the most important works of Galen and of Hippocrates soon became available. Dioscorides had been translated centuries earlier, but manuscripts both of him and of Pliny now began to be more common. The world was at last awakening from its long sleep.

There was another important movement in the thirteenth century that had its effect on Science. This was the great religious and artistic revival of which the wonderful cathedrals in the so-called 'Gothic' style are the best-known memorials. In certain departments this revival certainly had an effect upon biology. If we examine the illustrated manuscripts of the age, we shall find that, for the first time for many centuries, the artists do something more than copy. They are at last sometimes trying to represent things as they really are.

Now men cannot make their pictures like nature unless they study nature. This the illustrators of manuscripts at last began to do. Thus we meet with the interesting fact that the real revival of observation, the basic process of science, was the work not of men habitually occupied in scientific pursuits, but of artists. The process spread. It is to be found not only among the miniaturists, but also among the masons and minor craftsmen who decorated the cathedrals. These men loved to put their own observations into their work, and we often find good representations of animals and especially of plants in the architectural ornaments of cathedrals (Fig. 33, contrast 32).

§ 7. *Scholasticism*

But if this was the effect of the revival of art, literature exercised a less favourable influence. Reading, it is true, became much wider. The Universities, which came into great prominence at this period (Fig. 34), drew attention to the existence of a vast literature, partly sacred but to a great extent also philosophical and largely of Arabic origin. Some medical men were awakened by the trans-

lations of works of Hippocrates, Aristotle, and Galen from
Arabic into Latin. But it was the study of the texts of
these that absorbed their energies. They took almost as
little interest in nature outside their books as did the men
of the Dark Ages.

The system of thought that arose from the study and
comment on these texts is known as 'Scholasticism'. The

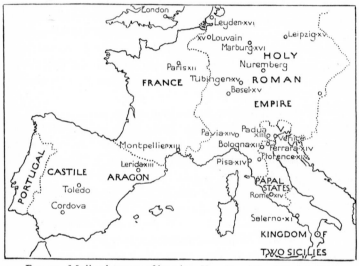

FIG. 34. Medieval centres of learning. V = Republic of Venice. For
universities, the century of foundation is given in Roman figures.

scholastics were characterized by their interest in words as
opposed to things. Very few exhibited any ability as
observers. The literature which they produced is enormous in bulk though low in quality. Yet the Scholastic
Age, even in scientific matters, had advanced greatly on
the Dark Ages. Some few of the ablest thinkers of the
time perceived the importance of going direct to the
original Greek texts instead of to the Arabic from which
their versions were derived. There was a yet smaller

group that also appreciated, if intermittently, the value of direct observation. Prominent among this select band were Robert Grosseteste (1168–1253), Albertus Magnus (1206–80), and Roger Bacon (1214–94). All were university teachers. All, despite appreciation of the value of Greek, had to depend on Latin translations of Arabic versions of Aristotle's biological works.

Grosseteste and Roger were in the relation of master and pupil. Long before there had been any appreciation of systematic observation of Nature, they perceived the value and interest of the process and dimly forecast some of the results to which it might lead. They studied specially optics which was the first science to which attention was directed in the middle ages. Both learned something of the action of lenses and Roger suggested their use as spectacles, a device which came into use in his time. Of the two Roger had the clearer vision of the possibilities of science. He made vigorous but ineffective attemps to draw to it the attention of the Papacy.

Interesting and fascinating as are the careers of Roger and Grosseteste, their scientific work was in departments other than biology. We note that the charge of practising magic, so freely made against Michael Scot, was no less freely made in after years against thememory of Roger, Grosseteste, and even of Albertus Magnus.

§ 8. *Albertus Magnus* (1206–80)

The Dominican Albrecht of Cologne, usually called Albertus Magnus, though less original as a thinker than Grosseteste or Roger, was far more influential on the age in which he lived. He exhibited amazing activity, alike as writer and teacher. He produced, in works of enormous bulk, a vast encyclopaedia of the science of his time based, however, almost entirely on the writings of Aristotle.

In his biological works Albertus follows Aristotle almost word for word, working always on Latin translations from Arabic. He gives a sentence or part of a sentence of Aristotle and then adds a few words in commentary. From our viewpoint it is difficult to imagine a worse method of dealing with a scientific topic. Nevertheless, investigation of the biological works of Albertus shows that they contain a considerable amount of personal observation. There can be no doubt that he was a naturalist of some ability. It is remarkable that so busy a man could observe quietly and accurately amidst his countless preoccupations, for he had onerous teaching and ecclesiastical duties to perform. Considering his circumstances, it is no less remarkable that he sometimes ventures to criticise Aristotle.

His most extensive biological effort, which follows closely the works circulated in his day in the name of Aristotle, is his treatise *On animals*. In it he says that

'between the mode of development of the eggs of birds and of fishes there is this difference: during the development of the fish the second of the two great veins which extend from the heart [as described by Aristotle in birds] does not exist. In fish, the vein which extends to the outer covering in the eggs of birds—wrongly called by some the *navel vein* because it carries the blood to the exterior parts[1]—is absent, but the vein that corresponds to the yolk-vein of birds—which carries to the embryo the nourishment by which its parts increase—is present. In fishes as in birds, channels extend from the heart first to the head and the eyes, and thus the upper parts develop first. As the growth of the young fish increases, the yolk decreases, being incorporated into the members of the young fish. It disappears entirely when development and formation are complete. The beating of the heart is conveyed to the lower part of the belly, carrying pulse and life to the inferior members. 'While the young fish are small and not yet fully developed, they

[1] Before the discovery of the circulation, veins were held to carry blood both *to* and *from* the heart.   (See Fig. 58, p. 107.)

have veins of great length which take the place of the navel-string, but as they grow and develop, these shorten and contract into the body towards the heart, as we have said of birds.' [*Abbreviated.*]

A glance at sketches of embryos of birds and of fishes will make Albert's meaning clear. In the embryos of birds, as of mammals, but not of fishes, there develops a membranous sac, the *allantois* (Greek 'sausage-shaped').

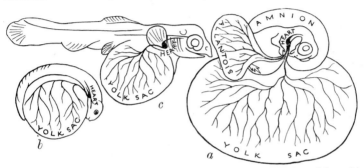

FIG. 35. Diagrams of a developing bird (*a*), and of earlier (*b*) and later (*c*) embryos of a fish, to illustrate the text of Albertus Magnus.

This sac is provided with a special blood-supply separate from the yolk sac (Fig. 35). It is to this special blood-supply that Albert is referring. We may add that the allantois remains free in the embryos of birds, but in those of mammals it takes part in the formation of the placenta. In this respect also the placenta of the placental dogfish of Aristotle differs from the placenta of mammals (Figs. 63–4).

The treatise of Albert *On plants* is one of his most learned works. It is, perhaps, the best work on natural history produced during the Middle Ages. The descriptions of the plants themselves are brief. When they are not first-hand, attention is always directed to the fact. Albert is, however, helpless in his attempt to draw up any general account of plants, since he reaches no satisfactory basis of classification and is equally ignorant both of their minute structure and of their true mode of reproduction.

A single example will illustrate the accuracy with which Albert observed plants. He describes the orange tree, which he had an opportunity of seeing on a visit to Italy. He says that

'the leaf is, as it were, in the form of two leaves, of which the greater is set upon the base of the lesser, and there is a definite mark

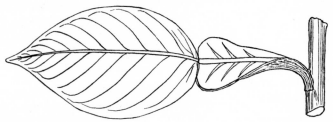

FIG. 36. Leaf of orange tree as described by Albertus Magnus.

at the place of junction of the two leaves. There are woody strands which run through the leaves. And it is characteristic of this tree that the strands, or veins, are so set that in the part near the stem they run towards the base and in the distal part they run towards the apex.'

In the leaf of the orange the petiole is winged, giving the appearance of a second leaf. The articulation between the blade and the winged petiole shows that it is really a compound leaf with a single terminal leaflet. The veins also run as Albert describes. Albert's meaning will be understood on glancing at an orange leaf (Fig. 36).

§ 9. *Medieval Anatomy*

In the great awakening of the thirteenth century the Universities came to exercise a profound effect on social, political, and intellectual conditions. In most of them medical faculties grew up. In some, anatomy came to be studied, at first simply from books. The atmosphere was still utterly scholastic, and the scholastic method devoted itself rather to sharpening wits than to training senses. There was no practical instruction.

FIG. 37. The earliest known representation of dissection (*c.* 1300). The operator is being admonished by a physician and a monk. The body, that of a female, has been opened. Kidneys, heart, lungs, stomach, &c., are strewn around. The operator holds a knife in his left hand and the liver in his right.

With the fourteenth century change set in. This first made itself felt at Bologna. The University there is the most ancient to which the term can be rightly attached. An organized medical faculty existed at Bologna as far back as the middle of the twelfth century. It was the law school, however, that was best known and most frequented by students. The medical faculty at Bologna was at first subject to the school of law. Dissection began there naturally as part of the process of legal investigation. There were cases of murder or of supposed murder in which an examination of the dead body was desirable. Thus human dissection was revived, after being in abeyance for more than a thousand years (Fig. 37).

The first practical anatomist of whom we hear was a teacher at Bologna. He bore the name Mondino (diminutive of Raymond; *c.* 1270–1326). He dissected a number of domestic animals, notably pigs and dogs, and a few human bodies. Although he observed first-hand, it cannot be said that he set out to make discoveries in the manner of a modern biologist. Like Albertus, he was still obsessed by the written word. Mondino had access to several Latin anatomical works translated from the Arabic. For these he had a very great regard, and his investigations were undertaken chiefly to verify them. He advised students to follow his example so that they might remember the better what the great anatomists had said. Yet it was something that he and his followers did really make an attempt to see things with their own eyes. It was, at least, an improvement on the mere book knowledge of the earlier scholastics.

There was, moreover, another department on which Mondino had a deep influence. All students of biology find as a preliminary difficulty the many terms for organs, structures, or processes. It is impossible to define these each time that they occur, and they are introduced as a convenient sort of shorthand. Terms invented from the dead languages, Greek and Latin, save much space. Now Mondino was engaged in introducing anew a study of the structure and functions of the body. He and his followers needed new words, and they took them, not as we do, from Latin and Greek, but mostly from Arabic and Hebrew transliterated and transformed into Latin forms.

III

REBIRTH OF INQUIRY

§ 1. *Naturalism in Art*

WITH the fourteenth century began a period of travel. Information concerning rare and strange creatures came in from overseas. Trade, especially with the East, was increasing, and drugs were brought from foreign countries. Along with commerce came also travellers' tales, both true and false. Those of Marco Polo (1254–1324) are specimens of the true, those bearing the name of Sir John Mandeville (*c.* 1370) of the false. Later, regular expeditions went forth to explore the unknown world. The most famous are the great journeys of Vasco da Gama (*c.* 1460–1524) to the East Indies and of Christopher Columbus (1446–1506) to the West Indies.

Curiosity was being aroused in other matters besides that of the greater and more distant world. Men were looking more keenly at the things immediately around them. In the thirteenth century attempts at naturalistic representations had been made (Fig. 33). These became more frequent and by the fifteenth century were much more successful.

One cause of this 'naturalism' was the 'rediscovery of antiquity'. Scholars were now investigating the great literary masterpieces of Greece and Rome. These, in their turn, aroused curiosity as to the material remains of ancient civilization. Men sought out and examined specimens of Greek and Roman statuary. By studying these masterpieces the great Italian artists learned to represent nature more accurately. The study reacted on the typical art of the period, that of Painting.

Another cause of this 'naturalism' was the invention

of scientific perspective, that is the way of representing perfectly in two dimensions objects which occur in nature in three. The discovery of this method was a true scientific process made about 1450 by the polymath architect Leone Battista Alberti (1404–72) and the learned artist, Piero della Francesca (1420?–92). Though usually discussed in relation to art, it was no whit less important for science and, not least, for the biological sciences. Without perspective the interrelations of the organs cannot be graphically represented nor can the forms of plants and animals and their parts be exactly portrayed. The immediate effects of the discovery were, however, more clearly perceived in the art than in the science of their time.

An examination of the works of the Italian masters of painting of the second half of the fifteenth century shows a very rapid improvement in the capacity for naturalistic interpretation. It affects alike the representation of scenery, the portrayal of the human and of the animal body, and the treatment of minute details of plants and flowers. A glance at the work of some of these great men will at once bring the process before the mind. We may select Botticelli and Leonardo da Vinci.

The charm of the work of Sandro Botticelli (1444–1510) is universally recognized. But his greatness was based on the fact that he was much more than a painter. He was also a man of extraordinary originality of mind and remarkable attainments.

Botticelli was born at Florence. From his earliest years he exhibited a desire to paint. He was apprenticed to Fra Lippo Lippi (1406–69), an exquisite artist whose work was but little touched by the naturalistic spirit. Botticelli's first independent picture was the *Adoration of the Magi* (1467), now in the National Gallery in London. At that time he had not yet come under the full influence

of naturalism. Very soon, however, he developed a peculiarly beautiful style of his own, which has earned him the
affection of all who love painting. With the general
character of his art we are not here concerned. What
matters to us is that he was an admirable observer and a

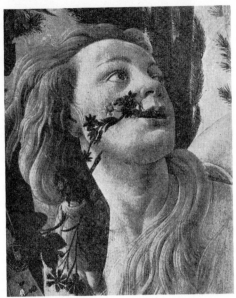

FIG. 38. Head of Flora from Botticelli's *Primavera* (1478).

lover of plant life. Nearly all his later pictures contain
very careful flower studies.

Perhaps the best known of all Botticelli's pictures is the
famous *Spring* at Florence. This entrancing picture,
painted in 1478, is typical of the spirit of the Renaissance.
Venus, with Cupid hanging over her, stands in a grove of
orange and myrtle. She welcomes the approach of Spring,
who enters with Flora and Zephyr. From the mouth
of Flora there flows a garland of flowers most accurately
and beautifully rendered (Fig. 38). The picture is, in fact,

largely a botanical study. It reveals Botticelli as the first painter of plants in modern times. Over thirty species can be detected in this one picture.

The greatest representative of the Renaissance artist-naturalists is Leonardo da Vinci. Not only did he excel in almost every kind of art—Painting, Sculpture, Design —but he excelled as an engineer and inventor, as an anatomist, and as an observer of nature. He occupies a place also in the history of philosophy and of mathematics. There are many other departments in which he added to knowledge. Here we are concerned only with his biological work.

(*a*) As human anatomist. Leonardo as an artist was particularly interested in the human form. Artists need some knowledge of anatomy, but they study for the most part the bones and the superficial muscles. Leonardo's insatiable desire for knowledge led him to much more searching investigations. His studies of human anatomy extended to all parts of the body and were most wonderfully illustrated by his own pencil. They were by far the best work of the kind up to his time, and were at least a century in advance of his age (Figs. 42–3 and 54).

FIG. 39. Drawing by Leonardo of the leaves and flowers of the marsh marigold, *Caltha palustris* (left), and wood anemone, *Anemone nemorosa* (right) (*c.* 1502).

(*b*) As investigator of animal bodies. Leonardo was led
to compare the structure of man with that of animals, the

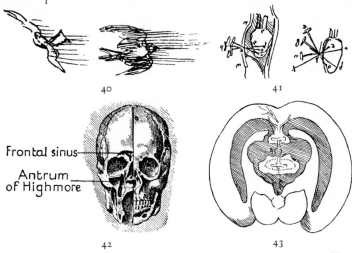

40   41

Frontal sinus

Antrum
of Highmore

42   43

FIG. 40. Drawings from Leonardo's note-book *On the flight of birds*. They
show the resistance of the air to the action of the wings at various angles and
in various positions (1505).

FIG. 41. Diagrams by Leonardo of experiment on heart. A needle through the
chest wall of a pig penetrates the heart and shows its movement externally
(*c.* 1500).

FIG. 42. The human skull. To the right it is shown complete. To the left it is
cut in a plane parallel to the paper. Above the orbit on this side is seen in the
thickness of the bone a large cavity known as the *frontal sinus*. Below the orbit
and in the upper jaw is an even larger cavity, sometimes spoken of as the *antrum
of Highmore*. Both these cavities are recorded by Leonardo for the first time.
Nathaniel Highmore, the 'discoverer' of the structure bearing his name, lived
1613–85. His discovery was 150 years after Leonardo's.

FIG. 43. Outline of human brain with cavities indicated by shading (*c.* 1514).
The outline of both brain and its cavities is more accurate than any drawing made
by previous anatomists. Leonardo suggests that the cavities of the brain may
be injected with wax, the first suggestion of injection with solid matter for the
examination of the form of bodily cavities. The method is now widely practised.

bodies of which he also dissected. He reached many
illuminating conclusions. Thus he brought out accurately
the relation of the bones in the human leg with those of

the horse's hind leg. He showed that the so-called 'hock' really corresponds to the ankle of man, while the 'stifle joint' corresponds to the knee. These facts were not generally recognized till long after the time of Leonardo. Here is but one example among many of his acute observation in comparative anatomy (Fig. 107, p. 225).

(c) As physiologist. Leonardo made many experiments which, had they been widely known, would have enormously advanced the knowledge of the working of the animal body. Thus he gave an account, in some respects singularly accurate, of the movements of the heart (Fig. 41). He described the action of the eye, particularly in relation to lenses. He investigated the mechanics of the various joints. He made striking embryological observations. His investigations of the flight of birds may be said to be the beginning of the scientific study of the mechanics of flight (Fig. 40).

(d) As student of plants. He prepared many drawings of plants which show him to have been a close student of their habits and habitats (Fig. 39). In this department he had only two serious rivals among artists, his older friend, the Italian Boticelli, and his younger contemporary the German Albrecht Dürer (1471–1528).

§ 2. *Humanism*

A very few of the scholastics—Grosseteste and Bacon among them—had attempted to reach the Greek originals behind the Arabic versions of their Latin scientific texts. Their efforts were all too rarely successful. But in later centuries the Near East, where men spoke and read Greek, became more accessible. Greek manuscripts came West in increasing numbers.

In the fifteenth century some of the acutest minds of the age—and indeed of any age—were busily occupied in seeking to retrieve the learning of antiquity. The libraries of old monasteries were ransacked; emissaries were sent to the East; Greeks were tempted to the West by princely payment; all with the hope of gleaning classical remains. The great energy and ability exerted in this task were at last rewarded. By the middle of the fifteenth century nearly all that was to be found of Greek learning had been brought to the West.

It happened, too, that a new invention came soon to the aid of these students of the classics. The art of printing was introduced about the middle of the fifteenth century. Printing helped to establish the Classics in the position in education that they occupied from that time until the nineteenth century. A landmark in our subject is the appearance at Venice in 1476 of the beautiful *editio princeps* of the biological works of Aristotle. These were translated into Latin from Greek by Theodore Gaza (1400–76) of Salonika, who had long resided in Italy.

The men who made a special study of the new classical material were very different from the old scholastics.

The schoolmen had been chiefly interested in theology and in a theological type of philosophy. They attached great importance to reasoning. The so-called 'disputations', or arguments in public, were the main exercises in the medieval university. The scholastics, moreover, gave very little attention to literary form. Medieval writings are generally tedious in presentation and are seldom attractive as literature.

All this was changed by the men of the new school. They studied the noble literary models of antiquity and they saw the beautiful ancient statuary which was being unearthed. They were conscious too of the creative spirit of the art of the age, which drew so largely on the past. Men under such influences felt themselves heirs to a great tradition. They were absorbed, like the great writers of old, in mankind, in humanity. 'I am a man and deem nothing human remote from me' was the dictum of one of their great models, Terence, the Latin poet. 'Consider the sorrows of thy fellows to be thine own', a yet greater, Menander, the Greek playwright, had said. The greatest and wisest of them all, Socrates, four hundred years before the birth of Christ, had lived a life and died a death that were to illustrate his own saying: 'As for me, it is not the Athenians nor even the Greeks that are my brothers, but all Mankind.' And Plato himself had written 'Man is the measure of all things'.

It was because they took to heart these sayings of the great writers of old, because they claimed to care for men as men, that those who studied antiquity became known as, and called themselves, *humanists*. The term *umanista* was invented by one of themselves, the poet Ariosto (1474–1533). Their studies were 'the humanities'. The humanists were informed by a lofty spirit which passed at times into ecstasy, though not seldom into arrogance.

The influence of the classical writings at their best has been summed up by a writer sufficiently removed from antiquity to feel its force as something outside his world, and yet sufficiently near to the Dark Ages to be instinct with a brooding fear that the days of spiritual freedom were gone for ever. Longinus, a philosopher of the third century of the Christian era, when active Greek thought was no more, had truly written that 'from the sublime spirit of the ancients there flow into the minds of those who emulate them, emanations like those from certain holy shrines. These inspire even the most ungifted with the enthusiasm and greatness of others.'

The humanists, moved by the spirit of antiquity, turned to the great scientific works of Greece. With loving care and with superb skill they prepared editions and translations of the classics. By the end of the first half of the sixteenth century Hippocrates, Aristotle, Theophrastus,

Dioscorides, Galen, and scores of others were available in texts which could probably be compared favourably with any accessible to the ancients in the period of classical decline.

But the humanist spirit was far from being all wisdom and love. Intoxicated by the beauty of the classics, the humanists developed a furious enmity against the scholastics. The Arabian texts, on which scholasticism had been nourished, aroused, in some of its enemies, an insensate anger which is very difficult for us now to understand, and impossible for us to share. The texts were 'castigated' and the very language was 'purified' by the humanists from the technical terms which had been derived from the Arabic. Words of Arabic origin were remorselessly hunted down and Greek or Latin terms were substituted.

Thus it has come about that our modern biological vocabulary is almost exclusively Greek or Latin. Nevertheless, an Arabic biological term has survived here and there, usually by accident. Sometimes such words have escaped suppression because they look like and were mistaken for Greek. They were, so to speak, protectively coloured against the attacks of their enemies. Perhaps the commonest of these Greek-looking words that are really Arabic is *nucha* (French *nuque*), a term for the 'back of the neck'. It survives in the anatomical term *ligamentum nuchae*. Many plant names, such as *sesame*, are also of Oriental origin. In the mathematical, chemical, and astronomical vocabulary, a larger number of Arabic terms has survived.

§ 3. *The German Fathers of Botany*

The group of movements which had thus come to flower by the beginning of the sixteenth century placed the student of nature in a peculiarly favourable position. He had now the works of antiquity on which to draw.

The craft of printing was at his disposal. Artists had learnt to represent details of nature effectively. And finally, the wood-cutter had so perfected his craft, that the figures of the artist could be effectively transferred to the printed page. Thus it came about that the first adequately illustrated botanical books were produced.

The new development began in Germany, the home of printing, where the practice had reached a very high standard. Otto Brunfels of Mainz (1489–1534) was the first to produce a work on plants, the figures of which rely wholly on observation (1530). The drawings are firm, sure, and faithful. It is very interesting to compare them with those of a good modern text-book. The text, however, is befogged by an error from which botanists took long to free themselves. Brunfels identifies his plants—gathered in the neighbourhood of Strasbourg—with those of Dioscorides, who worked on the shores of the Eastern Mediterranean.

A younger German botanist was Jerome Bock of Heiderbach (1498–1554), who escaped some of the errors of Brunfels. Bock's figures (1539) are not as good as those of Brunfels. Of interest for us, however, are his careful descriptions of plants and of their mode of occurrence, the first of the kind since Greek times. Only by collating a large number of such descriptions did botanists outgrow the habit of comparing all their plants with those of the ancients.

The most remarkable of the early German botanists was Leonard Fuchs (1501–66). His botanical work (1542), intended as a guide to the collection of medicinal plants, is a landmark in the history of natural knowledge. Fuchs had a good acquaintance with the Greek and Latin classics, and was, withal, an excellent observer, so that his identifications of plants are supported by adequate knowledge. His woodcuts are of extraordinary beauty and truth. They established a tradition of plant illustration which is traceable to the present day. Fuchs enjoys a verdant immortality in the beautiful group of American plants known as 'Fuchsias'.

Fuchs arranged his plants in alphabetical order. He gives us nothing of classification, hardly anything that can be called plant geography, little concerning the essential nature of plants or of their relation to other living things. His book, is in fact, a herbal pure and simple. Yet by close observation of details and by their accurate record on the printed page, it may claim a place among the pioneer works of modern science. He includes in his work an admirable glossary of botanical terms.

The works of Brunfels, Bock, and Fuchs will always be valued for the beauty and exactness of their figures. They had at their disposal the new instruments of perspective, printing, love of nature, the humanist tradition, and the improved art of the woodcutter. But they are almost devoid of new biological ideas and we may doubt if they can be regarded as representing great scientific advance. More important was the work of another German, Valerius Cordus (1515–44). He excelled in detail and exactness of description and he includes not only the medicinally used plants of Germany and Italy but numerous foreign species. He is the real father of scientific plant description. In addition his work contains the very first accounts of 66 'new' plants.

Valerius Cordus spent much of the years 1531–44 in exploring for plants in Italy and Germany, often in the wildest country. He is the first known to have undertaken dangerous journeys for this purpose and he died of an accident on such an expedition. His great work, *Historiae Plantarum* was brought out after his death by Conrad Gesner (pp. 94–6) in 1561. The wood engravings were added by Gesner himself with blocks from his own books or those of others.

§ 4. *The Naturalist Commentators*

The humanistic movement in its spread northward soon had further effects. Scholars, impressed by the new attitude to nature, occupied themselves in identifying the plants and animals mentioned by ancient writers and especially by Aristotle, Dioscorides, and Pliny. This gradually led them farther afield, as strange new animals and plants were coming to Europe from America and the East Indies.

One of the earliest and most typical of the scholar-naturalists was the Englishman, William Turner (*c.* 1510–68). While at Cambridge he published *Libellus de re herbaria* (1538). It treats of classical plant-names. He visited Conrad Gesner (pp. 94–6) at Zurich, and corresponded with him for many years. He travelled about the Continent. At Cologne he published a work on birds (1544) which attempts to determine those named by Aristotle and Pliny, adding notes, from his own observations, on many species. This is the first ornithological book in the modern scientific spirit. Turner was also a zealous botanist and showed sound judgment in botanical matters. His *Herbal* (best edition London 1568) marks the start of botany in England. Another service that he rendered to this country was the introduction of lucerne, which he called 'horned clover'.

In France the humanist movement was particularly active. In the works of Rabelais (1490–1553), some have discerned the new observational spirit. For our purpose, however, it is better displayed by two younger contemporaries, Belon and Rondelet.

Pierre Belon (1517–64) of Le Mans, after taking a medical degree, studied in Germany with Valerius Cordus.

The years 1546–9 he spent in travel in the Near East, during which he kept careful notes on natural history. Later he produced a book on fishes (1551) and another on birds (1555), both well illustrated. These, though based on Aristotle, show much original observation. It is interesting to note the clearness with which he grasps the general principles of comparative anatomy. Thus, for instance, he exhibits the skeleton of a bird placed by the side of a man, and compares them bone for bone (Fig. 44).

Belon was the first to figure the attachment of the

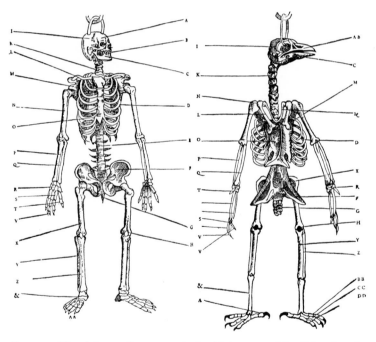

FIG. 44. The skeleton of a man and of a bird compared by Belon (1555). Corresponding letters point to parts which Belon regarded as homologous in the two cases. He is particularly successful in his treatment of the limb bones.

embryo 'cetacean' to its mother (Fig. 45). This established the group as mammals in confirmation of Aristotle (p. 31). The word 'mammal' (Latin *mamma*, an udder)

45

46

FIG. 45. A female killer-whale and her newly-born young. The young whale is linked to the placenta, which is in process of extrusion. From Belon (1551).

FIG. 46. A whale suckling her young and attacked by a 'killer-whale' *Orca gladiator*, the 'Belua truculenta dentibus' of Olaus Magnus. The breasts of the mother, the dorsal fin of the killer, and the blow-holes of all three animals are greatly emphasized. From Gesner after Olaus Magnus (1558).

The whale is much better represented by Belon than by Gesner. But Gesner, living at Zurich, saw nothing of marine animals, while Belon was quite familiar with them.

was introduced by Linnaeus (p. 191) in the eighteenth century and refers to the habit of suckling the young. The mammalian character of the cetaceans was, however, recognized by other sixteenth-century writers besides

Belon. Thus Gesner and several others emphasized it by figuring whales as equipped with a whole series of mammae (Fig. 46).

Belon wrote a systematic work on botany in the form of a short treatise on cone-bearing trees (1553), which is the first monograph of a plant group. He was a good draughtsman and in the course of his travels he sketched many interesting plants, several of which he was the first to figure (Fig. 98, p. 187).

More accurate than Belon as an observer, though less imbued with comparative principles, was Guillaume Ron-

FIG. 47. The placental dogfish of Aristotle (p. 32) with young still attached by navel-string. From Rondelet (1554).

delet of Montpellier (1507–66). Rondelet was a friend of Rabelais, by whom he was infected with the humanist spirit. Rondelet's great work is a painstaking and well-illustrated investigation of the fishes and other marine animals of the Mediterranean (1554). Its motive is largely to verify the views of Aristotle, but Rondelet describes many forms for the first time. He deserves great credit for having discerned the relation of embryo to mother in the placental dogfish (Figs. 47 and 17). Rondelet's illustration of the structure of a sea urchin is the earliest figure we have (1554) of a dissected invertebrate (Fig. 51). It exhibits the complex oral apparatus which has since become known as Aristotle's lantern (Fig. 22).

Another able naturalist of the time was Pietro Andrea

Mattioli of Siena (1500–77). He practised as a court physician and devoted his leisure to translating, annotating and illustrating the text of Dioscorides. He was a most skilful botanist and in constant correspondence with Gesner and Aldrovandi. The first edition of his *Commentaries on Dioscorides* appeared in Italian at Venice in 1544. It was translated into many languages, among them Latin, Bohemian, and French. Artistically somewhat inferior to the work of the German fathers, it is, especially in its later editions, one of the best works of descriptive natural history of the century. By reason of the extensive, accurate, and discerning scholarship that it displays, it is still in current use for the study of Dioscorides. Moreover, especially in the later editions, it contains descriptions of many plants of which Dioscorides had certainly no knowledge. Mattioli had access to the great Juliana Anicia manuscript (Figs. 26–7, p. 59) in the preparation of his work.

§ 5. *The Encyclopedic Naturalists*

Even during the sixteenth century, acute need was felt for some arrangement of the available biological knowledge. Thus arose the work of the 'Encyclopedists'. They made it their business to collect all known facts about living things. Their works are often enormously bulky and usually finely illustrated. Most of their figures are borrowed. Their huge volumes, however, are not only beautiful in themselves, but are interesting to the historian as giving a summary view of the biological knowledge of the day.

Conrad Gesner (1516–65), the great Swiss naturalist, is the best representative of this school. His learning was vast. The central position of Zurich, where he dwelt, enabled him to get news of scientific activities throughout Europe. The five folio volumes of his great *Historiae*

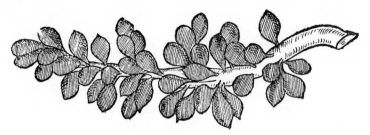

FIG. 48. Eggs of *Sepia* or cuttle-fish presenting the appearance of a bunch of grapes, from Rondelet (1554).

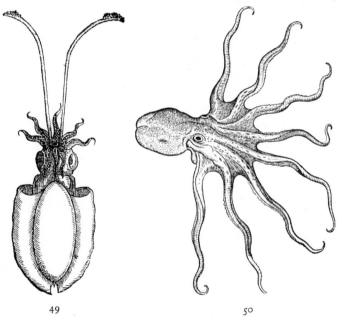

49   50

FIG. 49. The 'Sepia', *Sepia officinalis*, from Rondelet (1554).
FIG. 50. The 'Polypus' (French *poulpe*), *Octopus vulgaris*, from Gesner (1558) from a drawing sent him by Rondelet.

animalium (1551–1621) are really separate works on rather arbitrarily arranged animal groups. A number of the figures are original, and, like the descriptions, give evidence of careful observation. The account of fishes is perhaps the most valuable. It contains figures and descriptions of a number of invertebrates. Some of his figures of the commonest animals have a freshness and vigour seldom equalled in works on natural history (Fig. 52). The work of Gesner is regarded as the starting-point of modern zoology. To his contemporaries he was best known as a patron of learning and as a botanist, but his own most important botanical works were not published till two hundred years after his death (1751–71).

There is a special reason why Gesner should be remembered by nature lovers. It is strange that neither in Antiquity nor in the Middle Ages nor at the Renaissance of Learning was there any real appreciation of mountain scenery. Mountains were regarded with dread or even disgust. Gesner was among the first to voice a feeling for mountains. In a letter to a friend he speaks of the wonders of mountain scenery, and declares his intention of climbing at least one each year, not only to collect plants, but also for air and exercise. He wrote a description of Mount Pilatus which is probably the earliest work on mountaineering.

A writer worthy of mention was Ulissi Aldrovandi (1522–1605). Despite trouble with the Inquisition, he was appointed professor at the papal university of Bologna. The botanical garden founded there (1567) was among the earliest connected with a university, and of it he became the first director. He was a voluminous author. In 1599 he published three tomes on birds, and in 1602 a treatise, finely illustrated, on insects. This last was his best work. Although he exhibits no formal system of classification of animals, yet the arrangement of the figures displays an instinctive grasp of natural affinities. Others of Aldrovandi's monographs appeared after his death. He had designed the whole to form an enormous encyclopedia of living things.

51   52

FIG. 51. The Common Sea Urchin from Rondelet (1554). The animal has been opened by a section in the horizontal plane. The upper half, shown above, exhibits the intestine and the sexual gland. In the lower half can be seen the complex oral apparatus known as *Aristotle's Lantern* (see also Fig. 22). The arrangement on a 5-rayed basis is well seen. This is the earliest figure of a dissected invertebrate.   FIG. 52. Snails from Gesner (1558).

Among the encyclopedic naturalists is to be included the Londoner Thomas Moufet (1553–1604). Travelling extensively in Italy, Spain, France, Germany, Switzerland, and Denmark, he came to take much interest in insects. He kept copious notes of his observations and illustrated them by figures, for he was an accomplished draughtsman. He had considerable literary power, his master-piece being his *Theatre of Insects* (Latin 1634 English 1658). Both as regards text and figures it is the best work of its kind to its date, and it shows a new standard of exactness in the study of invertebrate animals. The manu-script still exists (Fig. 53).

Gesner, Aldrovandi, and Moufet represent an important phase in the history of biology. They were still tied to

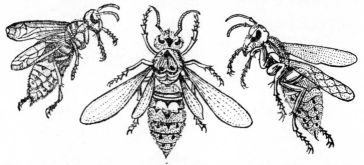

FIG. 53. Drawing of hornets from a manuscript prepared by Thomas Moufet about 1588.

classical traditions. But the biological knowledge of an-tiquity had now been fairly mastered. The significance of classical scholarship for science was on the wane, and the work of the later schools is conducted in a new spirit.

§ 6. *The Revival of Anatomy*

The beginnings of effective plant study have been traced to a fortunate combination of Humanistic Learn-ing, Renaissance Art, and the perfection of the Craft of Printing. The same is true of the study of the animal body. A beginning had been made by Leonardo, but his work remained hidden until our own time. The real father of modern Anatomy was Andreas Vesalius (1514–64).

From his childhood Vesalius accustomed himself to dissecting the bodies of animals. He studied first at

Louvain in his native Flanders and afterwards at Paris.
Now Louvain and Paris universities were both extremely
conservative.   The instruction was medieval in the
former and pinned to the texts of Galen in the latter.
Vesalius was highly successful as student and teacher,
and he became very learned in Galen.   Fortunately for
himself and for the world, Vesalius quarrelled with his
superiors and decided to seek his fortune elsewhere. He
determined on Italy, settled at Padua, and at the end of
1537, when twenty-three, was appointed to lecture there.
Immediately he showed great energy and introduced
sweeping reforms.

It may cause surprise that a Belgian new-comer should
be elected at an Italian University. But at that time there
was only one language of learning—Latin. The lectures
were in Latin. The native language of a lecturer was,
therefore, not of great importance provided that he spoke
and wrote Latin fluently. This essential accomplishment
Vesalius had early acquired.

In the old days of Mondino (p. 78) the professor had
dissected on his own account. The successors of Mondino had abandoned this difficult and tiring process.
They had been content to *read* their lectures from the
ancient texts, while a *demonstrator* (Latin *demonstro*, 'I
point out') indicated the parts to the students. Hence
our modern academic titles *Reader* and *Demonstrator*.
The basic reform of Vesalius was to do away with demon-
strators and other intermediaries between himself and
the object and, using his own words, 'to put his own
hand to the business.'

Such a change he had already sought to make at Paris,
where he was opposed by the conservative elements. At
Padua he succeeded, and was soon lecturing to large
audiences. His energy was irresistible. In five years he
had completed and printed the great work on which his

fame is based, and he was still only twenty-eight when that magnificent volume appeared. He did no further important work. His *De fabrica corporis humani* (Basel 1543) is both the first great modern work on anatomy and a foundation-stone of modern biology. It was repeatedly reprinted, abstracted, copied and pirated. The process long continued. It changed the whole course of anatomy and its influence can be traced right into the twentieth century.

§ 7. *Renaissance Art versus Modern Science*

If a modern man of science turn to the masterpiece of Vesalius, he meets a difficulty at the very outset. A modern biological work is divided into sections, each devoted to some special organ or function or process. But Vesalius does not abstract in quite this way. Though he speaks of the parts separately, he is always thinking of the body as a whole. This is in the background of his mind, even when he is speaking of some minor structure. He is, in fact, dealing with living anatomy, and tries to fit the part he is discussing into a living whole. For this reason his figures are seldom diagrammatic, but usually represent the parts of the body as in a living model (Fig. 55).

This habit of ever picturing the living figure, behind the organ or the structure, places Vesalius in a class quite apart from most modern biologists. His power of vision, in making him a great creative biologist, made him at the same time a great creative artist. To understand Vesalius we must try to rid ourselves of certain ideas which come to us from our education. We must think like Renaissance artists, and not like modern men of science.

The modern biologist's first task, like that of the anatomist of the time of Vesalius, is to describe carefully and accurately, the structures with which he deals. So far, and so far only, the methods of the two resemble each other. In the further application of their knowledge they differ greatly. The modern biologist treats his material comparatively, from the point of view of development and

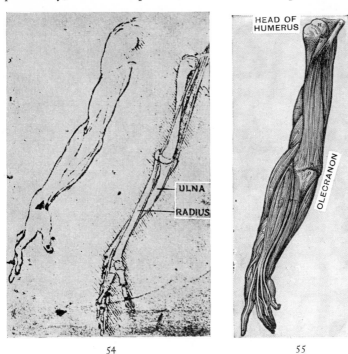

54   55

FIG. 54. Right arm and arm bones from Leonardo (about 1500). The limb is drawn from the front. The artist is portraying the effects of rotation of the *radius* about the *ulna*. The line of action of the muscles is indicated by cords attached to the bones.

FIG. 55. Dissection of left arm from Vesalius (1543). The limb is drawn from the back. The muscles are well portrayed. The bones come to the surface at *H* where the head of the *humerus* is shown and at *O* where the upper end or *olecranon* of the *ulna* is seen. The movement suggested is identical with that in Leonardo's drawings.

origins, and always with the idea of Evolution at the back of his mind. This was to some extent the case with a few Renaissance men of science, e.g. Belon (Fig. 44). But not

H.B.—5*

so with Vesalius and his followers. As the title of his great work implies, for him the body is always a *fabric*, that is a *piece of workmanship*, the artificer being the Great Craftsman, the Almighty Himself.

The Renaissance Latin *fabrica* is not, however, adequately translated by the English 'fabric'. *Fabrica* is not merely a thing wrought, but one which, being wrought, is doing that for which it was wrought. So we speak to-day of 'the workings' of a machine or the 'works' of a clock (compare German *Fabrik* = factory). The parts of it are meaningless without the whole. So these figures of bones or muscles by Vesalius are not placed in the symmetrical and diagrammatic positions as in a modern text-book, but are represented as parts of a living man.

Vesalius, a child of his age, could not help thinking of the end or object to which man is made. He considers the end, the modern biologist thinks of the origin. Though Vesalius had revolted against Galen, he was still looking with Galen's eyes. For Vesalius, as for Galen, Man is a work of art, God an Artist, men's bodies are the Artist's 'studies' for His Great Design.

There is another aspect of the vigorous mind of Vesalius. He was a very learned man. He was well acquainted with all the new-found wealth of antiquity which the humanists were making accessible to the reading public. He was, moreover, a somewhat boastful man and one given to display. Thus, for example, he claimed a knowledge of Arabic and Hebrew that he did not possess. But he was certainly well acquainted with the ancient anatomical texts, and he took an active part in that battle of books, the war of Humanists against Scholastics (pp. 84–7).

We must thus think of Vesalius as trebly equipped for his task. Firstly, by his native genius for dissection, developed at Paris, and stimulated by the freedom of teaching at Padua. Secondly, by the current attitude toward the human body, exalted by contact with Renaissance Art and sublimed by his own superb power of visual imagination. Thirdly, by an excellent education, according to the standards of his time, directed along humanist lines by some of the ablest writers of the day.

§ 8. *Vesalius on the 'Fabric of the Human Body'* (1543)

Since the work of Vesalius is so important for the history of our subject, and since it set a standard for after ages, we must give some account of it. The book opens with a description of the bones and joints, the general classification of which is from Galen. The first bone considered is the skull. It is astonishing to find here an examination in the modern manner of the different shapes of human skulls. Anthropologists attach great importance to these. Skulls are systematically measured and individuals and races are classed as broad-headed, long-headed, and round-headed. This is exactly what Vesalius does (Fig. 56). He follows this matter up farther by comparing the skull of man with that of certain animals, notably the dog.

Of all the anatomical subjects of which Vesalius treats, he is most successful with the muscles. His representations of these are actually superior in certain respects to

Fig. 56. From Vesalius (1543) illustrating types of skull. They correspond to the 'round' (left), the 'long' (middle), and the 'short' headed types of modern physical anthropology.

most modern anatomical figures. Vesalius, with an artist's eye, has succeeded in representing these muscles with their normal degree of contraction.[1] In other words, he has represented living figures as if their skins were transparent. This is a much more difficult task, and one involving more real knowledge, than any mere presentation of dead anatomy. Therefore naturalists still return to these figures and have something to learn from them, although they were prepared four hundred years ago (Figs. 55 and 156).

The account by Vesalius of the structure of the heart has a special interest. The workings of the heart and blood system had always been a puzzle. The current solution was that of Galen (Fig. 58).

Galen supposed that the basic principles of life were certain 'spirits' which dwelt in the blood. Blood was formed in the liver and there charged with *natural spirits*. It was distributed thence by the veins in which it ebbed and flowed. One main branch of the venous system was the right ventricle of the heart. For the blood that entered this there were two possible fates. The greater part, having parted with its impurities, which were carried off to the lung was returned to the venous system. But a small part of the venous blood trickled through minute channels in the septum between the ventricles to the left ventricle. There it met air brought from the lung and produced a higher type of spirit, the *vital spirits*. These spirits were distributed by the arteries, some of which went to the brain. There the blood became changed into a third spirit, the *animal spirits*. These were distributed by the nerves which were thought to be hollow. The conception has given rise to our modern term 'full of animal spirits' which means full of nervous energy.

---

[1] In fact most of the drawings are not by Vesalius himself, but there can be no doubt that he supervised them in every detail and determined the poses.

It will be seen that Galen's scheme depends on the supposed existence of pores in the septum between the ventricles.

Vesalius generally follows the physiological views of Galen. He gives a very fair description of the structure of the heart. When, however, he comes to the septum between the ventricles he is puzzled. He tells us that:

'The septum is formed from the very densest substance of the heart. It abounds on both sides with pits. Of these none, so far

FIG. 57. Skeleton prepared by Vesalius. In 1546 Vesalius passed through Basel. He had with him a skeleton on which he was invited to demonstrate. On leaving the town he presented it to the University, in whose charge it remains to this day. It is the oldest biological preparation made with a scientific motive that is now in existence.

as the senses can perceive, penetrate from the right to the left ventricle. We wonder at the art of the Creator which causes blood to pass from right to left ventricle through invisible pores.'

This exclamation shows that he was not quite satisfied with Galen's view. Twelve years after the appearance of the first edition of his great work Vesalius brought out a second edition. He has again examined the pits on the septum. This time he says:

'Although sometimes these pits are conspicuous, yet none, so far as the senses can perceive, passes from right to left ventricle.'

He goes on to say that:

'Not long ago I would not have dared to turn aside even a hair's breadth from Galen. But it seems to me that the septum of the heart is as thick, dense and compact as the rest of the heart. I do not see, therefore, how even the smallest particle can be transferred from the right to the left ventricle through the septum.'

This attitude to Galen makes it evident that we are on the eve of a biological revolution. Men are no longer satisfied with the traditions of the ancients. To grasp the situation the student is advised to give himself the delight of glancing over a reproduction of the series of figures of Vesalius.

§ 9. *Successors of Vesalius*

In his doubts as to the physiological views of Galen Vesalius was not alone. A fellow student with him in Paris had been a Spaniard, Miguel Servetus (1511–53). This Servetus was very fond of religious argument, and his writings were eqally unwelcome to both Catholics and Protestants. Early in his career the Spanish Inquisition made efforts to secure his person. His religious views are not of interest to us, but a passage in his *Restitution of Christianity* (1553) shows that he had abandoned Galen's view of the action of the heart. He was reaching out toward the idea of the blood moving in a circular manner. Servetus rightly considered that the arterial blood is 'produced by the mingling *in the lungs* of the inspired air with the blood which is communicated from the right ventricle'. He goes on to say that

'this communication does not take place, as is generally believed, through the septum of the heart, but by a remarkable device the *blood is driven from the right ventricle through a long passage in the lungs*. It is there rendered lighter in colour, and from the pulmonary artery is poured into the pulmonary vein. There it is mixed with the inspired air, and by expiration cleansed of impurities. At length, completely mingled with the air, it is drawn in by the left ventricle during its expansion, ready to become vital spirit.'

This is the first clear account of the pulmonary circulation. Similar views were expressed (1559) by Realdo Columbo (1516–59) of Padua. Servetus was burned at the stake at Geneva in 1553. The *Restitutio Christianismi* was burned with him. (It is right to add that an Arabic

writer of the thirteenth century had at least conceived
the circulation, but he did not influence later authors.)

Foremost as an anatomist among the contemporaries
and immediate successors of Vesalius was Bartolomeo

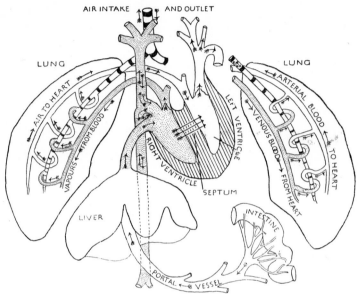

FIG. 58. Diagram of action of heart and blood-vessels according to Galen.
Inset on right is a diagram of the circulation in the lung according to Servetus.

Eustachi (1520–74). He is remembered for the 'Eustachian tubes' in the discovery of which he was, however,
anticipated by some two thousand years by Alcmaeon (p.
5). Among the investigations of Eustachi were his attempt
to study anatomical variations and his examination of the
organ of voice. Perhaps his greatest achievement was his
demonstration of the sympathetic nervous system. He
also investigated the nerve-supply of muscles. Eustachi
corrected many errors of Vesalius, especially on the blood
vessels and nerves.

The works of Eustachi remained for the most part unpublished during his life and were not given to the world until the eighteenth century (1714). Thus they had very little influence on his contemporaries.

In the second half of the sixteenth century, following on the stimulus of the work of Vesalius, there was much biological activity. Anatomy, Physiology, Botany, Zoo-

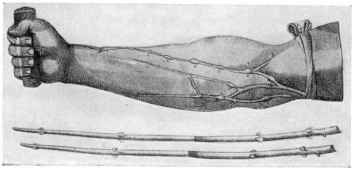

FIG. 59. From Fabricius ab Aquapendente (1603) to show valves in veins. The 'knots' (lettered *O O O O*) indicate the valves in the arm, which is bandaged above the elbow to bring them into greater prominence. Below are two veins that have been removed and turned inside out. On these the valves project.

logy, all made considerable progress. These subjects were taught especially in the Universities of North Italy. Nowhere were they prosecuted with such energy and ability as at Padua, the old school of Vesalius.

Directly in the tradition of Vesalius was his pupil Gabriele Fallopio (1523–62), who succeeded him at Padua. Fallopio made some important investigations of the nervous system and of the generative organs, both of which have parts to which anatomists still attach his name. He gave its scientific title to the 'placenta' or afterbirth and is himself commemorated by the 'Fallopian tubes' which link ovary and uterus.

Of the pupils of Fallopio the two ablest were the Hollander Volcher Coiter (1543–76?) and the Italian Hieronymo Fabrizzi. Coiter studied also with Eustachi and Rondelet (p. 93), and turned later to comparative anatomy. He was the first systematic exponent of this subject since Aristotle, but confined himself to the skeleton. Among his figures we note one of the skeleton of a monkey which he compares with that of man. Coiter also gave an account of the development of the chick, opening eggs day by day to trace the progress of the young animal. His work influenced Fabrizzi. He gave the best account since antiquity of the early development of the chick embryo.

Of all Paduan teachers the most influential, after Vesalius, was Hieronymo Fabrizzi (1537–1619), known as Fabricius of Aquapendente, from his native Tuscan village. He taught at Padua for sixty-four years, and was deeply learned and an admirable observer. Most of his writings have physiological bearing. He was the effective founder of the modern science of Embryology (p. 465). He also did important work in applying the principles of mechanics to muscular movement.

Late in the sixteenth century Fabricius made certain observations which he published in his book *On the Valves of the Veins* (1603). He describes

'thin little membranes distributed at intervals on the inside of the veins. Their mouths are directed towards the root of the veins, that is, towards the heart. In the other direction, that is away from the heart, they are closed.'

We know now that these valves hinder the flow except to the heart. Indeed, Fabricius shows this by an actual experiment (Fig. 59). He demonstrates that if an arm be lightly bandaged so as to compress the veins and thus prevent the flow of blood, the veins swell up. This is because blood can get into the veins from the arteries through the capillaries but cannot move farther on toward the heart, being stopped by the pressure of the bandage. The valves show up as swellings in the course of the veins. Fabricius draws attention to this. Yet his explanation is merely that the valves to a certain extent delay the blood and so prevent the whole of it flowing to the feet or the hands and collecting there.

He thus failed to recognize the true function of these valves. He was thinking on the old Galenic lines of the ebb and flow of blood in the veins. For him veins convey blood back and forth, carrying nutrient blood from the liver to the tissues, and bringing the exhausted blood from the tissues back to the liver.

§ 10. *Harvey* (1578–1657) *and the Circulation of the Blood*

As the sixteenth century was closing there came to Padua a young Englishman, William Harvey. He had been educated at Gonville and Caius College, Cambridge, which had already a link with Padua. Gonville Hall, as it had previously been called, had been refounded in 1557 by John Caius (1510–73) whose name it now bears. Caius met Vesalius in Padua as far back as 1539. Full of the stimulus of the teaching there, he returned to England and lectured on anatomy in London (1544–64) before he moved to Cambridge.

Harvey reached Padua in 1597 when Fabricius was at the very height of his powers. There still stands the curious theatre lined with carved oak, where the great teacher used to lecture by candle light (Fig. 100, p. 206).

In 1602 Harvey returned to England. In 1615 he was called to lecture on anatomy at the Royal College of Physicians in London. The manuscript notes of his first course show that he had already mastered the idea of the circular movement of the blood.

In 1618 Harvey was appointed physician to King James I, and became very busy. He continued, however, to lecture regularly at the Royal College of Physicians. These lectures embodied the most revolutionary doctrine delivered by the most sober and conservative of teachers. Their full importance was not appreciated for some years.

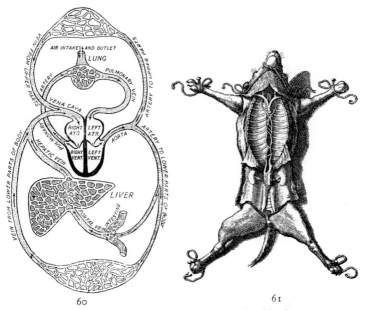

60   61

FIG. 60. Diagram of the circulation of the blood.

FIG. 61. From Pecquet (1651). A dog opened to show the main lymphatic channel or *thoracic duct*. This duct arises in the *receptaculum chyli*, a structure in the abdomen between the kidneys, into which chyle is brought from the intestines by the *lacteals*. From the *receptaculum chyli* the *thoracic duct* passes up through the chest or thorax along the backbone. It terminates by branches which empty into the venous system, usually in the *subclavian veins* (p. 207).

In 1628 there happened an event of primary importance for the progress of science, the publication at Frankfort of Harvey's great work. It is a miserably printed little Latin quarto of 72 pages. In this *Anatomical Dissertation concerning the Motion of the Heart and Blood* he refers to his examination of the heart's action in some forty species, including worms, insects, crustacea, and fish. It would to-day be regarded as a 'comparative' study. Since his time, physiologists have concentrated their attention almost exclusively on the workings of the body of man and of the higher animals which resemble him. During the last few years there has been a tendency to return again to the 'comparative' method on which the science of physiology was started by Harvey.

It is by no means difficult nowadays to understand the circulation of the blood (Fig. 60). If a frog be opened, the heart can be seen beating and the blood being propelled through the arteries and returned through the veins. With a microscope the capillary vessels can be discerned in the web of the frog's foot, and the blood can be watched as it courses through these minute vessels which connect arteries and veins. But Harvey had no microscope and he was brought up with a wrong view of the action of the heart.

He observed that the heart, if grasped, can be felt to harden during its active or contracted period. Thus it may be compared to the biceps muscle of the arm which does the same. He therefore regarded the heart as a hollow muscle.

Next he observed that the contraction of the heart is simultaneous with the expansion of the arteries. The expansion of the arteries can be felt at one's own wrist and is, in fact, the pulse. The determination of the pulse as simultaneous with the contraction of the heart was very important, since the pulse had been thought to be the *active expansion* of the artery. But as the *artery expands as the heart contracts*, it seemed to Harvey that the arteries must be expanding because blood is forced into them by the heart.

Harvey watched the hearts of cold-blooded animals.

These beat slowly, and their action can be the more easily followed. He saw that the atria contract first and that their contraction is followed by that of the ventricles.

Harvey now referred back to his knowledge of the structure of the heart. Blood, he observed, can enter the right atrium through the great vein, the *vena cava*, the opening of which is patent. When the right atrium contracts it drives the blood into the right ventricle, which is still dilated. The opening from right atrium into right ventricle, though guarded by valves, is then quite patent.

Having reached the right ventricle, the blood cannot return whence it came even when the right ventricle contracts, because valves between the atrium and ventricle prevent this. When the right ventricle contracts, the blood must therefore pass into the only exit open to it, namely the *pulmonary artery*. Again, it is prevented from returning to the ventricle by valves at the root of the pulmonary artery. Of these valves Harvey made a special study.

He then turned to examine the left side of the heart. He found there a state of things very similar to that in the right side. Blood can enter the left atrium from the pulmonary vein, and from the left auricle it can pass into the left ventricle. When the left ventricle contracts, the blood within it is stopped from returning into the left auricle by valves which lie between the left atrium and the left ventricle. It is thus forced into the only exit open to it, namely, the great artery or *aorta*. Return from there is also stopped, because the root of the aorta, like that of the pulmonary artery, is guarded by valves (Fig. 60).

Some of these observations had been made by Harvey's predecessors, Galen among them, but none had put them altogether in logical order. At this stage, however, Harvey introduces an entirely new idea. He insists that the flow of blood is not only in one direction, but is *continuously* in one direction. This leads him to his crucial discussion. Consider, he says, the capacity of the heart. Suppose the ventricle holds but two ounces. If the pulse beats seventy-two times in a minute, then in one hour the left ventricle will force into the aorta no less than $72 \times 60 \times 2 = 8,640$ of ounces = 540 pounds, i.e. three times the weight of a heavy man! Where can all this blood come from? Where can it all go to? The point is made with great force. There is no parallel to it in earlier literature.

Harvey reflected that blood sent out by the aorta can come only from the veins. This conclusion was reinforced by a very simple experiment. If an artery is cut, the animal bleeds to death. The bleeding gets slower and slower until finally it ceases as the blood is exhausted and death approaches. The reason must be that the blood that is lost does not reach the veins, and so cannot return to the arteries.

The great discovery now dawned on him.

'I began to think whether there might not be a *movement, as it were, in a circle*. I saw that the blood, forced by the action of the left ventricle into the arteries, was sent out to the body at large. In like manner the blood forced by the action of the right ventricle into the pulmonary artery is sent out to the lungs.'

It was now manifest why in dead animals there is usually so much blood in the veins and so little in the arteries, so much in the right ventricle and so little in the left.

'The true cause is that there is no passage to the arteries save through lungs and heart. When an animal ceases to breathe and the lungs to move, the blood in the pulmonary artery no longer passes therefrom into the pulmonary vein and thence to the left ventricle. But the heart, surviving for a while and contriving to pulsate, the left ventricle and the arteries continue to distribute their blood to the body at large, and to send it into the veins' [where it accumulates but whence it does not return]. [*Slightly abbreviated.*]

In the light of the new idea, Harvey repeated the simple experiments on the arm of a living man from which his teacher Fabricius had drawn his erroneous conclusion (pp. 109–10). By placing the fingers of one or both hands along the veins at different points, he showed that the

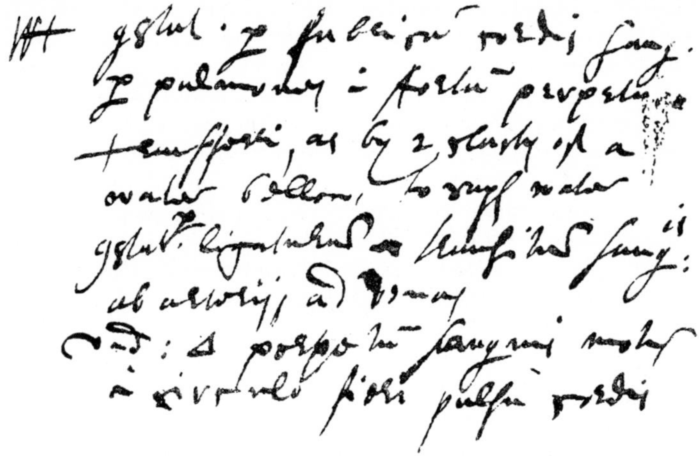

FIG. 62. Lines from autograph lecture notes of William Harvey, written in 1615 and containing the first reference to the circulation. *Transcription:*

> WH constat per fabricam cordis sanguinem
> per pulmones in Aortam perpetuo
> transferri, as by two clacks of a
> water bellows to rayse water
> constat per ligaturam transitum sanguinis
> ab arteriis ad venas
> unde Δ perpetuum sanguinis motum
> in circulo fieri pulso cordis.

Translation:—On account of the structure of the heart William Harvey holds that the blood is constantly passed through the lungs into the aorta, as by two clacks of a water bellows to raise water. Moreover, on account of the action of a bandage [on the vessels of the arm] he holds that there is a transit of blood from the arteries to the veins. It is thus demonstrated that a perpetual motion of the blood in a circle is caused by the heart beat.

Explanation of Terms: A *clack* is a valve, and a *water bellows* is a pump Δ is Harvey's shorthand sign for *it is demonstrated*, and his abbreviated initial WH is used to indicate his own views or observations. The *clacks* referred to are the two auriculo-ventricular—the so-called 'bicuspid'—valves.

flow of blood was always toward the knots and not away from them. Harvey's demonstration was now complete.

'All things, both argument and ocular demonstration, thus confirm that the blood passes through lungs and heart by the force of the ventricles and is driven thence and sent forth to all parts of the body. There it makes its way into the veins and pores of the flesh. It flows by the veins everywhere from the circumference to the centre, from the lesser to the greater veins. By them it is discharged into the vena cava and finally into the right atrium of the heart. The blood that is carried in one direction by the arteries, in the other direction by the veins, in so great a quantity that it cannot possibly be supplied all at once from the food that is taken into the body. It is therefore necessary to conclude that *the blood in animals is impelled in a circle, and is in a state of ceaseless movement.* It must be, moreover, that *this circulation is the act of function of the heart, which performs this act or function through the vessels.*'

§ 11. *Influence of the Discovery of the Circulation of the Blood*

Harvey's discovery was decisive in several directions.

Firstly, a word must be said of Harvey in relation to his predecessors and successors. Galen and those who followed him, including Mondino, Leonardo, Servetus, Vesalius, and Fabricius, knew well that blood went forth from the heart to the body by the great arteries. They knew, too, that there are valves at the roots of these vessels which prevent its return. Did they not then see that blood could not go on being pumped into the body for ever without returning whence it came?

This question has often been asked, and there is a complete answer to it. Galen and the others think and speak of the blood as *irrigating* the body. Now in irrigation water is led, by a conduit, to channels out into the field. Thence the water seeps slowly into the earth. From the earth it is absorbed into the substance of the corn or lost by evaporation. So thought the ancients of the body. The blood that went forth to it was building the tissues and was so used up, parts of it being lost in perspiration and excretion. Not so Harvey. He saw that the blood did not seep in slowly, but rushed through in a mighty torrent. Men had yet to see what a terrifically active thing life is. It took long to learn, and many of Harvey's followers failed to realize it. Thus, Harvey's contemporary, the philosopher Descartes (1596–1650), author of the first text-book of physiology (pp. 126–31), accepted the doctrine of the circulation but still spoke of it in the old terms of slow irrigation.

Next, we note that Harvey was the first to give a really adequate explanation, in physical terms, of any bodily process. From Harvey's time to our own the interpretation of bodily activities in terms of physics and chemistry has been one of the main tasks to which biologists have set themselves. Harvey's work is thus not only the starting-point of the modern science of physiology, but it is also the first milestone on the road to the modern rationalization of biological thought.

Lastly, we note, in those who follow Harvey, a change in attitude towards living processes. Until his day biological discussion had been full of vague terms intended to explain living activities. Such conceptions as 'innate heat', 'animal spirits', 'pneumatic force', often merely darkened counsel with words. Harvey's book rang the death knell of a whole vocabulary of such mysterious phrases. Though they survived for generations, the subsequent history of physiology is largely the history of the replacement of these vague entities by the simpler conceptions introduced into physics by Galileo (1564–1642) and into chemistry by Boyle (1627–91). In this sense Harvey initiates the modern period of biology.

And yet, in a very curious sense, Harvey belongs to the older rather than to the newer dispensation. The biologist of our time thinks of himself as advancing into new and untrodden tracts. This, however, was not how Harvey thought of himself. He was a man of very conservative temper. He was steeped in Galen. Above all, he proclaimed himself an adherent of Aristotle. In enunciating his great discovery Harvey almost disclaims originality and suggests that it is but a return to Aristotle!

Thus, although Harvey is beyond question one of the great initiators of modern biology, we can also treat him as the last of the classical tradition. We shall not meet others who work and think in his conservative spirit.

As we leave Harvey and his quaint quotations of an Aristotle misunderstood, we pass into a world not only of new discoveries but of a new outlook. Harvey stands just on the frontier. His gaze is turned in both directions. As the day breaks he salutes the dawn but ever and anon he casts a longing glance at the monstrous shadows that flee away and give place to the clearer outline of a grander landscape. He is in the new world and yet not of it. His experimental method is a reconstruction rather than a new creation. His stubborn heart is yearning for an impossible reunion with Hippocrates and Galen and Aristotle.

FIG. 63. Diagram of embryonic membranes and circulation in bird or reptile.
FIG. 64. Diagram of embryonic membranes and circulation in mammal.

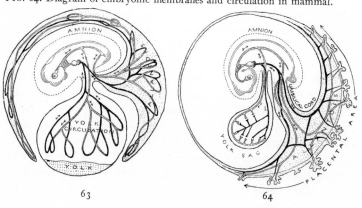

63   64

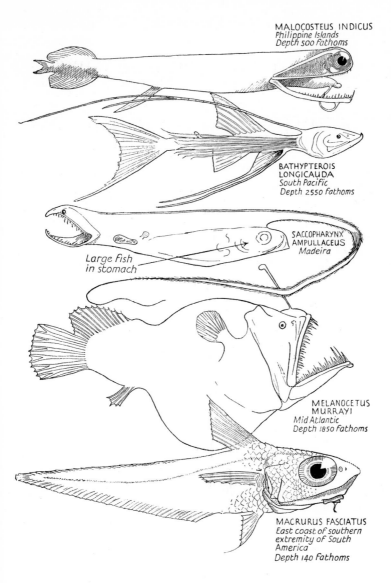

MALOCOSTEUS INDICUS
Philippine Islands
Depth 500 fathoms

BATHYPTEROIS
LONGICAUDA
South Pacific
Depth 2550 fathoms

SACCOPHARYNX
AMPULLACEUS
Madeira

Large fish
in stomach

MELANOCETUS
MURRAYI
Mid Atlantic
Depth 1850 fathoms

MACRURUS FASCIATUS
East coast of southern
extremity of South
America
Depth 140 fathoms

FIG. 65. Some fishes from the deep sea obtained by the *Challenger*.

PART II. THE HISTORICAL FOUNDATIONS OF MODERN BIOLOGY

IV

ON THE INDUCTIVE PHILOSOPHY AND SOME OF ITS INSTRUMENTS

§ 1. *The Change from Medieval to Modern Thought*

THE Revival of Learning, of Art, of Philosophy arose in Italy. It was intimately connected with the rediscovery of the ancient classics and with their more effective study. Of these ancient writings those on the sciences were late in arrival as compared to the literary, historical, and philosophical works and the last to be understood.

The movement known as the Renaissance spread northward from Italy to France, Switzerland, and Germany, and then to England and the Scandinavian countries. It was succeeded by the scientific wave which, as with the humane studies, having crossed the Alps, reached the other continental states before affecting England and the Baltic lands. At the end of the sixteenth century the northern countries had still produced but three men of first-rate scientific importance—the English physicist, William Gilbert (1540–1603), the Danish astronomer, Tycho Brahe (1546–1601), and the Flemish experimenter and mechanician, Simon Stevin (1548–1620). As regards biology there had been as yet no worker of the front rank north of the Alps. The first important exposition of biological science north of the Alps was the work of William Harvey, and he was trained by that true nurse of the sciences, free Padua. There, alone among the southern universities, religious tests were so applied as to be, in effect ,nugatory. Leyden in Holland had a free university from 1575.

It has been convenient to treat Harvey's main work as the last flower of Renaissance biology rather than the first of modern times. The turn of the sixteenth into the seventeenth century wrought, however, great changes in the philosophical outlook of the age—changes in which Harvey's conservative mind shared only imperfectly. A whole series of important thinkers had undermined the medieval system of thought and freed it from the trammels of traditional Aristotelianism. Men were beginning to look to an age when the nature of the physical world would be revealed by the new means of investigation. As is wont to happen at times of intellectual ferment, the new methods were actually in general application before their mode of action was fully understood or explicitly stated. Men of science were employing the inductive philosophy before the philosophers had adequately expounded its nature.

We cannot follow the long line of those who laid the foundations of the new philosophy. Four culminating figures of this great intellectual movement will, however, concern us. These are, in order of seniority, Francis Bacon (1561–1639), Fabri de Peiresc (1580–1637), Marin Mersenne (1588–1648), Pierre Gassendi (1592–1655), and, above all, René Descartes (1595–1650). With them Science passes to its modern stage.

§ 2. *Francis Bacon* (1561–1639)

How the modern world of scientific ideas differs from the old order we can, perhaps, best see by examining the philosophy of Francis Bacon. Was he truly a prophet of science? He erred in his view of the real nature of the scientific process, nor need we exalt the accuracy of his particular vision. But an examination even of his errors will help us, who stand upon his shoulders, to see more clearly things hidden from 'the greatest, wisest, meanest of mankind'.

The juristic eminence of Francis Bacon is universally conceded, the magic of his pen is abundantly evident, the ingenuity of his conceptions is generally allowed. But with all these powers he fails us in the first demand that we make upon a prophet—accuracy of vision. His view of the vast developments of science in the ages that were to follow was far from being close to the phenomena. Moreover, in applying what he believed to be the method of the new science, Bacon showed little skill and had no success. He thus failed in some measure as a scientific philosopher, and to a still greater extent as a practical man of science.

Let us consider Bacon's attitude toward the investigation of Nature. What was this new scientific process which he practised worse than he preached? The answer can now be given in a word. The process of scientific discovery is essentially an act of judgment. In ultimate analysis a scientific discovery is *a work of art*. This Bacon quite failed to envisage.

Bacon was for conducting his investigations by collecting all the facts. This done, he thought, the facts might be passed through a sort of automatic logical mill. The results would then emerge. But this method cannot be applied in practice, since facts, phenomena, are infinite in number. Therefore we must somehow choose from among them, though Bacon thought otherwise.

How then shall we choose our facts? Experience shows that they only choose profitably who have a knowledge of how their predecessors have succeeded or failed in their choosing. In other words, the process of choosing facts is *an act of judgment* on the part of a *learned chooser*, the man of science. So it is also with the process of choosing words on the part of the word-chooser whom we call a poet. The choice of the man of science, as of the poet, is controlled by knowledge of his art—of 'his subject' as we are wont to call it at the Universities or in the laboratories. The man of science, like the poet, exercises his judgment to select those things which bear a certain relation to each other. And yet no skill in reasoning, however deft, no knowledge of the nature of scientific method, however profound, no acquaintance with his science, however complete, will make a man a scientific discoverer. Nor, for that matter, will any learning in the lore of metre or of the nature and history of poetry make a man a poet. Men of science, like poets, can be shaped, but they cannot be made. They must be born with that incommunicable power of judgment.

The scientific man in the prosecution of his art of discovery has to practise three quite different mental processes. These may be distinguished as firstly, the choosing of his facts; secondly, the formation of an hypothesis that links them together; and thirdly, the testing of the truth or falsehood of the hypothesis. When his hypothesis answers numerous and repeated tests, he has made what is usually called a 'scientific discovery'. It is doubtless true that the three processes of choosing facts, drawing a hypothesis or conclusion, and testing the conclusion, are often confused, in his own thinking, by the man of science. Often, too, his demonstration of his discovery, that is the testing of his hypothesis, helps him, more or less unconsciously, to new acts of judgment, these to a new selection of facts, and so on in endless complexity. But essentially the three processes are distinct, and one might be largely developed while the others were in a state of relative arrest.

In this matter scientific articles, and especially scientific text-books—including text-books on the history of science—habitually give a false impression. These scientific works are composed to demonstrate the truth of certain views. In doing so they must needs obscure the process by which the investigator reached those views. That process consists, in effect, of a series of improvised judgments or 'working hypotheses', interspersed with imperfect and merely provisional demonstrations. Many hypotheses and many demonstrations have had to be discarded when submitted to a further process of testing. Thus an article or book, which tells nothing of these side issues, blind alleys, and false starts, is, in some sort, an attempt on the part of the investigator to conceal his tracks. For this reason, among others, science can never be learned from books, but only by contact with phenomena.

The distinction between the *process* of discovery and the *demonstration* of discovery was consistently missed during the Middle Ages. On this point, in which our thought is separated from that of the men of those times, Bacon remained in darkness. He succeeded, indeed, in emphasizing the importance of the operation of collection of facts. He failed to perceive how deeply the act of judgment must be involved in the effective collection of facts.

As an insurance against bias in the collection and error in the consideration of facts, Bacon warned men against his four famous *Idols*, false notions of things, erroneous ways of looking at Nature. There were the *Idols of the Tribe*, fallacies inherent in humankind in general, and notably man's proneness to suppose in nature greater order than is actually there. There were the *Idols of the Cave*, errors inherent in our individual constitution, our private and particular prejudices, as we may term them. There were the *Idols of the Theatre*, errors arising from received systems of thought. There were the *Idols of the Market-place*, errors arising from the influence of mere words over our minds (*Novum Organum*, 1620).

But did not Bacon himself fail to discern a fifth set

of idols? These we may term the *Idols of the Academy.*
Their worship involves the fallacy of supposing that a
blind though learned rule can take the place of judgment.
It was this that prevented Bacon from entering into the
promised land, of which but a Pisgah view was granted
him.

Yet despite Bacon's failure in the practical application
of his method, the world owes to him some conceptions
of high importance for the development of science.

(*a*) He set forth the widening intellectual breach which
separated his day from the Middle Ages. He perceived
the vices of the scholastic method. In the clarity and
vigour with which he denounced these vices, he stands
above those of his contemporaries who were striving
toward a new form of intellectual activity.

(*b*) He perceived, better than any of his day, the extreme
difficulty of ascertaining the facts of nature. He forecasted
the critical discussion that characterizes modern science.
He missed, however, the important point that the
delicate process of observation is so closely interlocked
with discussion that both must almost necessarily be
performed by the same worker.

(*c*) English writers of the later seventeenth century
concur in ascribing to the impetus of Bacon's writing the
foundation of the Royal Society. Thomas Sprat (1635–
1713), Bishop of Rochester, the first historian of the
Society, assures us of this (1677), as do Oldenburg and
Wilkins, its first secretaries. The opinion is fully confirmed
by Robert Boyle (1627–91), the most effective of its
founders, and by John Locke (1632–1704), the greatest
of English philosophers.

(*d*) It is, perhaps, in the department of psychological
speculation that the influence of Bacon has been most
direct. The basic principle of the philosophy of John
Locke is that all our ideas are ultimately the product of

sensation (*Essay concerning Human Understanding*, 1690). This conception is implicit in Bacon's great work, his *Novum Organum*. Through the 'practical' tendency of his philosophy and especially through Locke, Bacon was the father of certain characteristically English schools of thought in psychology and ethics. These have affected deeply the subsequent course of scientific development.

Whatever his scientific failures, we may thus accord to Bacon his own claim that 'he rang the bell which called the wits together'.

§ 3. *René Descartes* (1596–1650)

The role of René Descartes was very different from that of Francis Bacon. The great French thinker was the first in modern times to produce a complete and effective theory of the Universe (*Principia philosophiae*, Amsterdam, 1644). This became widely current. Though this scheme did not hold the field for long, yet the thought of Descartes exercised the most profound influence on the whole subsequent course of science. His general view of the universe covered also the living things therein. To the physical and mathematical sciences Descartes made important contributions that are of biological application. Notably, the so-called *Cartesian co-ordinates*, in constant biological use for graphic methods, were introduced by him (*La Géometrie*, Leyden, 1637) and still bear his name. He exercised direct influence on physiological thought, through his *Traité de l'homme* (Paris, 1664; Latin version, Leyden, 1662. This book, though not published till after its author's death was complete by 1637 and was thus the first important work to accept the circulation of the blood.

Descartes is usually regarded as the founder of modern philosophy. In his fascinating *Discourse on Method* (Leyden, 1637) he expounds the procedure that he proposed to adopt in his own investigations. The substance of this is still recommended to the scientific worker, though often with less succinctness than characterized Descartes. He resolved:

(*a*) 'Never to accept anything as true which I do not clearly know to be such, avoiding precipitancy and prejudice and comprising nothing more in my judgment than is absolutely clear and distinct in my mind.

(*b*) 'To divide each difficulty under examination into as many parts as possible.

(*c*) 'To proceed in my thoughts always from the simplest and easiest to the more complex, assigning in thought a certain order even to those objects which in their nature do not stand in a relation of antecedence and sequence—i.e., to seek relation everywhere.

(*d*) 'To make enumerations so complete and reviews so general that I might be assured that nothing was omitted.'

Descartes believed that truth is ascertainable solely by the application of these principles. His field of ascertainable truth includes that occupied by religion. This point is important for the subsequent history of science. If he be right, revealed religion is superfluous, since the data of religion should be ascertainable by the same process as the data of science. This was widely held, especially in the eighteenth century. Those who took this view came to be known as *Deists*. The school to which they belong has exercised great influence on the development of science.

An aspect of the thought of Descartes that has set its mark upon modern philosophy is his criterion of truth. For him the fundamental test of truth is the clearness with which we can apprehend it. *Cogito ergo sum*, 'I think, therefore I am', is the most clearly apprehended of all truths. Therefore our thinking, at least, can be no illusion. The cogency of our thought may be denied, but that we do think, and that we therefore exist, is sure. In some of his further deductions he applies the same principle less cogently. Thus he holds that the conception of the soul as separate from the body is clear and even obvious. He maintains that it must be accepted as a reality on that account.

The general conception of the nature of the material universe set forth by Descartes places him among the moderns, and separates him from his predecessors and even from Francis Bacon. We cannot discuss his view of the universe, but one of his theses we must consider. Descartes, unlike Bacon and unlike the philosophers of the Middle Ages, regarded the Universe as infinite. He did not hold, as did Bacon, that the earth is at the centre of the solar system. This change of view is clearly of fundamental importance for astronomy and for any general conception of the physical universe. It is, however, less widely recognized that it exerted of necessity an equally revolutionary influence upon other sciences, including biology. The point needs some elucidation.

The old Aristotelian philosophy of the Middle Ages treated the universe as limited by the sphere of the fixed stars. In such a finite world—with earth as centre—science, that is the observation of Nature, must needs occupy a subordinate place. The general scheme of the world being known, the sole aim of science, in such a world, was to fill in the details. This was doubtless a difficult task, but it was terminable. Thus accounts of the universe could, on this view, be made so complete that at last a man could know all that there was to be known about Nature. The undertaking was comparable, let us say, to a statement of the possibilities of the game of chess. The intricacies of that game are very difficult to master, but their complete mastery could be accomplished. If a sufficient number of men of sufficient ability were to apply themselves systematically to chess problems for a sufficient time, the subject could be explored to the very end. The number of possible combinations and positions in chess, though very great, is yet limited and is calculable.

Such was the view of Nature held by the medieval thinker. For this reason scientific effort often took the 'encyclopedic' form during the Middle Ages and Renaissance. Such attempts to compass all knowledge did not then seem by any means foredoomed to failure. The task was hard, but why should it be thought insurmountable if the universe be finite?

Now at the end of the sixteenth and at the beginning of the seventeenth century there came a change in this attitude. Copernicus (1473–1543) had set forth a scheme of the universe, in which the earth did not occupy the central position (*De revolutionibus orbium celestium*, Nuremberg, 1543). Giordano Bruno (1548–1600) had represented the universe as infinite (*De l'infinito universo*, London, 1584). The Englishman William Gilbert (1540–1603) was the first important disciple of Bruno, and probably met him in London. The views of Bruno and Gilbert, cloaked under the name of Copernicus, were supported by Galileo (1564–1642) and by Kepler (1571–1630). The details of these men's work lie outside our discussion, but their conception of the endless extension of the universe and the acceptance of that view by Descartes necessarily affected all the sciences.

· The biological sciences were not in any way exempt. For with an infinite universe as a postulate, does not the adventure of science take on an entirely new character? There can now be no hope of complete comprehension. Is not the field boundless? Science is no longer a sea to be explored and mapped. Seas, even oceans, can be crossed, but for science there is no further shore. The task before the medieval encyclopedist had been regarded as vast but susceptible of ultimate accomplishment. Complete knowledge could be attained by *extension* of the old method. The stress now came to be laid on *intensiveness*. Specialism became possible, and, though its day was not yet, its advent was inevitable. The 'infinite universe' rang the death knell of the encyclopedist. But more, it rang in a new era not for religion and philosophy alone but also for science. Of that era Descartes was a herald.

There is much evidence that we are now living at the very end of the scientific era which Descartes initiated, and that radically new conceptions are dawning for science. It is at least probable that we shall need, in the near future, to revise our views as to the scope and end of science. It is certain that scientific methods are being subjected to more searching analysis than ever before.

For the completion of the system of Descartes, it was necessary for him to include the phenomena presented by living things. He was not wholly unsuccessful in this, though we must, for the moment, defer their consideration (p. 358). But a very important point for us here is that, by their attitude toward science, Descartes and Bacon became the progenitors of the associations known as the 'Scientific Societies' or 'Academies' (p. 136). The essence of a scientific academy may be summed up in two words 'organization' and 'specialization'. The organization of scientific effort was explicit in the system of Bacon, the specialization of scientific effort was implicit in the philosophy of Descartes.

For the extension of natural knowledge, science has come to depend, more and more, upon co-operative effort. *Patrons of learning* played an important part in the earlier period. They were the progenitors of the *Academies*. Very helpful to scientific advance have been *Collections* of living things in botanical and zoological gardens, *Museums* where specimens can be studied and inspected, and scientific *Journals* in which the results of research can be set forth. Apart from these means of communicating and preserving knowledge, we must note that a new attitude toward nature came in with the advent of the *microscope*.

Fig. 66. The University of Padua about 1600. In the foreground are several groups of students. One group is headed by a professor in academic dress. Over the door is written *Gymnasium omnium disciplinarum*, 'The University of all departments of learning', and above this *Sic ingredere ut te ipso quotidie doctior evadas*, 'Enter in such mood that you daily come out wiser'. The front ground floor is occupied by shops. In this little building was given the most important biological teaching of the sixteenth and seventeenth centuries. It still stands almost unaltered.

This is the biological instrument *par excellence*. We shall devote some attention to its introduction (p. 146)

§ 4. *Early Collections of Plants and Animals*

The first step in the process of making science is the systematic collection of facts. In biology this is specially aided by botanical and zoological gardens. The habit of forming these is of great antiquity. We hear of them from Pliny. The monasteries of the Middle Ages often had their herb gardens. The plans of several have survived, together with more than one work on their care. Exotic animals, too, were occasionally brought to medieval courts. A well-known instance was the elephant sent to Charlemagne about 800 A.D. The curiosity and luxury of the

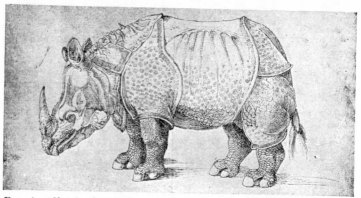

FIG. 67.   Sketch of an Indian rhinoceros by Albrecht Dürer. It was made from a drawing sent by a friend from Lisbon in 1515.

Renaissance were, however, specially conducive to the formation of such collections.

Toward the end of the fifteenth century, collections of live animals were founded by many princes. This was notably the case in Italy. The enlightened Duke Ferrante (1433–94) of Naples owned the most famous of contemporary animal parks. It contained a giraffe and a zebra, two creatures not previously seen in Europe, which were gifts from the Caliph of Bagdad. Many animals were collected at Florence by Lorenzo the Magnificent (1449–92). There was also a collection at Lisbon where, for the first time, a rhinoceros was brought in 1513. The artist, Albrecht Dürer, has left a sketch of it (Fig. 67).

Since the sixteenth century the practice of keeping botanical and zoological gardens has been continuous. The sixteenth century saw at Padua the establishment of the first botanical garden attached to a university (1545). Other Italian universities, notably Pisa (1547) and Bologna (1567), soon did the same. Leyden (1577) and Montpellier (1598) and Oxford (1621) followed. Historically a very important place has been occupied by the *Jardin des Plantes* founded at Paris by Cardinal Richelieu (1585–1642) in 1626. Dried collections of plants—Herbaria—began to be formed by private collectors in the sixteenth century.

The exploration of the world soon began to make itself felt in biological literature. Voyages of exploration, and notably the opening up of America and of the East Indies, resulted in the recognition of many foreign plants and animals. The *Coloquios dos simples e drogas e cousas medicinais da India* ('Discourses on simples, drugs and *materia medica* of India', Goa, 1563) of the Portuguese Jew Garcia del Huerto (1490–1570) was the first book printed in India. It contains the first description of many Indian plants, among them being the coco-nut. The Spaniard Nicolas Monardes (1493–1588) performed somewhat the same service for America. His work (Seville, 1596) was soon Englished under the cheerful title *Joyfull Newes out of the Newe founde Worlde* (London, 1577). It contains descriptions of the armadillo, of tobacco, and of many other plants and animals. The ingenious Venetian traveller, Prospero Alpini (1533–1617), issued his finely illustrated work, *De plantis Aegyptii*, in 1591. It contains numerous good figures and descriptions including the first account of coffee. The most remarkable early faunistic and floristic work was produced as a combined effort by the Academy of the Lynx (p. 136), *Plantarum, animalium, mineralium mexi-*

canorum historia (complete edition, Rome 1651). It gives a fully illustrated account of all the living things known at the time in Mexico.

In the sixteenth century many attempts were made to acclimatize exotic plants. One of the best-known instances is that of tobacco. A pioneer of its culture was Jean Nicot (1530–1600), French ambassador to Portugal, whence he brought seeds which he gave to Catherine de Medici (1519–89). After Nicot the plant has been named *Nicotiana*, whence the alkaloid *nicotine*. Other plants acclimatized in the sixteenth and seventeenth centuries were the Jerusalem artichoke, arrowroot, tomato, maize, pumpkin, and acacia. Very remarkable is the case of the agave which, though now dominant in the scenery of the Mediterranean area, was introduced from Mexico about the middle of the sixteenth century. The introduction of the potato and the turnip were of the highest economic importance. The one provided an almost dangerously cheap carbohydrate food. The other, by making it possible to keep cattle through the winter, has made fresh meat an accessible commodity the year round.

§ 5. *Early Patrons of Science*

In the later sixteenth century it became generally recognized that science was a study whose devotees were somewhat separated from the rest of mankind. They soon began to speak of themselves as the *curiosi rerum naturae*, 'inquirers into Nature's ways', 'the curious', or 'the *virtuosi*'. Their peculiar interests and character called for some means of intercommunication.

This need was but imperfectly supplied by correspondence when posts were still uncertain, slow, and costly. Thus arose a class of eminent and wealthy patrons of science. These *amateurs*—a noble title, too often now debased[1]—sought to bring together their less fortunate colleagues in their houses, or to store their letters for subsequent discussion or distribution.

Such a function was performed in England, in modest fashion, by William Gilbert (1544–1603), physician to Queen Elizabeth. He was the first Englishman to occupy a position of high importance among the 'curiosi rerum naturae'.

A more influential 'amateur' was the wealthy Frenchman, Nicholas Fabri de Peiresc (1580–1637). This most industrious and learned man acted for many years as a general agent for the exchange of scientific knowledge and ideas. He made it his business to know every one with any scientific attainments, whatever his nationality. His enthusiasm was boundless and tireless. He was intimate with Galileo, and bought forty telescopes that he might verify the discoveries of the great Florentine. He followed closely the investigations of Aselli on the lacteals and of Harvey on the circulation. He won the philosopher Pierre Gassendi (1592–1655) to the study of the works of Galileo and Kepler, and thus made a most important recruit for science.

Peiresc left a vast correspondence. It yields a pleasing picture of the lives and activities of men of science in the early part of the seventeenth century. Other wealthy patrons of learning of the time followed a course similar to that of Peiresc.

A very important amateur was the French Minorite friar, Marin Mersenne (1588–1648). He was a physical investigator. For us his chief importance is for the place he occupied as a correspondent and friend of men of science. He was intimate with Descartes and was the

---

[1] The word appears in the higher sense in sixteenth-century French. Thus Rabelais (died 1553) speaks of an 'amateur non seulement des lettres, mais aussi des gens lettrés'.

main agent by whom Descartes communicated with other learned men. Mersenne was a skilful writer. He translated important scientific treatises of Galileo and others into French, thus acting as a popularizer of science. In his cell near Paris the most learned men in France habitually gathered. His circle was the origin of the first scientific societies in France and England.

§ 6. *The First Scientific Societies*

Somewhat like Peiresc in his outlook was the brilliant and wealthy young Italian, Federigo Cesi (1585–1630), Duke of Aquasparta. His early death robbed science of one of its most forceful advocates. As a youth Cesi drew to himself a number of others dedicated to science. These young men, despite much opposition, formally banded themselves into a scientific society (1609). They named it the *Academy of the Lynx* ('Accademia dei Lincei'). Soon they were joined by Galileo, Peiresc, and others. This first scientific academy took the lynx as emblem because of its supposedly piercing sight. It was thus natural that the improvement in vision given by the microscope should make a special appeal to the 'Lincei'.

In the event, the Academy of the Lynx made the first effective contribution to microscopy (p. 145). The word *microscope* was itself invented by one of the original members, Johannes Faber of Bamberg (1574–1629). The 'Lincei' also prepared the first important monograph on

Description of Fig. 68.

Louis XIV enters, escorted by Colbert. All the faces are portraits. Different members exhibit specimens and apparatus. On the table to the left is a microscope and also the air-pump recently improved by Hooke (1660). In the background are ranged skeletons of a man, an antelope, and a lion, together with chemical apparatus, growing plants, and engineering models. To the right anatomical diagrams, maps and plans are exhibited, and, in the foreground, a telescope, a great concave reflector, a pot of exotic plants, an armillary sphere, and a stuffed civet.

FIG. 68. *See note opposite.* A meeting of the Académie des Sciences in the Royal Library at Versailles in 1671. From the superb *Mémoires pour servir à l'histoire des animaux.*

the Natural History of America. However with the death of Cesi in 1630 the society ceased its activities, though it was revived in a later generation. Unfortunately much of the co-operative work of the Lincei and most of that of Cesi has disappeared.

The group of savants around Marin Mersenne was more successful in perpetuating itself. It grew in power, number, and organization. Without a fixed home, it met at the houses of its more influential members. Foreigners often visited Mersenne and his circle. Among them were men who later became fellows of the English *Royal Society*. Such were the father of vital statistics, Sir William Petty (1623–87), and the first secretary of the Royal Society, Henry Oldenburg (pp. 140–1).

Among the more powerful members of the French group was Melchisedec Thevenot (1620–92, pp. 140, 162), the patron of Swammerdam. It was Thevenot who secured the manuscript of that unfortunate Hollander's *Bible of Nature* (p. 164), and so preserved from loss, perhaps, the ablest series of biological investigations of the seventeenth century. Stensen (pp. 406, 463) experimented in Thevenot's house, through which there passed a constant stream of men of science. Jean Baptiste Colbert (1619–83), the minister of the Grand Monarque, Louis XIV, became the patron of this society (Fig. 69). In 1668 Colbert gained it an official recognition as the *Académie des Sciences*. The early biological work of the Academy, from 1666 onwards, was issued chiefly as anatomical descriptions of various animals, mostly the work of Claude Perrault (1613–88, p. 208). The result was the most sumptuously reproduced of all biological works and was distributed as a royal gift. (*Mémoires pour servir à l'histoire des animaux*, Paris *à l'imprimerie royale*, 1771–6 (Fig. 68).)

The English *Royal Society* began much as the *Académie*

des Sciences. A little later in origin as an informal gathering, it was somewhat earlier in official recognition. The Society began in London about 1645. The members referred to it as the 'Invisible College'. The early gatherings were often held at Gresham College. Sir Thomas Gresham (1519–79) had intended his foundation to be a university for London, and some of its 'professors' were among the earliest members of the Invisible College.

In 1649 many members of the Invisible College left for Oxford where meetings of the same kind soon began to be held. It was in 1660, the year of the Restoration of the British Monarchy, that those assembled at Gresham College constituted themselves into a formal society. It earned the King's approval and was incorporated as the *Royal Society* in 1662.

Before long similar associations were formed in Italy, Germany, and Denmark. Scientific societies increased steadily though slowly in the eighteenth century. In the nineteenth century they began to specialize and their name became legion.

§ 7. *The Advent of Scientific Journals*

One of the most familiar manifestations of science in our own time is the scientific journal. There are now thousands of these in various languages, devoted to various branches and aspects of science.

Early in the second half of the seventeenth century an important patron of learning was the Parisian, Denys de Sallo (1626–69). In the course of his self-imposed task this learned man employed copyists to extract for him what he regarded as the most remarkable passages that he encountered. He was of Colbert's circle (p. 136), and suggested to that Minister of State the publication of such extracts at regular intervals. Thus in 1665 was born the first scientific periodical, the *Journal des Sçavans*.

The journal contained much outside the department of science. From the first however, it undertook to publish the scientific discoveries of the day. It was, in the beginning, rather what would now be called a 'review' than a record of original observations. The original element gradually became more prominent. Very soon the *Journal des Sçavans* was imitated in England, Italy, Germany, Switzerland, and Holland. Before long, moreover, the *Académie des Sciences* issued separate volumes in addition to its journal.

The *Philosophical Transactions of the Royal Society* had a somewhat similar origin to that of the *Journal des Sçavans*. At first it was a private venture of one of the secretaries of the Society, and it has continued publication, with a short break, until this day. Its early history is of considerable interest.

In 1662 when Charles II bestowed its charter, the Royal Society made a very fortunate choice of two secretaries.

One was the Reverend Dr. John Wilkins (1614–72), an old parliamentarian, at whose house in Oxford scientific men had been accustomed to meet after the break up of the 'Invisible College' in London (p. 139). Wilkins, who afterwards became a bishop, occupies an important place as a philosopher of science.

The other secretary was Henry Oldenburg (1615–77), a native of Bremen, then one of the league of Hansa towns. Oldenburg had come to England as diplomatic agent of Bremen. Moving to Oxford, he came in contact with the chemist Robert Boyle (1627–91), and the other virtuosi of that city. Although not himself an original worker, Oldenburg had boundless enthusiasm for the new science. He conducted on behalf of the Society a vast correspondence with foreign men of science. 'Foreign members' were early elected, among them being Malpighi (p. 152) and Leeuwenhoek (p. 166).

In 1665, three months after the appearance of the *Journal des Sçavans*, the energetic Oldenburg began on his own account the monthly *Philosophical Transactions*. At first they consisted largely of reviews. Their character gradually changed until only original contributions were included. Later, the responsibility for publication was accepted by the Society itself.

During the seventeenth century the Royal Society published also a number of independent scientific treatises. Among them were biological monographs by Malpighi, Hooke, and Grew. None would have seen the light in anything like so effective a form had the Society not been ready to issue publications independent of its *Transactions*.

The object of these scientific journals has been well set forth by Oldenburg himself:

'Whereas there is nothing more necessary for the improvement of philosophical matters, than the communicating to such as apply their studies that way, such things as are discovered by others; it is therefore fit to employ the press to gratifie those whose delight in profitable discoveries doth entitle them to the knowledge of what this kingdom, or other parts of the world do afford, as well as of the progress of the studies, labours and attempts of the curious and learned in things of this kind. Such productions being clearly and truly communicated, desires after solid and usefull knowledge may be further entertained, ingenious endeavours and undertakings cherished, and those conversant in such matters encouraged to search out new things, impart their knowledge to one another, and contribute to the grand design of improving natural knowledge.' [*Abbreviated.*]

The Italian *Academia de Cimento* ('Academy of Experiment') proceeded along somewhat different lines from those of the French and English societies, in that its publications were anonymous and were set out as the combined work of the society as a whole. The *Cimento* corresponded with Oldenburg, Thevenot (p. 138), and many other scientists but continued for only ten years. It was the inspiration for a German society, but it was nearly a generation before the German publication became of any scientific importance.

The other scientific journals of the seventeenth and eighteenth centuries were formed on the French and English model. Those intended for wider reading usually imitated the *Journal des Sçavans*; those that restricted their contributions to the severer type of original contribution followed the *Philosophical Transactions*.

The need for specialist journals was not felt until the end of the eighteenth century. Since then the number has become enormous. One of the earliest, in the biological series, was *The Botanical Magazine*, which is still running and has appeared under varying titles since 1777. Among the editors have been W. J. Hooker (1785–1865) and his son, Sir J. D. Hooker (1817–1911), who between them were responsible for it for seventy-six years.

Almost coincident with the appearance of specialist journals was the founding of specialist societies. Of these the *Linnean Society* began to publish its *Transactions* in 1791, under the glamour of the name of the great naturalist. The Geological Society, founded in 1807, began to publish its *Transactions* four years later. Both the Linnean and the Geological Society's *Transactions* have appeared continuously ever since.

In France the earliest important biological journal was the *Annales du Muséum d'Histoire naturelle*. This began in 1802 and has continued with various changes of title to the present day. Cuvier and de Candolle contributed largely to its early numbers.

The German-speaking countries have surpassed all others in the number of their biological journals. Worthy of special commemoration is the *Archiv für die Physiologie*, begun in 1795 and edited for a time by Johannes Müller (p. 393), and the *Zeitschrift für Wissenschaftliche Zoologie*, founded by von Siebold (pp. 323, 341) and Kölliker (p. 310) in 1848. An influential botanical journal, *Flora*, appeared first at Ratisbon in 1818. All these were in existence till the outbreak of the Second World War.

In the course of the nineteenth century the whole of the ever-extending field of science became covered by the formation of Societies and by the publication of their journals. In the twentieth century the needs of those who are specially interested in the History of Science have been met both by associations and by periodical publications.

§ 8. *Early Museums*

The origin of our term 'museum' is of some interest. In its Greek form it is an adjectival noun referring to the Muses. Hence Plato uses the word for 'a temple of the Muses'. Thence it came to be applied to schools for the arts and specially for philosophy. The 'Museum' *par excellence* of antiquity was the great school and library founded at Alexandria in the third century B.C. (p. 54). To it the word without qualification was customarily applied. Among the Latins the word museum was sometimes used for a library. There were no collections in antiquity of the kind that we should now call a museum. The word itself disappeared with the classical civilization, to be revived in the seventeenth century.

There must, however, always have been a tendency to hoard natural curiosities. Even during the Middle Ages skeletons and dried specimens formed part of the traditional stock in trade of the alchemist and the apothecary. In the fifteenth century, with the revival of learning, the collecting of coins and other antique objects came into vogue. Natural curiosities were soon added. The German

scholar, Georg Agricola (1494–1555), father of minera-
logy and one of the founders of geology, put together
such a collection. Vesalius, too, had the elements of an
anatomical museum at his disposal. Among those who
made scientific collections in the sixteenth century were
Gesner (pp. 94–6) and Belon (p. 91–2). Collections of dried
plants—'herbaria'—were made by Cesalpino (p. 176) and
Aldrovandi (p. 96), remains of which are still in existence.
At Copenhagen the anatomist, Ole Worm (1588–1654),
had a regular museum of natural curiosities of which an
excellent descriptive catalogue was printed by his son.

In England the earliest museum of which we have a full
account was put together by the gardeners John Tradescant, father and son (1567?–1637; 1608–62). They are
commemorated in the American plant *Tradescantia*. The
younger printed in 1656 a catalogue as *Museum Trade-
scantium: or a collection of rarieties preserved at South Lam-
beth near London*. This collection passed in 1694 to Elias
Ashmole (1617–92) who installed it at Oxford. Remnants
of this collection have been permitted to survive there.
Among them are a couple of fragments of what the

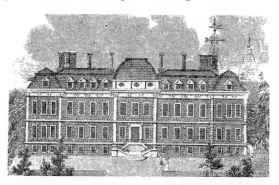

Fig. 69. Montague House, the first home of the British Museum as it appeared
toward the end of the eighteenth century. A house on the site was built in 1676,
Robert Hooke (p. 169) being its architect.

Tradescant catalogue called a 'Dodar from the Island of Mauritius; it is not able to flie being so big'. This was the Dodo, now extinct.

The first scientific museum in England of real educational value was that of the Royal Society. Of this a good catalogue was printed by Grew (p. 157) in 1681. The collection was transferred to the British Museum in 1781. Very few of the specimens in Grew's catalogue can now be traced. A famous collection of the period was that of Sir Hans Sloane (1660–1753), whose vast hoards were acquired by the nation (1759) and formed the nucleus of the British Museum (Fig. 69).

In early days it was very difficult to preserve biological specimens. They had of necessity to be dry, and thus were often deformed and unrecognizable. Three inventions greatly extended museum possibilities. The discovery of alcohol as a preservative, the introduction of flint glass, and the devising of methods of injection, were all contributions of the seventeenth century. The preservation of dried plants presented, from the first, comparatively little difficulty.

The use of alcohol as a preservative was suggested in 1663 by Robert Boyle, who was himself a collector. The idea was soon taken up in other collections, including that of the Royal Society. No method has been so helpful for the preservation of specimens intended for exact scientific investigation.

Hardly less important has been the introduction of flint glass. The older type of glass was unsuitable for specimens intended for exhibition. Until the introduction of suitable glass in the seventeenth century, full advantage could not be taken of alcohol as a preservative. The price of flint glass, however, was one of the serious items of museum expenditure.

On account of the high cost of spirit and of glass, there

was ample room for new methods of preserving dried animal specimens. Of these the most effective were injections of solidifying substances into the vessels. The technique of this process was greatly elaborated in the seventeenth and eighteenth centuries, notably by the three Dutch investigators Jan Swammerdam (1637–80), Regnier de Graaf (1641–73) and Frederick Ruysch (1638–1731).

The development of the biological museum entered on its modern stage in the eighteenth century with the formation of John Hunter's collection (p. 212). From his time, museums have been among the main instruments of biological advance. They have become linked up not only with teaching but also with every form of scientific research.

§ 9. *Introduction of the Microscope*

Until the seventeenth century, naturalists depended upon their unaided senses. We now, however, enter upon a stage in which the senses become refined and sharpened by instrumental aids and especially by the microscope. Lenses have played no small part in ushering in the new inductive philosophy. Robert Hooke says in his great *Micrographia* (1665) that, by the use of such instruments as the microscope, 'the power of considering, comparing, altering, assisting, and improving' the works of nature can be 'so far advanced by the helps of Arts and Experience, as to make some men excel others in their Observations and Deductions, almost as much as they do the Beasts.'

Lenses, consisting of segments of glass spheres, were known in antiquity and were used by the Arabian mathematicians. In Europe lenses were constructed from the thirteenth century onward. By the end of the fifteenth century spectacles with convex and concave lenses were familiar. At the beginning of the seventeenth century a spectacle-maker in Holland happened to put together a convex and a concave lens in a tube. The combination formed what physicists now call a 'Galilean telescope' when looked through from one end, and a 'Galilean microscope' when looked through from the other. A vague rumour of this successful experiment came to Galileo. At once he set about constructing a telescope of his own. In 1610 he published his great work, *Sidereus nuncius* ('The Starry Messenger'). In it he records for the first time the appearance of the mountains in the moon, the rings of Saturn, and the four satellites of Jupiter.

Galileo was thus the effective inventor of the microscope. Its optical properties were worked out first by Johannes Kepler (1571–1630) and by Christiaan Huygens (1629–95). A host of lens-makers in the seventeenth century improved technical details in the construction of the instrument. No one gave a better account of one of these early microscopes than Robert Hooke (Fig. 75). The development of the microscope and of microscopic technique has had incalculable effects on subsequent biological investigation.

Galileo was not himself a biologist. Nevertheless, the earliest biological observation with the microscope was made by him. An Englishman travelling in Italy in 1610 wrote that 'I heard Galileo himself narrate how he distinguished perfectly with his optic glass the organs of motion and of sense in the smaller animals. Especially he observed in a certain insect that each eye is covered by a thick membrane, perforated with holes, thus affording passage to the images of visible things'. Galileo was describing the compound eye of an insect.

The first systematic investigation of living things with the new instrument was made by the 'Academy of the Lynx' (p. 136). A member of it wrote in 1628 that 'our prince (i.e. Federigo Cesi) commissioned an artist to make draughts for him of numerous plants hitherto regarded by botanists as seedless. But the microscope clearly showed them to be teeming with seeds. Such is the wonderful and minutely fine dust adherent to the back of fronds of ferns, and seen as big as peppercorns.'

The dark structures seen on the fronds of ferns are the 'indusia'. The spores that emerge from these are not, however, of the nature of seeds, but give rise to the *prothalli* or *sexual generation*, from which in turn the structures that we call ferns—really the *asexual generation*—are produced. The knowledge of this was deferred for more than two hundred years (pp. 525 ff.). Sporangia of ferns were, however, well figured in the seventeenth century by Swammerdam (Fig. 182, p. 526).

Continuing his narrative, our Lincean author tells that 'with this microscope Francesco Stelluti, the companion of the Lynx, has marvellously set forth the external anatomy of the bee. And he has caused to be engraved the eyes, tongue, antennae, head, legs, digits, and other parts of this little animal.'

Description of Figs. 70–3.

FIGS. 70–1 are engravings of two bees. These originally appeared in a printed sheet prepared by the first 'Academy of the Lynx' (p. 136) in 1625. A single complete copy of this entitled *Melissographia* has survived and is in the Lancisian Library at Rome. The figures were re-engraved and printed more clearly by Francesco Stelluti (1577–1653) in an Italian translation of the Latin poems of Persius published at Rome in 1630. Our figures are taken from this. The magnification of the bees themselves is 5 diameters. See also Fig. 96.

FIG. 72. Between the bees is a drawing of the mouth parts of the insect. It is magnified about 10 diameters and should be compared with the figure of the same structures by Swammerdam some sixty years later (Fig. 88).

FIG. 73 is a representation of a weevil, magnified 10 diameters. This appeared in Stelluti's *Persio* for the first time. The long beak-like rostrum, the form of the antennae, the joints of the legs, and many other minutiae are accurately recorded. A more enlarged detail of the rostrum, magnified about 20 diameters, is also shown. The animal is drawn to its natural size in the top right-hand corner.

In both bee and weevil, the compound eyes have impressed the artist.

FIGS. 70–3. *See note opposite*. The earliest published figures prepared with the aid of the microscope.

We have a good record of this investigation. The figures of the bee and its parts thus set forth by Stelluti are the first illustrations prepared with a microscope that were set forth in a printed book. In 1630 Stelluti wrote that 'I have used the Microscope to examine bees and all their parts. I have also figured separately all members thus discovered by me, to my no less joy than marvel, since they are unknown to Aristotle and to every other

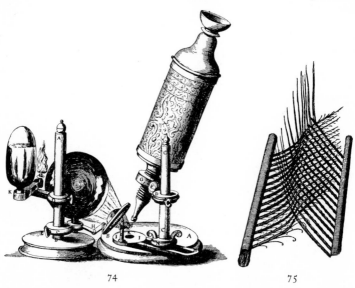

74   75

FIG. 74. Microscope used by Hooke (1665). It had a tube of about six inches long, which was provided with further draw tubes by which it could be lengthened. Focusing was brought about by a screw on the nose of the tube. This moved in a ring attached to the stand. The objects, fixed on a pin attached to the base, were examined by reflected light obtained from a lamp to which a spherical condenser was attached.

Hooke also used minute simple lenses fixed in a hole in a metal plate, much like the microscopes of Leeuwenhoek.

FIG. 75. Portion of a feather, magnified, from Hooke (1665). The barbules and their hooklets are well seen. It is these hooklets that give the feather its peculiar contexture.

naturalist. And I have caused them to be engraved here in such detail as was revealed by the glasses of the Microscope' (Figs. 70–2, 96). These insect drawings stood unrivalled in accuracy of detail until the appearance of the work of the 'classical microscopists.' (See below.) With the collapse of the Academy of the Lynx on the death of its president in 1630 (p. 136) systematic microscopical work fell into abeyance. Such microscopic observations as were recorded between 1630 and 1660 were desultory and of no great consequence.

One writer of this period, however, deserves some commemoration. Henry Power (1623–68) was an intimate friend of Sir Thomas Browne (1605–82). His *Experimental Philosophy* (London, 1663), contains a section 'Microscopical Observations'. The softly glowing music of the *Religio Medici* (1643) has infected it, and Power's English has something of the magic cadence of the great stylist. The *Experimental Philosophy* is among the most beautifully written of English scientific books. It contains many interesting, though hasty, observations. Writing to Sir Thomas Browne in 1649, twelve years before Malpighi's communication of 1661 (p. 153), Power had spoken of 'the minute and capillary channels' between the veins and arteries. He has thus some claim to have preceded Malpighi in the completion of Harvey's discovery of the circulation of the blood (pp. 110 ff.). Power's early death robbed English science of her best spokesman.

In the last forty years of the seventeenth century there arose a series of great investigators who may be described as the *Classical Microscopists*. Of these, two, Robert Hooke and Nehemiah Grew, were English; two, Antony van Leeuwenhoek and Jan Swammerdam, were Dutch; and one, Marcello Malpighi, was Italian. The more important work of all these men, except Swammerdam, was first published in England.

§ 10. *Malpighi* (1628–94)

Marcello Malpighi studied at Bologna, and became professor of Medicine at the University, where most of his life was passed. He began to work with the micro-scope when about thirty and acquired great technical

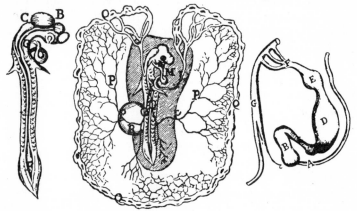

FIG. 76. Malpighi's figures of the developing chick at the end of the second day of incubation (1673). In the centre is the vascular embryonic area. Enlarged about five diameters. From the protruberant and coiled heart, arteries PP carry blood to the heart while the venous system QQ brings it back. To the left is the entire embryo, further enlarged. The coiled heart can be more dis-tinctly seen. The three vesicles that are the rudiments of the brain are prominent with the eye below. The body, narrower than the head, is segmented and the neural groove is open below. To the right is the heart yet further enlarged. The ventricle D gives rise to the *truncus arteriosus* E which branches at F, as in the gill arches of a fish, to unite in forming the dorsal aorta G. The atrium B receives the main venous trunk A.

skill. In 1667 the Royal Society wrote to him suggesting that he should send them his scientific communications. Most of his discoveries appeared in London in volumes issued under the auspices of the Society. He was a fine observer and draughtsman. His Latin style is involved, and rendered more obscure by vague theoretical explan-ations. He was, moreover, neither apt at devising experiments nor skilful in interpreting their results. As an observer, however, he has seldom been excelled. He used a simple microscope.

Malpighi's first important work was an extension of that of Harvey. The Englishman had shown that the blood leaves the heart by the arteries to return by the veins. He had not, however, seen the 'capillary vessels' which connect arteries with veins. In 1660 Malpighi demonstrated this capillary system in the lung of a frog. He hit on the idea of examining the lung by injecting water into the pulmonary artery and seeing it issue from the pulmonary vein. This washed the blood out of the lung and made it more transparent. Thus he could examine it easily under the microscope. When he did so, the network of capillaries presented itself. He followed up the clue and demonstrated capillaries in other parts of the body. He wrote to a friend about this time:

'While the heart is still beating two movements, contrary in direction, are seen in the vessels, so that the circulation of the blood is clearly laid bare. The same may be even more readily recognized in the mesentery. By the impulse of the heart the blood is showered down in minute streams through the arteries. By repeated division these *lose their red colour* until they approach the branches of the veins.'

This loss of colour is a mere optical effect. Under high magnification the pigment of the blood, which is confined to the red blood-corpuscles, becomes less and less obvious as the artery divides, until in the capillaries the red colour is very faint. It is seen to be darker again as the capillaries join together to form small veins.

Malpighi greatly extended the work of Fabricius on the development of the embryo. Fabricius had already investigated the embryonic history of a number of animals, notably the chick. Being unprovided with a microscope, he had seen and depicted only naked-eye characters. He had concentrated on the later stages, paying little attention to the very minute earlier stages. Malpighi now stepped in with his microscope and made many important embryological discoveries (p. 467).

Malpighi observed that in the chick, at an early stage,

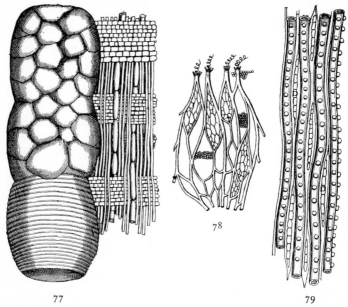

77   78   79

From Malpighi (1675), showing various plant vessels in section. All highly magnified. FIG. 77. Large tyloses and medullary rays from chestnut tree. FIG. 78. Spiral vessels from fig tree. FIG. 79. Pitted vessels from pine tree.

a series of vessels, given off by the aorta as it leaves the heart, encircle the gullet (Fig. 160B). These disappear or become modified in a later stage of the development of the animal. Malpighi was unaware of the nature and meaning of these vessels. Since his time (pp. 477–9) they have played an important part in the progress of biological ideas. They represent, in fact, the vessels of the gills of a fish. Birds, like mammals, are descended from remote fish-like ancestors. In their embryonic development they still bear traces of this, and among the traces are these vessels of the ancient gill-slits.

Among the most striking of Malpighi's researches was that on the anatomy of the silkworm. It was still believed that such minute creatures are devoid of internal organs. Malpighi dissected the silkworm under the microscope with wonderful skill. He was astonished to observe that its structure is no less intricate than that of larger animals. His treatise on the silkworm is the first monograph on an invertebrate.

Insects have no lungs, but breathe by a complicated system of tubes (*tracheae*) which are distributed like our blood-vessels to every part of the body. The tubes open at a series of breathing-holes ranged along the sides of the insect's body. Malpighi was the first to observe these tracheae and 'spiracles', and he arrived at a correct conclusion as to their function. He also demonstrated many other important structures in insect anatomy, and certain of these still bear his name.

The bulkiest of Malpighi's contributions are on the anatomy of plants. His interest in this subject was stimulated by a peculiar observation. Walking one day in a wood, his eye fell on the broken bough of a chestnut tree. He noticed very fine threads projecting from the fractured surface. Taking a lens he found that they presented a spiral appearance (Fig. 78). Struck with their resemblance to the air-tubes of insects, Malpighi wrongly regarded them as subserving the function of breathing (cf. Fig. 89). These vessels impelled him to investigate plant structure. Many of Malpighi's accurate figures of plant anatomy remained unintelligible to botanists until the structures that they portray were redescribed during the last century.

Malpighi's figures show that he attained to a good general notion of the anatomical structure of the stems of the higher plants. They exhibit a clear distinction between the herbaceous and the woody dicotyledonous stem on the one hand, and between the dicotyledons and monocotyledons on the other. We see clearly the scattered bundles of monocotyledons, the annual rings, medullary rays, and rearrangement of fibres at the nodes of the

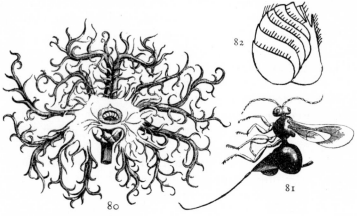

From Malpighi, *On galls*, 1679. FIG. 80. Oak gall cut open. The grub lies within. FIG. 81. Magnified view of gall insect, showing the long ovipositor. FIG. 82. The outline of the abdomen of the insect shows it to be a species of *Rhodites*.

dicotyledons. Wood fibres, resin passages, pitted vessels, and tyloses are all well represented (Figs. 77–9).

Malpighi was familiar with the walls of cells, but in this matter he was preceded by Hooke (pp. 171, 330). He was the first to see and figure the stomata on the undersides of leaves, but he could make nothing of their function. He has good descriptions of the parts of the flower, but was in the dark as to their sexual nature. He also made an interesting research on the relation of the mistletoe to its host (Figs. 139–140, p. 325).

In the study of the development of plants, as with the development of animals, Malpighi was a pioneer. He gives many figures of the embryo sac and endosperm. He has an admirable and classical account of the germination of the bean, laurel, and date-palm. His descriptions of the seedling of the pea and of the wheat are no less good. Very interesting are his admirable distinctions of the different modes of germination of monocotyledons and dicotyledons (Figs. 19–20). It is worthy of remark that on the rootlets of the bean he figures small tubercles. These we now know to be bacterial in origin, and to have a special function in fixing the nitrogen of the atmosphere (pp. 384 ff.). Malpighi, as was inevitable in his day, misinterprets their nature.

Malpighi devoted a special work to galls. Like nearly all of his age, he believed that these extraordinary growths are spontaneously produced. He showed, however, that some galls contain a grub. In some cases he traced the grub back to an egg, and onward to a hymenopterous insect. He described the long and peculiar egg-laying apparatus ('ovipositor') of the gall flies (Cynipidae, Fig. 81) in a way that should enable us to identify the species. The formation of some galls he ascribed to the injection of a fluid by the insect at the time of oviposition. In this he was probably wrong, but it was a useful hypothesis which long held its ground.

§ 11. *Grew* (1641–1712)

Nehemiah Grew was educated at Cambridge and Leyden. He practised as a physician in London, and was one of the early fellows of the Royal Society, of which he became secretary in 1677. Grew was a pious man and he tells us that he was led to the study of vegetable anatomy because both plants and animals 'came at first out of the same Hand and are therefore the contrivances of the same

Wisdom'. Being both parts of God's design, he thinks that plants and animals will present some similarity in structure. He is therefore ever looking for animal analogies in plants. Not infrequently he finds what he seeks.

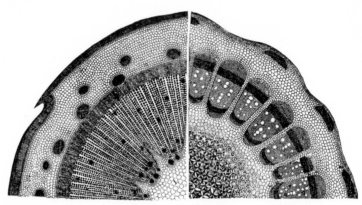

FIG. 83. Section of pine branch magnified (at left).
FIG. 84. Section of thistle stalk magnified. Both from Grew (1682).

FIG. 85. Fibres of wood of fir magnified. From Grew (1682).

'In the woody parts of plants, which are their bones, the principles are so compounded, as to make them flexible without joynts, and also elastick. That so their roots may yield to stones and their trunks to the wind, or other force, with a power of restitution. Whereas the bones of animals being joynted are made inflexible.'

Grew gives admirable and elaborate sections of plant stems and roots, which bring out clearly the difference in structure of the main plant types (Figs. 83–4). Of spiral vessels he learned from Malpighi. He followed them in a variety of plants and called attention to the fact that they never branch. He saw and traced the vessels of seed-lings (Fig. 86).

Grew was well aware that the tissues of plants are porous, sponge-like, or as we should now say 'cellular', in structure. He frequently represents the outlines of the walls of the 'cells', but his view of their nature seems very strange. In his great *Anatomy of Plants* (London, 1682) he wrote:

'The most unfeigned and proper resemblance we can at present make of the whole body of a plant, is to a piece of fine bone lace, when the women are working it upon the cushion; for the pith, insertions and parenchyma of the barque, are all extream fine and perfect lace work; the fibres of the pith running horizontally, as do the threads in a piece of lace; and bounding the several bladders of the pith and barque, as the threds do the several holes of the lace; and making up the insertions without bladders, or with very small ones, as the same threds likewise do the close parts of the lace, which they call the clothwork. And lastly, both the lignous and aer vessels, stand all perpendicular, and so cross to the horizontal fibres of all the said parenchymatous parts; even as in a piece of lace upon the cushion, the pins do to the threds. The pins being also con-ceived to be tubular, and prolonged to any length; and the same lace work to be wrought many thousands of times over and over again, to any thickness or hight, according to the hight of any plant. And this is the true texture of a plant; and the general composure, not only of a branch but of all other parts from the seed to the seed.'

Perhaps the most striking of all Grew's contributions is his guess that flowers are the sexual organs of plants. He distinguishes the calyx, corolla, and stamens—to which he gives the curious name of *attire*—and pistils. He tells that the anthers when they open scatter 'a congeries of many perfect globes or globulets, sometimes of other figures, but always regular'. These 'globes' are the grains of pollen, and Grew figures them for a number of plants.

Grew tells us that 'in discourse hereof with our learned

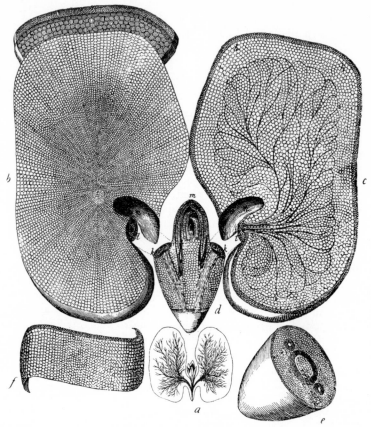

FIG. 86. A bean dissected, from Grew (1682).

a, The entire bean, natural size, opened and with the surfaces of the cotyledons pared away to display their vascular system. This system Grew calls the 'seminal root' because he thought the plant drew nourishment through it as the older plant does through its root. He distinguished the seminal root from the true root or 'radicle', which is also here shown. *b*, Cut surface of cotyledon. Part of the coats adhere to the top. The cotyledon is cut off from the stem at *l*. *c*, Cotyledon with the surface layers dissected off to expose the 'seminal root'. The cotyledon is cut off at *l* as in *b*. *d*, The young remainder of the plant dissected. The 'plume or bud' is at *m*, the root is below. The cotyledons have been cut off at *k*. *e*, Shows the top of the true root cut off and seen in section. *f*, A piece of the surface layer of the cotyledon, displaying the cellular structure.

Sedleian Professor, Sir Thomas Millington, he told me he conceived that the *attire* doth serve as male for the generation of the seed. I immediately replied that I was of the same opinion; and gave him some reasons for it, and answered some objections which might oppose them.'

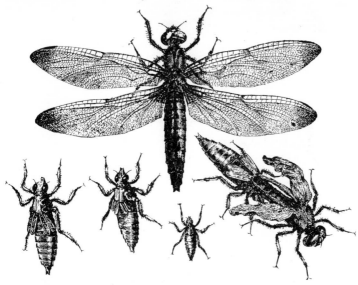

Fig. 87. Series showing the development of a Dragon-fly, from Swammerdam (1685). Below and to the right the complete insect is seen emerging from its last moult with the wings not yet fully unfolded. Swammerdam has beautiful descriptions of the expansion of the wings in several species of insects.

He notices that pollen grains 'are that body which bees gather and carry upon their thighs, and is commonly called their bread. For the wax they carry in little flakes in their chaps, but the bread is a kind of powder, yet somewhat moist, as are the little particles of *attire*.'

§ 12. *Swammerdam* (1637–80)

Jan Swammerdam was the son of an Amsterdam

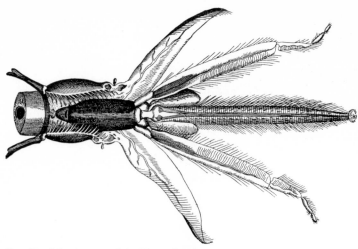

FIG. 88. Mouth parts of the Honey-bee from Swammerdam's *Biblia Naturae*, published in 1737 long after his death. The drawing is excellent and, with some minor changes, would serve for a modern scientific treatise.

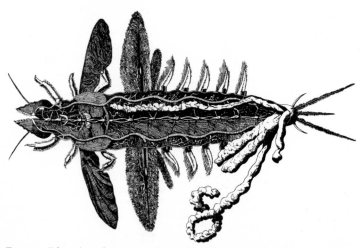

FIG. 89. Dissection of the larva of the May-fly from Swammerdam (1675). It is the best early representation of dissection of an insect. Running in a wavy line along each side is a long tracheal respiratory tube. In the mid line is the double nervous system united at a series of segmentally arranged 'ganglia'.

druggist who had collected a museum of curiosities. From his youth he had a passion for natural history. He helped his father with his museum and did not go to Leyden University till he was twenty-four, when he began to study Medicine. At this time Leyden was the best scientific school in Europe, and was fast surpassing Padua (pp. 108–110).

Swammerdam visited Paris. There he drew the attention of men of science and especially of the 'amateur' Thevenot (p. 138) by his marvellous skill as a dissector. It was realized, even at this time, that he was a man of unstable mind. On his return he published his *Algemeene Verhandeling von bloedloose diertjens* ('General account of bloodless animalculae', Utrecht, 1669, French and Latin versions, 1685). It deals with the modes of transformation of insects, and brings out the different manner of development of the different types of insects. Text and figures are equally good, and the book at once obtained popular recognition. An essay in the same direction had been made a few years previously (1662, Fig. 169) by his fellow-countryman Jan Goedart of Middelburg (1620–68). Swammerdam, however, far surpassed Goedart in the minuteness of his observations and in the intelligence with which they were prosecuted.

Swammerdam came under the influence of a half-mad woman who was a religious fanatic. During the remainder of his life he suffered from fits of depression. While in a state of mental disturbance he produced his *Ephemerae vita* ('Life of the Ephemera', i.e., the May-fly, Amsterdam, 1675). This contains some very remarkable pieces of minute anatomy. Among them is a figure of the dissected larva which shows the series of nodules or ganglia characteristic of the nervous system of insects, and the tracheal system and the musculature as well as the series of breathing processes or gills (Fig. 89). The minuteness of the work can perhaps be realized from the fact that the insect itself is less than a quarter of an inch long.

Swammerdam's mind was now frequently clouded. He died at 43 in a condition of great mental distress, having burnt much of his work. He bequeathed his manuscripts and drawings to Thevenot. On the death of Thevenot the papers passed to his heirs, who sold them. Many years later they were published under the title of *Biblia Naturae* (Leyden, 1737–8) in two magnificent folios with many plates at the expense of Boerhaave (p. 366).

The *Bible of Nature* is the finest collection of microscopical observations ever produced by one worker. It is astonishing that one with so few years to work and who was so often mentally incapacitated, could have accomplished so much. Contemporary accounts describe him as working with the concentration of a madman. The book that is the product of this intense industry is consulted by naturalists to this day. Some of the figures have never been excelled.

Among the most remarkable of the contents of the *Bible of Nature* is the description of the anatomy of the bee. The account of the internal organs of this creature is truly wonderful. Swammerdam's descriptions of the development of gnats and of dragon-flies are astonishing in accuracy and modernness (Figs. 87, 90–2). He traced the development of the frog, and made fine dissections of the tadpole. He figures for the first time the spore cases, now known as *sporangia*, of ferns (Fig. 182). These studies place him in the front rank of biological observers.

Nor is it only as an observer that Swammerdam exhibited skill. He was also a delicate and judicious experimenter, and did something to advance physiological knowledge. His very earliest publication (*Tractatus physico-anatomico-medicus de respiratione usuque pulmonum*, Leyden, 1667) threw some light on the respiratory pro-

cesses. He improved the technique of injection (p. 166).
He demonstrated, also, by a simple but excellently devised
experiment, that muscles, though they alter in *form* do

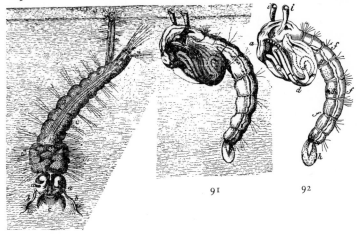

91   92

90

FIG. 90.   Larva of *Culex nemorosus* from Swammerdam.   It hangs from the
surface film by the respiratory syphon, which contains two air tubes.   These pass
into two other tubes which traverse the whole length of the body.   They supply
every part with air.

FIG. 91.   Pupa of *Culex nemorosus* from Swammerdam.   The animal here
breathes from two respiratory trumpets which arise from the thorax.   These
trumpets suspend the creature from the surface film.

FIG. 92.   The same pupa drawn in outline to show internal structure.   *a*, one
of the eyes.   *b*, one of the antennae 'divided into black joints'.   *d d d d*, the legs,
'the hinder ones coiled up in a very surprising manner and lying for the most
part under the wings through which, however, those belonging to one side show
themselves in this figure'.   *e e*, one of the wings.   *f f f*, 'eight rings of the body'.
g g, 'a beautiful edging belonging to the belly'.   *h*, 'the tail, hanging down with
its rowing fins'.   *i i*, respiratory trumpets.

not alter in *size* when they 'contract'. This is evidence
that nothing material passes into the muscle on contraction. The fact is now well recognized, but in Swammerdam's day it ran counter to the current physiological
views.

H.B.—7*

§ 13. *Leeuwenhoek* (1632–1723)

Of all the classical microscopists Antony van Leeuwenhoek made the most impression on his contemporaries. He was born at Delft in 1632, and he lived there until his death in his ninetieth year. He never learned any language but his own Dutch. At the age of sixteen he went to Amsterdam as a shop assistant. He returned soon after to Delft where he owned a draper's shop and spent all his spare time on microscopy. He was entirely without thought of worldly advancement.

Leeuwenhoek always constructed his own microscopes, lenses and all. These were sought after, and he sometimes gave one away, but he never sold them. He did not show his best instruments, however, and in one of his communications he observed that he kept some microscopes absolutely for his own studies, and that 'through these no man living hath looked, save only myself'.

Leeuwenhoek must have had exceptionally acute vision for he worked exclusively with simple lenses. At that time compound microscopes always had what is now called 'chromatic aberration'. A simple lens has no such drawback but, on the other hand, if the power of such a lens is at all high, the area of clear field becomes very small. The general principle of construction of Leeuwenhoek's microscopes is quite simple. The method of using them, on the other hand, was a matter of very great difficulty.

From an early date Leeuwenhoek sent his principal communications to the Royal Society, a practice to which he was introduced by De Graaf (pp. 465, 472). The Society published them in English or in Latin translations from the original Dutch. He went on making observations almost to the end. In one of his later communications he says:

'A certain gentleman entreated me to go on in making observations, adding that the fruit which ripened in autumn was the most lasting. This is now the autumn of my life, I being arrived to the age of 88.'

Leeuwenhoek was a very great observer but certainly not a solver of problems. He allowed himself to wander through Nature with his microscope, finding new wonders at every turn. Many of his observations show great acumen, but it is impossible to do justice to such a desultory mass.

Leeuwenhoek extended the knowledge of the capillary circulation which Malpighi had discovered, describing capillaries in a variety of situations. He followed this by a description of the blood-corpuscles, and he truly observed that the blood-corpuscles of fishes and frogs are oval, and those of man and mammals round. His figure of the nucleus of the red corpuscles of fish is certainly the first of its kind. He also described the blood-corpuscles of some invertebrates. He is the founder of the science of *Histology*, that is, the study of the microscopic structure of tissues, In particular, he gave excellent accounts of the microscopic structure of the muscles (Fig. 93), of the lens of the eye, of the teeth, the skin, and other structures.

Very remarkable is Leeuwenhoek's work on the so-called 'compound eyes' of insects. These, as Galileo had observed (pp. 147–9), differ from ours in that each is provided with a number of lenses. Leeuwenhoek showed that these lenses form numerous inverted images. He believed that compound eyes endow the insect with quickness of sight. As proof he relates how he watched a swallow—the swiftest of birds—chasing a dragon-fly over a pond. The swallow was baffled by the rapid and unexpected turns of the insect. Any one who has tried to catch dragon-flies will have earned a respect for the resource of these creatures.

Certain peculiarities in the life-history of the Aphis insects that infest plants are of great interest to biologists.

To these Leeuwenhoek was the first to draw attention. He opened the bodies of several species of aphides in search of their eggs. To his surprise he found no eggs,

FIG. 93. Branching muscle-fibres of the heart, highly magnified. From Leeuwenhoek.

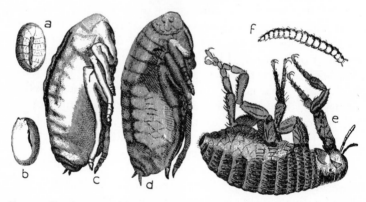

FIG. 94. Development of flea, from Leeuwenhoek (1693), *a*, Egg. *b*, Egg-shell after escape of larva. *c* and *d*, Stage of pupa, *e*, Young complete insect. *f*, Larva.

but young aphides resembling the parents though naturally much smaller. An aphis of a fortnight old might contain as many as sixty young. The birth of these was observed. He found no males and he rightly concluded that the female aphis reproduces *parthenogenetically*, that is without intervention of a male (Greek *parthenos*, 'virgin', *genesis*, 'birth'. See pp. 544 ff.)

Among the best of Leeuwenhoek's observations are those on the development of the ant, on the insect nature of cochineal, on the spinning and poison apparatus of spiders, and on the development of mussels. In his account of the metamorphosis of the flea (Fig. 94), Leeuwenhoek described a mite parasitic upon its larvae. This and similar observations gave rise to the lines of the satirist Jonathan Swift (1667–1745), on which there have since been many variants:

> So naturalists observe, a flea
> Has smaller fleas that on him prey;
> And these have smaller still to bite 'em;
> And so proceed *ad infinitum*.
>
> (*Poetry, a Rhapsody*, 1733)

Very beautiful communications of Leeuwenhoek are the papers on minute freshwater creatures, such as *Rotifers*, *Hydra*, and *Volvox*, objects now well known to every explorer of pond life. He and Hooke were the first to describe specimens of the group of unicellular organisms now known as 'Protozoa'. Probably Leeuwenhoek's most remarkable achievement is that he caught a glimpse of Bacteria (1683). There can be no doubt, both from his figures and descriptions, that he did perform this feat—extraordinary with a simple lens. No one but Leeuwenhoek recorded bacteria until, in the nineteenth century, the improvement of the microscope made it possible for others to see them. And Leeuwenhoek was working with a lens made by himself!

§ 14. *Hooke* (1635–1703)

Intellectually Robert Hooke was unquestionably the most distinguished of the classical microscopists. He was, however, primarily a physical experimenter, and most of his best work lies outside our field. Sickly from childhood, his health prevented him from receiving a normal education. He was, however, a precocious and rapid worker. At Oxford he attracted the attention of Robert Boyle. When the Royal Society was founded, he entered its service as a salaried 'curator of instruments'. This country has produced no more brilliant, ingenious, and inventive experimenter, and in certain important

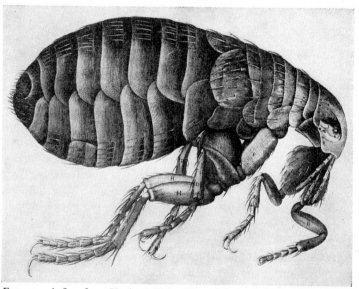

FIG. 95. A flea, from Hooke (1665). The original figure is nearly eighteen inches long.

matters he anticipated Newton. He was an acidulous controversialist, with a spirit warped by congenital infirmities of body and temper.

Hooke's *Micrographia*, published in London in 1665, opens with a description and figure of his microscope. This account is a valuable landmark in the history of the subject and contains many acute observations. Their chief biological importance is in the accuracy and beauty of his figures, which formed a standard for generations. There is evidence that some, at least, of the figures in this great book are by Sir Christopher Wren.

Hooke has a figure of the microscopic structure of cork, showing the walls bounding the cells (Fig. 142). He refers to the walls as *cells*. That word in modern biological nomenclature comes from him. He shows also cells on the surface of a stinging nettle leaf, and gives a good account of the stinging apparatus of the plant (Fig. 143). An important botanical observation by him is on a leaf fungus, the development of which is well displayed. He also gives accounts of the development of a mould, of the structure of a moss, and of experiments on the sensitive plant.

Among his figures taken from the animal kindgom are many of historical importance. Thus he depicted a Polyzoon for the first time. He discerned the beautiful minute markings on fish scales, the structure of the bee's sting, the foot of the fly, and the 'tongues' or 'radulae' of molluscs. All of these have since become classical subjects of microscopical study. He made a fine investigation of the structure of feathers (Fig. 76), which was not improved until the nineteenth century. His representations of minute insects, spiders, and mites had no rivals for a century except those of Swammerdam. Really wonderful are his figures of a gnat and its larva, of the compound eyes of a fly, and two perfectly gigantic pictures of a flea (Fig. 95) and of a louse.

§ 15. *Influence of the Classical Microscopists*

A perusal of the writings of the classical microscopists gives a first impression that their work, except in the case of Grew, is without system or definite objective. It must be remembered that these men were entering an unexplored region, a field of investigation so novel that its very existence had not been suspected.

The infinite complexity of living things thus revealed was as philosophically disturbing as the ordered majesty of the astronomical world which Galileo (1564–1642) and Kepler (1571–1630) had unveiled to the previous generation, though it took far longer for its implications to sink into men's minds. Considering the smallness of their numbers and the greatness and novelty of the topics that they were developing, it is subject for remark that the classical microscopists hit upon so many observations of fundamental importance, and that the views they propounded were, on the whole, so rational and coherent.

One cannot fail to be struck by the isolation of these men. As it happens, they were all of eccentric character. They form a group almost entirely cut off from other workers. They had little intercourse with each other, they founded no schools, they had no pupils, and they were wellnigh without imitators. Thus there is no one of their contemporaries that may be placed with them, even in the second class. They remained without effective followers until the nineteenth century. This is a remarkable fact and one that awaits adequate explanation.

But despite the inferiority of their contemporaries and successors, the work of the classical microscopists was far from being without influence. Their writings were not forgotten, but were constantly studied and their observations not infrequently verified. The knowledge of the infinitely minute complexity of the structure of living things gave a new and more philosophical trend to biological thought. Thus the general tone of the biological writing that followed them is very different from that which precedes them. Variety and complexity now begin to overawe the naturalist. Amidst the multiplicity of phenomena, order must be sought if knowledge is not to lose itself in detail. So it is that in the age that follows, the importance of *classification* becomes greatly emphasized.

Intensive research with the compound microscope was hardly taken up again until the nineteenth century. Not till then were any great practical improvements made in the instrument. 'Achromatic' instruments appeared about 1830. An immersion system was introduced by Amici (pp. 514–5) in 1840, but it was long before it became available to naturalists generally. The modern microscope really dates from 1878 when the models of E. Abbe (1840–1908) were first constructed in the workshops of Messrs. Zeiss at Jena.

Soon after, a form of substage illumination was designed by Abbe for Koch (pp. 542–5). The placing on the market of the new substage condenser by the firm of Zeiss—of which Abbe became proprietor in 1888—has had great influence on microscopical research. It was the last radical improvement in the ordinary laboratory instrument.

FIG. 96. Heads of bees, from Stelluti (1630).

V

RISE OF CLASSIFICATORY SYSTEMS

§ 1. *Absence of System in the Early Naturalists*

THE sixteenth-century 'fathers of botany' hardly attempted to arrange their plants in any logical order. They contented themselves with figures and brief descriptions (pp. 88–9). They recognized kinds or 'species' of plants, and saw that some kinds resembled each other more than they did other kinds.

With the exploration of foreign lands, it became more and more evident that each country has plants and animals peculiar to itself. The description of these taxed the powers of naturalists. Bewildering confusion ensued. Deriving from Aristotle, there was a fairly definite conception of the nature of species, but no satisfactory manner of arranging the vast numbers of them. The embarrassing character of this accumulation may be illustrated by a single handy example. The *Theatre of Insects*, written about 1590 by Thomas Moufet (pp. 67–8), sets forth the sum of contemporary knowledge of the subject. It shows a naturalist overwhelmed by the wealth of material collected from many countries. His description of grasshoppers and locusts runs thus:

'Some are green, some black, some blue. Some fly with one pair of wings, others with more; those that have no wings they leap, those that cannot either fly or leap, they walk; some have longer shanks, some shorter. Some there are that sing, others are silent. And as there are many kinds of them in nature, so their names were almost infinite, which through the neglect of Naturalists are grown out of use. Now all Locusts are either winged or without wings. Of the winged some are more common and ordinary, some more rare; of the common sort, we have seen six kindes all green, and the lesser of many colours.

'The first of the bigger, hath as it were a grass cowle or hood which covers the head, neck and almost half the body; the wings come from the neck underneath, of a greenish colour, speckled with a few small spots, the back green, the belly dusk coloured, the tail or stem at the end blackish; it hath a great mouth, and strong big teeth, excellently made to devour the fruits withall. The second seems to be like this, but that the hood is fastened to the neck; the nose also and mouth are more red, and it hath greater spots in the wings. The third is of a green countenance, the shanks whitish, the tail blackish, the wings beset with greater store of spots, and about the edges of a pale red [*Abbreviated from English translation of* 1658].

Moufet collected most industriously on his travels, and many specimens were sent him. These but befogged him the more. He is at his wits' end for descriptive terms. Thus:

'I procured one from Barbary that was brought out of Affrick with some cost to us, slender, five inches long, hooded, the head pyramidal, very long, out of which almost at the top came forth two little broad cornicles about an inch long, much like that Turbant which the Turkish Janizaries use with two feathers in it.'

Moufet's contemporary, Aldrovandi, treats the group of grasshoppers in much the same manner. Belon, Rondelet, and Gesner are in like case. To remedy such confusion some system was demanded. The first help came from the botanists.

§ 2. *First Attempts at Formal Classification*

An early attempt to arrange plants in accord with their structure was made by the Fleming, Matthias de l'Obel (1538–1616). He came to England in his youth and dedicated his first book (1570) to Queen Elizabeth. Later he was botanist to King James I. The plant known as *Lobelia* is named after him. His most important work appeared in 1576.

De l'Obel takes the form of the leaf as a basis for grouping. As a natural result, he places together many quite unrelated plants. He begins well with the Grasses, which have narrow, long, simple, pointed leaves. He goes on to plants with broader, though still simple, leaves, placing together Lilies and Orchids, both of which have lance-like leaves with parallel veins characteristic of Monocotyledons. But from them he passes to a miscellaneous group which contains certain Dicotyledons together with Ferns! They are thus approximated because the highly subdivided fronds of ferns bear a superficial resemblance to the 'compound' leaves of such forms as Tansy or Hemlock.

Andrea Cesalpino (1519–1603), Professor at Pisa, made a more effective attempt. He would arrange plants according to their flowers and fruit. On this basis he sketched a complete scheme of classification (*De Plantis*, Florence, 1583). An abortive scheme, following somewhat the same line, was made by Prince Federigo Cesi (published posthumously 1630). Part of Cesalpino's scheme was, however, absorbed into the important and influential work of Bauhin. It exerted influence also on Jung (pp. 181–2), and through him on Ray (pp. 183–6) and Linnaeus (pp. 188–195).

The Swiss, Kaspar Bauhin (1560–1624), was a pupil of Fabricius (pp. 109–10) at Padua. He went thence on many botanizing expeditions throughout Italy. Returning to his native Basel, he spent forty years arranging his botanical book, which appeared in the year before his death.

The great work of Bauhin contains descriptions of about six thousand species of plants. In general outline it is distinctly inferior to that of Cesalpino. Its merit is that it distinguishes rather more clearly between the idea of a *genus* and of a *species* of plant.

§ 3. *What is a Genus? What is a Species?*

The reader will here ask what is meant by these words, *genus* and *species?* It is disappointing to be forced to reply that no complete definition can yet be given. This answer will naturally raise a second question, 'How then can the introduction of such conceptions be a scientific advance?' The fullest answer available would lead us into a very intricate and unsatisfactory philosophical discussion. The preparation of a complete answer is the object of biological science as a whole and especially of genetics. But the reader cannot await this consummation, and he will naturally demand at least a provisional reply. He must rest in the partial satisfaction that there is a partial answer. The question, in fact, lays bare one of the weak points in biological science. Genetics is slowly working out an answer.

Living things, as we meet them in the world, exist as separate individuals. (There are exceptions to this statement, but they need not detain us.) If we examine a large and miscellaneous collection of individuals, we find that they fall naturally into separate groups, sharply marked off from other groups. Any horse can be distinguished from any ass. All the sheep can be separated from all the goats. It is true that occasional *mules* are encountered. These are the result of the crossing or interbreeding of two species. Such products are, however, rare, and since they do not usually propagate their kind, they tend to disappear and can, in the first instance, be disregarded.

These groups into which living things can naturally be placed are known as *species*. So far as direct experience goes, species always breed true, in the sense that they never produce individuals of other than their own species. Horses may produce peculiar, deformed, or monstrous offspring, but their foals, however unusual, are always recognizable by naturalists as of the horse species. They can never be mistaken for asses or zebras or any other kind of similar creature.

Thus, much had been known to naturalists since the time of Aristotle. The recognition of the general characters of the various *species* is one of the first demands of biology as a science.

The number of species known was rapidly accumulating in the seventeenth century. If any general survey of the knowledge of living things was to be obtained, it was essential to find some method, some general formula, by which these species could be grouped.

Now it is precisely by the finding of such general formulae that science advances. The word *science* means, of course, *knowledge*, but that is only its first meaning. Science means a particular kind of knowledge. It does not mean the picking up of bits and scraps of information, here and there, just as they happen to turn up. It means, on the contrary, arranged, classified, orderly knowledge. Another way of saying this is that science means knowledge for which general formulae can be found. The progress of science is to be measured by the success men have in applying these general formulae to the knowledge they have collected.

Let us apply these principles, as Bauhin applied them, to the thousands of species of known plants which presented themselves to him. He had the species fairly well defined. The puzzle was how to arrange them. What general formulae could he find for them?

It may be said at once that he reached no useful scheme of classification for all the species with which he had to deal. He took, however, a step towards this end in recognizing more clearly than his predecessors that species fall naturally into small groups—and that these groups are more distinct from one another than are the species of which they are composed. Such groups we now call *genera*.

It is here convenient to interpolate a piece of knowledge concerning genera that was hidden from Bauhin. It has been mentioned that we can occasionally cross individuals of different species of the same genus. The products are mules or 'hybrids'. But it is far more seldom possible to obtain progeny from a cross between individuals belonging to different genera. We can cross horse and ass, ass and zebra, or zebra and horse. All are species of the genus *Equus*. But we cannot cross any of these with their next nearest living representatives, the tapirs, animals of about the same size and build. Tapirs form a genus of their own, far removed from horses, asses, or zebras. There are many animals and plants of different genera that are very much nearer to each other than horse and tapir but from which it is impossible to obtain progeny by crossing. The difficulties of crossing are rendered partially intelligible by the chromosome theory, the discussion of which is a little beyond our field.

So far as plants are concerned, Bauhin reached, in some cases at least, a workable division into species and genera, and a fairly satisfactory method of describing some of these groups. He thus made possible the systematic treatment of vegetable forms. It is not that he attained to an exact delimitation of the meaning of *genus* and *species*. No man has done that. But Bauhin did foreshadow a method by which the recognized *genera* and *species* could be designated, and the distinction between different *genera* and *species* indicated. In fact, he adumbrated a system of binary classification which led ultimately to the 'binomial nomenclature.'

We may end this unsatisfactory discussion by saying that hybrids are much commoner and more likely to be fertile among plants than among animals.

§ 4. *The Binomial Nomenclature*

To every known species of animal and plant, naturalists now attach a scientific name. This name is always in Latin and is always double. Thus the common English primrose is *Primula vulgaris*, the first name *Primula* being the name of the *genus*, the second *vulgaris* that of the *species*.

But the primrose of the hedgerows is not the only species of *Primula*. All over England, in meadows and pastures, is to be found another species of plant which it resembles in many ways. It is usually known as the *cowslip*. The leaves and flowers of the cowslip are very like those of the primrose. The plant of the cowslip is, however, a little less hairy, the flowers are a little smaller and darker than those of the primrose, and, unlike the primrose flowers, they are grouped at the end of a longish stalk. Despite resemblances, cowslip and primrose do not form a permanent cross. Naturalists never have any difficulty in distinguishing the one from the other. The cowslip is, therefore, regarded as a separate species of *Primula*. The name *Primula veris* (i.e. 'Primula of spring') has been invented for it.

Moreover, in the north of England, though not in the south, a third *Primula* is sometimes seen. Its flowers issue from the stem much as in the cowslip, but they are smaller and of a lilac colour. The leaves too are much smaller and their under sides are covered with a white mealy down. This is a separate species, called in ordinary speech 'bird's-eye primrose', and known to botanists as *Primula farinosa* (i.e. 'mealy primrose').

Primula farinosa is also a native of the United States, where *Primula vulgaris* and *Primula veris* are not found in the wild state. In other countries there are many other species of Primula. They are found all over Europe and Central Asia, including the Alps, which yield their own peculiar species. There are other species in China and in the Himalaya Mountains, and yet others in North America and one in South America. The *genus*, in fact, is almost world-wide.

The use of the double Latin name to describe every species is known as the 'binomial nomenclature'. By it the arrangement and easy presentation of a vast amount of knowledge concerning species is rendered feasible.

The binomial nomenclature is usually ascribed to Linnaeus (pp. 188–95). He was the effective introducer of the system and he applied it skilfully, but the idea itself is in the books of Bauhin (p. 176) and even of earlier writers. And although Bauhin applied the method falteringly, fitfully, and inconsistently, we yet look upon him as one of the founders of that 'systematic' description of living forms which is fundamental to biological progress. Bauhin was the first of the *systematists*. His tradition was carried on by Jung, Ray, and Tournefort to Linnaeus.

§ 5. *Jung* (1587–1657)

A peculiar place in the history of biology is taken by Joachim Jung. His real merits have been adequately appreciated only during the last half century, but his influence was exerted on his immediate successors.

Jung was born in the free town of Lübeck. Abandoning mathematics for medicine, which he studied at Padua, he came under the botanical tradition of Cesalpino. Returning north, he held chairs at several of the Hansa towns. For the last twenty years of his life he was director of a school at Hamburg (1629–57).

Jung published little, perhaps because of the suspicion of heresy under which he suffered. Long after his death, certain of his pupils, who had carefully preserved his manuscripts, printed them in pious duty (*Doxoscopiae*,

1662, *Isagoge phytoscopica*, 1679). These printed works became excessively rare and were little noticed. Fortunately in 1660 a manuscript of his writings fell into the hands of John Ray, who was deeply impressed by them. It was through Ray that the ideas of Jung reached Linnaeus.

Jung's literary form is concise, and suggests well-arranged lecture notes. His two little works exhibit not only great biological insight but a real genius for classification. They undoubtedly approach nearer to the modern point of view than any other systematic work of his century. This comes out in his terminology. Thus we note that in discussing leaves he distinguishes *simple* and *compound, pinnate* and *digitate, paripinnate* and *imparipinnate, opposite* and *alternate.* These modern terms were coined by him. Jung treated the kinds and parts of stems and roots with equal skill. He introduced *petiole* for the leaf stalk. The technical terms *perianth, stamen,* and *style,* in their modern signification, are his invention. He discerned the true nature of the flowers of the Compositae, distinguishing disk florets from ray florets. The whole composite 'flower' he rightly regarded as an inflorescence or *capitulum.* Perhaps Jung's greatest intellectual achievement is his clear division of the departments of botanical study into what we should now describe as morphology (pp. 218–20), physiology, systematics or taxonomy, and ecology (pp. 283–4).

Jung knew nothing of the sexual nature of flowers, but he used flowers as the foundation of classification. He distinguishes clearly such groups as the Compositae, the Labiatae, the Leguminosae, by the form of their flowers.

His manner of naming plants approaches extraordinarily near to a formal binomial system. The plants in his alphabetical list are almost always given two names. The first is a noun and in effect a generic name, the second is an adjective and in effect a descriptive specific name. The method is developed from Bauhin and leads on through Ray and Tournefort to Linnaeus. The great rarity of Jung's writings has militated against adequate reference being made to them.

§ 6. *Ray* (1627–1705)

The English naturalist, John Ray, was the son of a blacksmith. At Cambridge he acquired varied learning and took to making natural history excursions. Thus he rode through the Midland and Welsh counties, taking notes on plants and collecting dried specimens. In 1660 he published a scientific description of all the species of plant that grow in the neighbourhood of Cambridge. There had been earlier accounts of local plants, but this was the first real *Flora*, i.e. a systematic and searching catalogue of the plants of a given locality. In the year of its appearance, Ray took orders in the Church of England.

Soon after this, Ray came in contact with a younger enthusiast, Francis Willughby (1635–72). The two resolved to prepare a systematic description of the whole organic world. Willughby would finance the venture and undertake the animals, while Ray was to deal with the plants. The friends travelled together on the Continent and in England to enlarge their knowledge of natural history. Willughby died early and left an annuity to Ray, who edited his friend's work on fishes.

Ray's next work was a handy *Catalogus* of British plants (1667). This was a complete Flora of the British Isles, which became the pocket companion of every botanist of this country for generations. It is dedicated to Willughby.

Ray is, with Linnaeus, the chief founder of the science of 'systematic biology', that is, the science that aims at the orderly arrangement of the species of animals and plants. After his catalogue of British plants, his next attempt was on birds, and was based on Willughby (1676). Next followed a booklet, *Methodus plantarum nova* (London, 1682), which shows the influence of Jung. It demonstrates the true nature of buds, and uses the now familiar division of flowering plants, into Dicotyledons and Monocotyledons. These words were invented by Ray but did not appear in print until the issue of the second edition (1703). Ray based his system chiefly upon the fruit but also upon the leaf and, following Cesalpino and Jung, upon the flower.

Thus Ray succeeded in indicating many of the larger groupings of plants which botanists now call *Natural Orders*, and thereby took the first decided step towards a *natural system* of classification. In a 'natural system' all resemblances are taken into account, a value being ascribed to them according to their presumed importance. A system is said to be 'artificial' in so far as resemblances and differences in some few particulars only are taken into account, independently of all others. Nevertheless in the particular case of the flowering plants, as Ray well knew, accumulating experience had shown that the first consideration must be given to the structure and development of the flower. Almost any type of leaf stem or root may occur in group after group, but the flower is the compass for the explorer of the plant world.

Ray's greatest work is his *Historia generalis plantarum* (3 volumes, London, 1686–1704). This huge work contains a most able account of everything then known of the structure, physiology, distribution and habits of plants. It does justice, moreover, to the memory of the men from whom Ray derived many of his ideas. All the plants made known by his predecessors and contemporaries, about 18,600 in number, are described methodically and clearly. The species are arranged in 125 'sections' of which a number are still recognized as 'Natural Orders'. The work earned the special admiration of Cuvier (p. 227).

Ray's writings on animals were nearly as important as, though less exclusive than, his botanical works. His *Historia Piscium* (1686), part of which is by Willughby, and his *Synopsis methodica animalium quadreupedum et serpentini generis* (1693) were both of great value. The latter contains the first truly systematic arrangement of animals. This is based primarily upon the toes and teeth. Abundant traces of his general arrangement of animals survive in modern systems of classification. With much acuteness Ray perceived that such forms as the armadillo, the hedgehog, and the mole could not be made to fit into his scheme, which may be set forth thus for mammals:

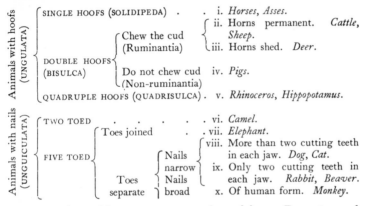

Apart from his great systematic writings, Ray showed insight and wisdom in other departments of science, as in his private life. In his *Widsom of God Manifested in the Works of the Creation* (1691), he set forth the true nature of fossils, as petrified remains of species now extinct. This was a great advance on current views. He also accepted the sexual character of flowers. Toward the end of his life he devoted himself to the study of insects. When a certain lady of Exeter was judged insane because she collected insects, Ray appeared as a witness to her sanity.

Ray's work on insects was not published till after his death. It combined the system of Aristotle with that of Swammerdam (p. 163) and prepared the way for Linnaeus. It was regarded by Cuvier (p. 231) with the greatest respect.

§ 7. *Tournefort* (1656–1708) *and Artedi* (1705–35)

In France an authoritative position somewhat similar to that of Ray was occupied by the botanist, Joseph Pitton de Tournefort.   From his earliest youth Tournefort exhibited unrestrainable enthusiasm for the study of nature.   He travelled much throughout Europe and the Near East, diligently collecting plants. From 1683 till his death he was a professor at the Jardin des Plantes (p. 196).

Tournefort was inferior to Ray in philosophic grasp as well as in the naturalist's acumen. His descriptions of plants are good. His classification, though convenient, is artificial, being based on an inadequate and hollow investigation of the form of the flower. He is important as a predecessor of Linnaeus in developing both the binomial nomenclature and the method of describing plants.

Tournefort denied the sexuality of plants, and classed those with small flowers along with those without flowers. He thus missed a quite fundamental distinction.   His chief work, the *Institutiones rei herbariae* (1700) has considerable significance as foreshadowing the rigid systematic spirit that coloured the biology of the eighteenth century. It is beautifully illustrated. The third (1719) and subsequent editions were improved by Antoine de Jussieu senior (1686–1728; see p. 196).

Tournefort, following Bauhin, lays great stress on *genera*. His chief contribution to systematic botany lay in his method of distinguishing these. Bauhin had given only a name to each genus, while the terms that he applied to the species were, like those of Jung, purely descriptive.

Tournefort adds descriptions to the names of the genus. The names for the species and varieties, however, are almost devoid of description save such as is provided by the admirable figures. The method of Linnaeus might be briefly described as adaptation of Bauhin's method for genera *plus* Tournefort's for species *plus* a descriptive method largely his own.

The Swede, Peter Artedi (1705–35) was a few years senior to Linnaeus and had already taken to biology

97

98

FIG. 97. Linnaeus with equipment for exploration. From a mezzotint.
FIG. 98. An oriental species of spurge, *Euphorbia apios*, of Linnaeus depicted for the first time by Belon (p. 93) in 1555. He called it by its native Greek name 'Apios'.

when the two met as students at Upsala university. They had similar tastes and aims and became fast friends. After Linnaeus left for Lapland (see below) Artedi came to England (1734) to study fishes and on to Holland where he became curator to a wealthy collector. Next year he died in an accident but his work *Ichthyologia* was in manuscript. It was published by Linnaeus in 1738 and is the first major treatise on systematic zoology. It is divided into five parts; (a) a bibliography of ichthyology; (b) a 'philosophy' of ichthyology, that is a general consideration of the features of fishes on which his divisions of them are based, and this has proved to be the most important section of the work; (c) a description of the genera; (d) synonyma; (e) a detailed account of the species. The Latin used by Artedi is of the condensed and intelligible form later made familiar by Linnaeus who adopted Artedi's system for fishes unaltered. It is remarkable that Artedi still classes the Cetacea among fishes, being on this point, behind Aristotle.

§ 8. *Linnaeus* (1707–78)

The Swede, Karl Linnaeus, as a boy was backward and destined to be an artisan. His father, a poor clergyman, was reluctantly persuaded to send the lad to Lund, and afterward to Upsala as a student of medicine. It soon became evident that, despite literary defects, he had extraordinary application and great aptitude in certain directions.

While a student at Upsala, certain French botanical works set Linnaeus examining the sexual parts of flowers. Convinced of their importance for the life of plants, he conceived the idea of basing a system of classification on them. He wrote a short treatise on sex in plants which was far from original. He was asked to act as deputy to the professor of Botany on the strength of it.

In 1732 Linnaeus was chosen by the Upsala Academy to visit Lapland as a collector. He at once set out alone and travelled more than 4,600 miles at a total cost of £25. In five months he explored Lapland, and much of northern Sweden and Norway. Of his equipment he says:

'My clothes were a light coat of linsey-woolsey, leather breeches, a round wig, a green leather cap and a pair of half-boots. I carried a small leather bag containing one shirt, two pair of false sleeves, two vests, an inkstand, pen-case, microscope and telescope, a gauze cap to protect from gnats, a comb, my journal and a parcel of paper for drying plants, my manuscript Ornithology, *Flora Uplandica* and *Characteres generici*. I wore a hanger at my side, and carried a small fowling-piece, as well as an octagonal stock graduated for measuring.' (Fig. 97.)

Accompanied by two Laps, he crossed the peninsula on foot to the Arctic Ocean—itself no mean feat—and returned again by a parallel route. After incredible fatigues and hardships, climbing precipices, crossing torrents, suffering extreme heat and cold and hunger and thirst, and constantly pestered by mosquitoes, Linnaeus reached again the Gulf of Bothnia, and so home. He had observed many wild animals and discovered a hundred new species of plants.

Linnaeus next visited Germany, Holland, England, and France. During these travels he made the preliminary draft of his famous work the *Systema Naturae*. It was first published in Holland in 1735. At the same period he also issued several botanical works. The most important were the *Fundamenta botanica* (1736), the *Genera Plantarum* (1737), and the *Classes Plantarum* (1738).

On his return to Sweden Linnaeus was appointed Professor of Natural History at Upsala. He became an extraordinarily popular teacher. Students crowded to him. Much of his best work was done by sending them out on expeditions. These pupils explored a large part of the known world in search of new animal and plant forms. Many, it is said one-third, died on these expeditions. Many more are commemorated in the *Systema Naturae*. One, Solander (p. 240), accompanied the English explorer, Captain Cook, and was long resident in London.

§ 9. *The 'Systema Naturae'* (1735–58)

Linnaeus had a passion for classification. Not only did he draw up classified lists of plants and animals, but he also classified minerals and even diseases. To his work on plants and animals naturalists still constantly refer.

It would be a mistake to suppose that it is only in classification and nomenclature that Linnaeus has left his mark. He also developed a method of formal description or organisms which is essentially similar to that still in current use. He took the parts of a plant or an animal in regular sequence and described them according to a recognized rule. This brief and graphic method of description, though it tended to become mechanical, was yet a great improvement on the verbose accounts common till his time.

Linnaeus succeeded in assigning to every known animal and plant a position in his system. This involved placing it first in a *Class*, then in an *Order*, then in a *Genus*, then in a *Species*. Orders were large divisions, each usually containing a number of Genera. Classes were yet larger divisions, each containing a number of Orders. Organisms from two different Orders in the same Class, e.g. an orchid and a hyacinth, differ from one another less than either differs from a daisy, which is in an Order belonging to a different Class from either.

The Linnaean Classes and Orders of Plants were primarily those set forth by Tournefort (p. 186). They were based on the parts in the flower. The number of 'stamens', or free male parts, in the flower was the founda-

tion of this division into classes. Thus Linnaeus grouped
plants with one stamen in the Class *Monandria*, plants
with two in the Class *Diandria*, plants with three in
the Class *Triandria*, and so on. Each Class was divided
into orders, according to the number of 'styles' or free
female parts in the flower. Thus, the Class *Monandria*
was divided into the Orders *Monandria Monogynia* with
one style, *Monandria Digynia* with two, and *Monandria
Trigynia* with three; and so on with the other Classes.

To take a random example, the spurge *Euphorbia Apios*
(Fig. 98, p. 187) which he named, he placed in the Class
Dodecandria as having its male parts in systems of twelve,
and in the Order Trigynia as having its female parts in
systems of three. Of this Order he recognized three
genera. Of the genus *Euphorbia* he recognized over a
hundred species from various parts of the world, the
Mediterranean species *Apios* coming to him from the
island of Crete.

As to animals, Linnaeus distinguished the Classes of
Mammals, *Birds*, *Reptiles*, *Fishes*, *Insects*, and *Vermes*. The
first four classes had already been grouped together by
Aristotle as 'Animals with red Blood' (p. 42) or as we
now call them 'Vertebrata'. The remaining Classes of
Insects and *Vermes* contain, bundled together, all the
groups of animals without vertebrae. Here Linnaeus was
behind Aristotle, who had broken up these groups (p. 43).

This basis of arrangement that Linnaeus adopted for
the Animal Kingdom may be thus diagrammatically set
forth:

Heart with 1 or 2 ventricles and 2 atria Blood warm and red.	{ 1. Viviparous. MAMMALS. 2. Oviparous. BIRDS.

Heart with 1 ventricle and 1 or 2 atria;
Blood cold and red.
	{ 3. Breathing by lungs.
	      REPTILES.
	  4. Breathing by gills. FISHES.

Heart with 1 ventricle and no atrium;
Blood cold and colourless.

{ 5. With antennae. INSECTS.
{ 6. With tentacles. VERMES.

Some of these distinctions have been revised by more modern researches. Thus, there are some Mammals, such as the Australian duck-billed Platypus, that are not viviparous, but that lay eggs. That animal, however, was unknown during the lifetime of Linnaeus. Some of the Linnaean Class of Reptiles, which include our 'Amphibians', breathe with gills; and only by forcing the term can they be said to have but one atrium. Not all of the group that Linnaeus called 'Vermes', a group broken up by Lamarck and now abandoned (p. 199), have colourless blood, nor do all possess tentacles. Nevertheless, the scheme is workable and gave naturalists something on which to build.

The contribution of Linnaeus to biology that has lasted best is his division of living things into *genera* and *species*, with his development of the binomial system. Bauhin, Jung, Ray, Tournefort, had already attached short descriptive titles, often of two words, to all their species. Moreover, a Leipzig botanist and physician, August Quirinus Rivinus (1652–1723), made a definitive suggestion (1690) that no plant-name should contain more than two words. Both Ray and Tournefort were quite aware of the work of Rivinus. Thus the idea cannot be placed wholly to the credit of Linnaeus. But it was Linnaeus who secured the general adoption of a system in which every species has two names definitely allotted to it. The first work in which he undertakes this on an extensive scale is his *Species Plantarum* of 1753. About 7,300 species are there provided with names, most of which are still in use.

Linnaeus was accustomed to give a rigid definition, in his terse Latin, to each genus and species. This method has been followed by his successors. His classification has been repeatedly revised by more modern naturalists. Little is now left of his division into *Classes* and *Orders*. A number of his definitions of *Genera* and of *Species* are, however, still accepted.

When a plant or animal retains the titles given by Linnaeus, it is usual to add to them an abbreviation of his name. Thus, for example, naturalists refer to the common daisy as *Bellis perennis*, Linn., or to the common frog as *Rana temporaria*, Linn. In general, indeed, it has become the custom to add to the Latin terms for any species an abbreviated form of the name of the man who first bestowed it, especially in difficult or doubtful cases.

The great work of Linnaeus is his *Systema Naturae*. It was first drafted in 1735. He modified it and amplified it a good deal as it went through its many editions. Of these, biologists have selected the tenth, which appeared in 1758, as their basis. If a species is given its Linnaean name by a modern naturalist, it means that adopted in this tenth edition.

Linnaeus attached a generic and specific name even to man, who is thus known to naturalists as '*Homo sapiens*, Linn.' The designation may be translated 'Man, the reasoner'. Linnaeus included in the genus *Homo* another species, the Orang-outang. This creature he designated *Homo troglodytes*, 'Man, the cave-dweller'. The decision of Linnaeus to unite man and orang in one genus has long since been reversed by his modern successors. Naturalists now consider that there is only one living species of genus *Homo*, though the species *Homo sapiens* may be divided into several *varieties*.

The terms *subspecies*, *variety*, and *race* are applied to divisions of species of which the peculiar characters are less marked, and usually less constant, than those which separate species. The discussion of these distinctions we must defer, but we note that subspecies, varieties, and races of the same species are always fertile with each other. Admittedly these terms are loosely used.

Of the varieties of man, Linnaeus knew the black or Negro, the yellow or Mongol, the white or European, and the red or American. Since his day, forms of *Homo* have been described in a fossil state which differ from any

FIG. 99. A Swedish warship in pursuit of the vessel containing the Linnaean collections in 1784 (R. J. Thornton, 1797). The incident is of more than doubtful authenticity.

of the living varieties of man far more than any of these varieties differ from each other. They form, therefore, separate species of *Homo*. The best known fossil species (?subspecies) of man is the so-called *Homo Neanderthalensis* which once inhabited Europe and the Near East (p. 315).

The general ideas of Linnaeus concerning the nature of species are of great historical importance. He held that species are constant and invariable, a view in which he differed from John Ray. 'There are just as many species as there were created in the beginning,' said Linnaeus, and again, 'There is no such thing as a new species.' In this respect we have departed completely from his standpoint.

When Linnaeus died in 1778 his collections and books were bought by a wealthy young English naturalist on the suggestion of Banks (pp. 240–3). It is said that the King of Sweden sent a war vessel after the ship that bore the Linnaean specimens to England. The British proved the faster, however, and reached England without mishap (Fig. 99). The purchaser of the collection became the first president of the English biological association named the *Linnean Society*, which still owns his collections and library. The pursuit of the collection by the Swedish navy is just one of those picturesque legends that cannot be killed! (Fig. 99).

§ 10. *The Successors of Linnaeus*

Linnaeus focused biological interest on classification, especially by external parts. He thus withdrew attention from the intimate structure and workings of the living organism. The search for new genera and species became, for generations, the chief aim of most naturalists, to the neglect both of anatomical and of physiological studies. The tendency was especially marked among botanists, who were esteemed in proportion to the number of flowering plants whose characters they could memorize.

But the Linnaean influence showed itself also in another and more attractive manner. Linnaeus had a passionate love of wild nature, and the close examination of species in a living state provided a stimulus toward that type of study to which the term 'Natural History' is sometimes restricted. The eighteenth century and the early part of the nineteenth furnished examples of nature-lovers of high ability and great literary power. Thus, the *Natural History and Antiquities of Selbourne* (1789) by the Rev. Gilbert White (1720–93) and the *Wanderings* (1825) of

Charles Waterton (1782–1865) are classics both of the English tongue and of biological expression. In other languages there are contemporary works comparable to these.

More in the direct line of the Linnaean tradition were the activities of a group of botanists in France, Progress in that country was aided by several organized botanic gardens and by the devotion of two famous families, de Jussieu and de Candolle.

The family of de Jussieu came from Lyons. Three sons of an apothecary of that town all became distinguished for their studies of plants. For more than a century and a half the stock continued to produce eminent botanists. The names of no less than six de Jussieu recur constantly in scientific literature. Bernard (1699–1777), the most famous, worked first at the Jardin des Plantes and from 1759 at the garden of Le Trianon at Versailles. There he made it his chief task to elaborate a 'natural system'. He sought to arrange living plants after the Linnaean manner, though his method was based on somewhat different principles. He published nothing, but his nephew and assistant, Antoine Laurent de Jussieu (1748–1836), developed his system and gave it to the world in his *Genera plantarum secundum ordines naturales disposita* (1774).

The terms 'Monocotyledons' and 'Dicotyledons', introduced by Ray (1703) for the two primary sub-divisions of flowering plants (p. 184), were popularized by Antoine Laurent de Jussieu. He distributed the classes on a principle taken partly from Tournefort. This further division depended partly on the number and arrangement of petals in the flower and partly on the position of the stamens with reference to the ovary (*Hypogynous* = below the ovary, *Perigynous* = around the ovary, *Epigynous* = above the ovary). Thus:

			Classes
Acotyledons	. . . .		I
Monocotyledons	Hypogynous		II
	Perigynous		III
	Epigynous		IV
Dicotyledons	Apetalous	Hypogynous . .	V
		Perigynous . .	VI
		Epigynous . .	VII
	Monopetalous	Hypogynous . . .	VIII
		Perigynous . . .	IX
		Epigynous . . .	X, XI
	Polypetalous	Hypogynous . . .	XII
		Perigynous . . .	XIII
		Epigynous . . .	XIV
	Irregular		XV

Of these fifteen classes the 'Acotyledons' provide but one. This group, which we now know to be no less miscellaneous than vast, includes Fungi, Algae, Liverworts, Mosses, and Ferns. It will be seen how overwhelming is the stress on flowering forms—Monocotyledons and Dicotyledons. The artificially symmetrical element in this classification, which fits these forms tolerably, is as prominent as in the Linnaean system. The Classes were broken up into Orders, Genera, and Species much as the *Systema Naturae*.

Augustin Pyramus de Candolle (1778–1841) was the most dinstinguished member of a wealthy Genevan family which, like the de Jussieu, produced a number of eminent botanists. He exhibited life-long industry and unsurpassed enthusiasm in botanical research. Most of his investigations were made in France at a period of the most intense scientific activity in that country. The width, depth, and philosophic insight of his studies on plants made him no unworthy colleague of Cuvier (pp. 227–37)—also by birth a Swiss—who devoted his energies to animal forms.

In his *Elementary Theory of Botany* (1813 and many later editions), de Candolle set forth his general views. There are many parts of this fine work which can still be read with profit. The classification which de Candolle adopted was far more natural and more deeply based on anatomical structure than any yet attempted, nor does it exhibit the straining after symmetry of Linnaeus and de Jussieu. The classification of de Candolle may be abstracted thus (*Prodromus systematis naturalis*, Paris, 1824–70)[1]:

I. Structure 'cellular'. Scarcely proper seeds. Propagation by spores.

    6 families, including Fungi, Lichens, Mosses, and Liverworts.

    [A miscellaneous group of non-flowering plants.]

II. Spiral vessels. True seeds. Sexual parts not double.

    5 families, including the Ferns and Club mosses.

    [A fairly natural group, wrongly described.]

III. Spiral vessels scattered in bundles throughout stem. Sexual parts obvious. Embryo buried in albuminous substance. Arrangement in threes common.

    18 families, including Cypresses, Grasses, Palms, Irises, Orchids, being gymnosperms and monocotyledons.

    [Two natural groups linked as one.]

IV. Spiral vessels in concentric rings. Sexual parts obvious. Embryo not buried in albuminous substance. Numerical proportions various.

    85 families, all dicotyledonous. [A natural group.]

This system, with all its faults and errors, is a real attempt at a 'natural' classification in that it takes many different parts into consideration and in this respect is thoroughly modern. It may be placed beside that set

---

[1] In sixteen volumes completed by the author's son Alphonse de Candolle.

forth for animals by Cuvier in *Le règne animal* (pp. 331–35), which is based on the whole findings of comparative anatomy and physiology and not on the appearance of arbitrarily selected parts.

Independent of Cuvier, however, and in his own time, there was one writer who, though better remembered for his evolutionary speculations, has left a deep impression on classificatory schemes. This was Lamarck (pp. 296–9). In his *Philosophie zoologique* of 1809 he adopted Cuvier's useful term *invertebrates* ('invertébrés'). He arranged the animal kingdom in fourteen classes thus:

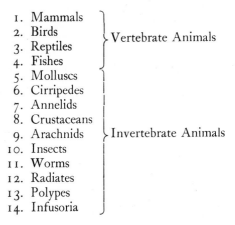

1.	Mammals	⎫
2.	Birds	⎬ Vertebrate Animals
3.	Reptiles	
4.	Fishes	⎭
5.	Molluscs	⎫
6.	Cirripedes	
7.	Annelids	
8.	Crustaceans	
9.	Arachnids	⎬ Invertebrate Animals
10.	Insects	
11.	Worms	
12.	Radiates	
13.	Polypes	
14.	Infusoria	⎭

The arrangement is in the form of a 'scale' or ladder and thus contrasts both in principle and in appearance with that adopted by his great rival, Cuvier.

§ 11. *Modern Systems of Classification*

Since the time of Linnaeus almost every important biological movement has left its mark on the system of classification current in its day. The classification of living things adopted by a biological writer may often be treated as an epitome of his views on many important biological problems, and especially on comparative studies, to which we shall presently turn. This was notably the case with the system of Cuvier.

We would stress the fact that, from the time of Linnaeus to our own, a weak point in biological science has been the absence of any quantitative meaning in our classificatory terms. What is a Class, and does Class *A* differ from Class *B* and much as Class *C* differs from Class *D?* The question can be put for the other classificatory grades, such as Order, Family, Genus, and Species. In no case can it be answered fully, and in most cases it cannot be answered at all. Again the geneticist of our century are seeking the keys that may unlock these secret things of Nature.

Until some adequate reply can be given to such questions as these, our classificatory schemes can never be satisfactory or 'natural'. They can be little better than mnemonics—mere skeletons or frames on which we hang somewhat disconnected fragments of knowledge. Evolutionary doctrine, which has been at the back of all classificatory systems of the last century, has provided no real answer to these difficulties. Geology has given a fragmentary answer here and there. But to sketch the manner in which the various groups of living things arose is a very different thing from ascribing any quantitative value to those groups.

It is not possible to discuss all the systems of classification that have attracted naturalists since Lamarck, Cuvier, and A. P. de Candolle. In bulk, systematic works form the main part of biological literature. It is convenient, however, to set down the chief features of a scheme generally acceptable to modern naturalists, based largely on the process of development and inspired by evolutionary teaching. It will be appropriate to discuss its limitations.

Firstly, the basic division is into a plant series and an animal series. No such absolute distinction can, however, be rigidly maintained. There exist many organisms which cannot be left definitely in either the animal or plant kingdom. Most of these forms of uncertain position are characterized, in some stage of their existence, by the possession of a whip-like organ, the 'flagellum'. We are, therefore, forced to include the group FLAGELLATA, for example, in both tables.

Secondly, modern investigation has shown that there are many groups of living things which have no clear relation, either in structure or in mode of development, to any other group. This is even more the case among animals than among plants. Such major groups are known as PHYLA. Many PHYLA contain only a very few species, but may not be the less interesting on that account. In our table on pp. 202 and 203 we have only been able to include the larger PHYLA.

Thirdly, it will be noted that while we divide the animal kingdom into two main sub-kingdoms, according as the organism is composed of one or of many cells, no such attempt has been made with the plant kingdom. In fact this mode of division is not satisfactory even for animals, and there are animals concerning which we might discuss whether they are unicellular or multicellular. For plants, however, it will not serve at all. One reason for this is fairly apparent. Most animals have a definite form with definite body shape and definite number of organs. But plants present not so much *form* as *pattern*. A plant repeats itself over and over again and will do so as long as it lives. These patterns which we call plants arise ultimately from a single cell. Now in all the first three groups of plants which we enumerate, as well perhaps as in others, there are forms which may be either one-celled or many-celled according to conditions of life.

SOME IMPORTANT PLANT PHYLA

1. FLAGELLATA. Usually unicellular, with one or more lash-like organs of motion (flagella). Many are on the border-line between plants and animals, e.g. Peridinians (p. 263).

2. SCHIZOPHYTA. Include the Bacteria, the so-called 'Blue-Green Algae', and the Desmids.

3. THALLOPHYTA. *a.* ALGAE. Some are unicellular, some multi-cellular. Include Brown, Green, and Red Algae and Diatoms (263).

    b. FUNGI.

4. BRYOPHYTA. Mosses and Liverworts (p. 530).

5. PTERIDOPHYTA. *a.* PSILOPHYTALES. Found only as fossils (p. 280).

    b. LYCOPODIALES. Chiefly extinct. Include Lepidodendrons (pp. 280–1), modern Club-mosses, &c.

    c. EQUISETALES. Chiefly extinct. Include Calamariae (p. 281) and modern Horse-tails.

    d. FILICALES. True ferns (pp. 280–2).

6. SPERMOPHYTA ('Seed Plants').

    a. PTERIDOSPERMAE 'Seed Ferns' (pp. 282). Found only as fossils.

    b. GYMNOSPERMAE. With naked seeds. Include the Cone-bearing plants (p. 282), the Cycads, the 'living fossil' *Ginkgo*, and the fossil *Cordaites*.

    c. ANGIOSPERMAE. Seeds enclosed in a peri-carp or seed vessel. Flowering plants, divided as Monocotyledons and Dicotyledons.

SOME IMPORTANT ANIMAL PHYLA

SUB-KINGDOM **Protozoa**. Unicellular Organisms

1. FLAGELLATA, as above.

2. INFUSORIA or CILIATA (pp. 331–4). Nearly all have minute hair-like cilia covering a large part of the body, e.g. *Paramecium*.

3. SARCODINA. Send out lobe-like processes of protoplasm, e.g. *Amoeba*.

　　SUB-KINGDOM **Metazoa**. Multicellular Organisms.

　　　a. **Acoelomata**. Without definite body-cavity (p. 479).

4. SPONGIARIA. Sponges.

5. CNIDARIA. *Hydra*, Jellyfish (pp. 490–1), Sea-anemones and Corals.

6. PLATYHELMINTHES. Flatworms. Include many parasitic forms, e.g. Tapeworms, Liver-flukes, etc.

　　　b. **Coelomata**. With definite body-cavity (p. 491).

7. MOLLUSCA. Unsegmented forms with small body cavity.

8. ANNELIDA. Round segmented worms, e.g. Earthworms.

9. ARTHROPODA. Jointed organisms with external skeleton. Include Insects, Crustacea, Spiders, &c.

10. BRACHIOPODA. An isolated group of extreme antiquity. Shells superficially like those of bivalve molluscs.

11. ECHINODERMATA. Five-rayed forms including Starfish, Sea-urchins, &c.

12. NEMATHELMINTHES. Round unsegmented worms. Many are parasitic. Free living forms of great importance in soil.

13. CHORDATA include the vertebrates (pp. 486–7).

In leaving the subject of classification, we would call attention to the fact that the modern concentration on the chromosomes and their behaviour (chaps. xiv and xv) has not yet had time to react upon the general scheme of classification. There can be no doubt that this must happen in due course.

VI

RISE OF COMPARATIVE METHOD

§ 1. *Comparative Studies in the Seventeenth Century*

RESEMBLANCES and differences in intimate structure drew the attention of naturalists earlier in the
case of animals than of plants. For most of their course,
studies of this order have been prosecuted mainly on the
higher backboned animals. This has had a narrowing
effect from which biological science has shown signs of
recovery only during the last hundred years.

It is necessary also to remember that, with John
Hunter (1728–93), the analytical study of forms had for
a time as its object the exposition of function. Thus
comparative anatomy was designed to lead up to what we
should now call 'comparative physiology'. With the
death of Hunter this movement waned and with
the general acceptance of doctrines of common descent
the attitude changed. Naturalists concentrated on demonstrating the historical relations of organisms with less
regard to the functions of organs. Thus the evolutionary
bias divorced physiology from anatomy.

Man was the first animal whose anatomy was adequately
explored. With Vesalius that study became exact (pp. 99–
104). He frequently compared human structures with
those found in animals, and to this day much of the
nomenclature of comparative anatomy is strictly appropriate only to man.

During the sixteenth century many besides Vesalius
made dissections of animals and compared them with man.
At the end of the century a lawyer of Bologna, Carlo
Ruini, undertook a detailed investigation of the anatomy
of the horse. His treatise is the first devoted exclusively to

the structure of a single species other than man. The monograph of Ruini of the horse (Bologna, 1598) is the rival, in exactness and beauty, to that of Vesalius of man.

At Padua, Fabricius ab Aquapendente transmitted the tradition of Vesalius to his pupils, William Harvey of Folkestone (pp. 110–18, 465–8) and Giulio Casserio of Piacenza (1561–1616). The latter succeeded to the chair of his master in 1604. Casserio had as strong a comparative bias as Fabricius. In his extensive investigations of the sense-organs, he habitually followed a structure through a long series of widely different species.

The successor of Casserio at Padua was the Belgian Adrian Spigelius (1578–1625), who has left his name on an important structure in the liver. He was a very exact student of the anatomy of man, but he did a signal disservice to science by his formal separation of human from comparative anatomy. Curiously enough, the founder of the great Paduan anatomical tradition, Vesalius, and its last great representative, Spigelius, were both natives of Brussels.

The evil example of Spigelius was followed for two centuries. 'Anatomy'—that is human anatomy—ceased to be a true science and became a mere medical discipline. The historian of biology can afford to pass lightly over this restricted study until it begins to be informed by a new spirit in the late nineteenth century. We shall, however, constantly need to refer to the greater human anatomists who rose above their medical atmosphere.

Contemporary with Casserio and Spigelius at Padua, Gasparo Aselli (1581–1626) professed anatomy at Pavia. His great discovery (published 1627) was that of *lacteal vessels*. These gather and convey the fatty material of the food from the intestines. The lacteal vessels converge to the so-called *thoracic duct*, which runs by the side of the spine to open into the left subclavian vein. The thoracic

FIG. 100. Anatomical theatre at Padua built for Fabricius in 1594 and still intact. Harvey, Casserio, and Spigelius all attended lectures in it.
From a stipple engraving of 1844 taken from an older drawing.

duct itself was first clearly seen twenty years later by the Frenchman, Jean Pecquet (1622–74), at Montpellier in 1647 (Fig. 61). Both Aselli and Pecquet worked on dogs. Both would have extended their researches to other animals, but for the intervention of an early death in the case of the first, and an even earlier addiction to alcohol in the case of the second.

The thoracic duct, before it opens into the blood system, also receives tributaries from a system of vessels, the *lymphatics*, which ramify throughout the body. The lymphatic system was first demonstrated in 1653 by Thomas Bartholin (1616–80) of Copenhagen. He was a pupil of Severino (see below) and a member of a family that long monopolized anatomical posts in Denmark. During the eighteenth century an immense amount of detailed work was done on the lymphatics, without, however, adding to the conception of their role in the animal economy.

Marco Aurelio Severino of Calabria (1580–1656), long professor at Naples, earned the suspicious attentions of the Inquisition, perhaps by reason of the anti-Aristotelian bias which he shared with his contemporary, Galileo. In 1645 he issued his *Zootomia democritaea, id est anatomia generalis totius animantium opificii* ('Democritean Zoology, or a general anatomy of the whole animal creation'). Democritus, we recall, was the philosopher to whom Aristotle was most opposed. Nuremberg, where this volume was published, was out of reach of the Inquisition.

In this book Severino sought to trace analogies of construction in corresponding parts of various animals. He had dissected many forms, both vertebrate and invertebrate, but he exhibits no deep understanding of their structure. He was, however, convinced that microscopic research would throw light on comparative anatomy.

The classical microscopists began to prove him right. Swammerdam investigated the anatomy of the may-fly, of the bee, of the snail, and of many other forms. Malpighi worked on the silkworm, and Leeuwenhoek studied innumerable small creatures. Thus by the later seventeenth century, partly through the diffused influence of the Paduan school, partly under the stimulus of the classical microscopists, the comparative method of investigating animal structure caught fast fold.

Anatomical monographs of various vertebrates now began to appear. Worthy of notice is a series of descriptions of dissections by Claude Perrault sumptuously issued, from 1671 onwards, by the French Académie des Sciences (p. 137). This included monographs of the beaver, the dromedary, the bear, the eagle, the tortoise, and many other forms. Comparable to these are treatises of the English naturalist Edward Tyson (1650–1708) on the porpoise (1680), the rattlesnake (1683), the opossum (1698), and the chimpanzee or 'pigmy' (1699). In the same category are works on the ostrich (1712) and the chameleon (1715) by the accomplished Paduan, Antonio Vallisnieri (1661–1730), who is commemorated in the familiar water-weed, *Vallisneria spiralis.*

One of the most remarkable comparative studies of the seventeenth century is Grew's *Comparative Anatomy of the Stomach and Guts* (1681) based on the study of some thirty-five species. Grew had himself introduced the term 'comparative anatomy' in his *Comparative Anatomy of Trunks* (1675).

The Fleming, Gérard Blaes (Blasius, 1646–82), who became professor of medicine at Amsterdam, devoted many years to a comparative investigation of the structure of vertebrate animals. His great *Anatome animalium* (Amsterdam, 1681) is a compendium of previous writings on the same subject, but contains some original observations. Each animal is treated, organ by organ, in an orderly manner, and the work is entitled to respect as the first general systematic treatise on comparative anatomy of animals. It throws into clear relief the community of structure of the larger groups of vertebrate animals, such as birds, rodents, and carnivorous mammals.

The common elements in the structure of the great botanical groups are less obvious than those of comparable zoological divisions. An excellent start was made, however, by Malpighi and Grew. The latter author especially attained a view of the structural distinction between roots and shoots, and the differences between the systems of 'vessels' in the two. Grew and Malpighi displayed the distinction between dicotyledonous and monocotyledonous stems, and between both and those of such plants as the pines and cypresses, now classed as *Gymnosperms* (Greek, 'with naked seeds').

Grew showed remarkable grasp of the relationship of the various organs of plants throughout different groups. Thus, he traced stems through their modifications, and perceived that thorns, for example, are modified branches. Of subterranean bulbs he says, truly, that 'the *Strings* (i.e. the adventitious rootlets) only are *Roots*; the *Bulb* actually containing those *Parts* which springing up make the *Leaves* or Body, and is, as it were a Great Bud under ground'.

Grew perceived that the growing zone of a stem lies near the surface and that 'the young *Vessels* and *Parenchymatous Parts*' (pp. 157–61) are formed each year 'betwixt the *Wood* and the *Barque*'. Thus, 'every year the *Barque* of a *Tree* is divided into Two Parts, and distributed two contrary ways. The outer part falleth towards the *Skin*; and at length becomes the *Skin* itself. The inmost portion of the *Barque* is annually distributed and added to the Wood, the *Parenchymatous Part* thereof making a new addition to the *Insertions* within the *Wood*.'

In France, worthy successors of Perrault were J. G.

Duverney (1648–1730), his pupil, J. B. Winslow (1669–1760), L. J. M. Daubenton (1716–1800), and Vicq d'Azyr (1748–94). Duverney wrote a fine monograph of the organ of hearing (1683), did good work on the comparative anatomy of the heart and great vessels (1701) and on the structure of fish. Winslow carried further the muscular mechanics of Borelli (p. 360). Daubenton anatomized many species and provided Buffon (pp. 293–5) with exact comparative knowledge.

Vicq d'Azyr, a busy Parisian practitioner, was the best exponent of the comparative method in France before Cuvier. His work, largely based on the study of function, leads naturally to Hunter. His intensive study of a large variety of forms led him to apply the principle of correlation (pp. 228–31) and links him to Cuvier (pp. 227 f.). He made extensive investigations of the human brain and consistently compared the anatomy of man to that of other mammals. His suggestions of a detailed comparison of the parts of the fore and hind limb were of great service for subsequent comparative studies.

In England in the earlier eighteenth century, the only names worth mention are James Douglas (1676–1742) who wrote well and learnedly on myology (1707) and John Hill (1716–75) a verbose and quarrelsome busybody whose voluminous works conceal some real advances towards a comparative anatomy of plants.

§ 2. *Some Eighteenth-century Conceptions of Living Nature*

Following the great comparative investigations of the later seventeenth century, there was a lull in progress. Certain philosophical prepossessions tended to obscure good work. Among these was the conception of the 'Ladder of Nature'. Aristotle had been content with the formal projection of the idea (p. 40). He did not erect it as a rigid framework into which all observations were to be fitted. This, however, was the policy of many eighteenth-century naturalists.

Prominent among those who thus approached nature with preconceived ideas was the Genevan, Charles Bon et (1720–93). It was an age when 'free-thinkers' and 'deists' (p. 126) on the one hand, and Christians on the other, inhabited sharply-divided camps. It was the century of Voltaire (1694–1778), the scoffer, whose works everybody read, and of Archdeacon William Paley (1743–1803), whose unphilosophical *Evidences of Christianity* (1785) hypnotized until this century the University where Newton had taught.

Bonnet was very influential on the Christian side and was, in some sense, a predecessor of Paley. He raised the doctrine of preformation, or rather of *emboîtement*, to the rank of a dogma, and both it and the process of partheno-genesis which he rediscovered (p. 168) he made to serve religious ends. Moreover, he stamped upon the comparative anatomy of his age a rigid interpretation of the Aristotelian ladder of nature. Passing from the most subtle of the elements, fire and air, through water and earth, to the minerals, it ascended through crystals to living things, proceeding via moulds, plants, insects, and worms to fish, birds, mammals, and finally to man. Man is the type by which other forms must be tested. 'All beings', wrote one of Bonnet's followers, 'have been conceived and formed on one single plan, of which they are the endlessly graded variants. This prototype is man, whose stages of development are so many steps toward the highest form of being.' 'Man is the measure of all things' said Plato, two thousand years before him.

Such views drew much from the philosophy of Leibniz (1646–1716). They pass insensibly into the attitude known as *Naturphilosophie*, which became very popular in Germany (pp. 215–27). This temper has, at times, served well the cause of scientific advance and, surprisingly enough, has often stimulated comparative studies.

§ 3. *Hunter* (1728–93)

A remarkable and isolated position is occupied by the great English biologist John Hunter. Dull at school, he

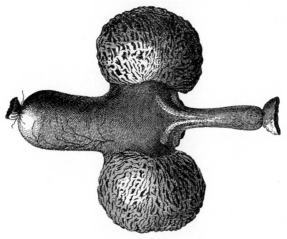

Fig. 101. The crop from a male pigeon, while the female was breeding. It is turned inside out to show the development of the mucous membrane on its internal surface. This secretes a substance which nourishes the young. (From Hunter, 1786.)

was sent to assist a Glasgow cabinet-maker, but persuaded his brother William, an eminent anatomical teacher in London, to accept him as his assistant. His real genius now declared itself, and he became a very great exponent of the comparative method. He distinguished himself as a practical surgeon, but his earnings and his powers he devoted entirely to science.

Hunter accumulated a wonderful museum; he paid and kept assistants and draftsmen; he even set up his own printing-press and acted as his own publisher. He performed an enormous number of physiological experiments, kept his own menagerie, and never missed an

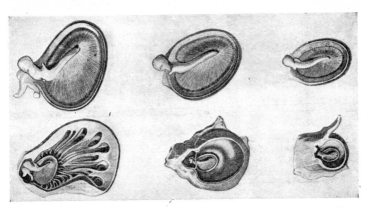

FIG. 102. Ear membranes and their apparatus, from drawings by Clift of specimens in Hunter's collection. From left to right are the structures in the cow, deer, and hare. In the tier below is shown the whole apparatus in these animals. In the tier above the membranes and their ossicles only.

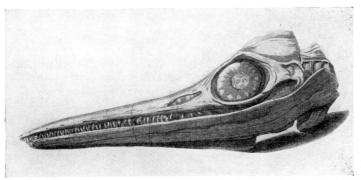

FIG. 103. Skull of Ichthyosaurus from a drawing by Clift in the collection of Hunter.

opportunity of securing the body of a rare beast for dissection. He was interested in fossil no less than living forms (Fig. 103), in physiology as much as in anatomy.

His energy in the pursuit of science was inexhaustible. He was ever fearful that he would die with his museum and his papers in disorder. In fact, of the great mass of notes that he left behind, some were published in the name of another and others were wantonly destroyed.

Hunter willed that on his death his collections should be offered to the British Government for purchase at a low price. The Prime Minister, the younger Pitt, on being approached, answered, 'What! buy preparations? Why, I have not enough money to buy gunpowder!' Fortunately an old servant was more generous. He devoted himself to these despised preparations, though he had to live on seven shillings a week. The price of bread rose enormously owing to the war, but this zealous man never relaxed his ward. Hunter died in 1793. In 1795 the Government decided to reconsider the matter. The guardian of the collection was almost starving, but he held on. In 1799, having taken six years to make up its mind, the Government bought the collection and handed it over to the Royal College of Surgeons of London. Hunter's old servant was appointed curator. At the end of his vigil, the specimens were in a better state than on his master's death. His name was William Clift (1775–1849). Though an admirable assistant, he was devoid of initiative and made no important discovery, but he has an honoured place in the history of science.

A spirit informs the Hunterian Museum which is essentially different from the 'magpie instinct' which has been the motive of many great collections. Here every object has its place and its reason for being included. Hunter had anatomized over 500 species of animals— many a great number of times—as well as numerous species of plants. He designed to trace systematically the different phases of life as exhibited by the organs, the structures, and the activities of both animals and plants.

A few of the biological observations which occupied
Hunter appeared in the *Philosophical Transactions*. They
show the range of his studies. He was interested in
monstrosities, and was the founder of their experimental
investigation. Animal behaviour attracted him, and he investigated, for example, that of bees. He made many
experiments on the temperature both of plants and of
animals. Anything related to animal mechanism particularly appealed to him, and his researches on the air-sacs
of birds, on the electric organs of fish, and on the structure
of whales and sea-cows added greatly to knowledge.

No great naturalist is so refractory to literary treatment
as Hunter. He was himself lacking in literary power and
could not put outside himself the thoughts surging in his
mind. Often we can reach his meaning only by piecing
together hints and notes. His few systematic treatises
fail to do him justice. His main work is his museum
and the influence that it has exerted. And yet his real
monument is not so much the museum itself as his ideal
for a Museum. There were many museums before his
(p. 143), but they were seldom more than collections of
curiosities, exhibitions to illustrate human anatomy, or, at
best, attempts to set forth the superficial characteristics
of a number of species. Hunter created the conception of
a collection to illustrate the varieties of structure and function right through the organic series. His monument is
therefore the British Museum of Natural History of
which his son-in-law Richard Owen (p. 287) was director.

§ 4. *The Naturphilosophen; Kant* (1724–1804), *Goethe*
(1749–1832), *and Oken* (1779–1851)

Hunter was averse from all abstract discussion and had
no feeling for what is usually called 'philosophy'. But the
thought of his age was being given new direction by the
Königsberg philosopher, Immanuel Kant. The change

was inaugurated by this writer's famous *Critique of pure reason* (1771).

Kant was primarily a mathematician and physicist. His mind was slow in developing to its full powers, and his philosophical interest emerged only gradually from his treatment of scientific problems. Beginning with a world of phenomena, of nature, of experience—the determinate world of the man of science—he gradually passed into the world of the intelligible, of freedom, of ends.

To most, the two worlds seem still to confront one another. Scientific men of our own time still affirm this when they say that 'teleology is the enemy of science', meaning that the study of purpose, of ends, is inconsistent with the adequate description of phenomena. It was Kant's thought that the two attitudes are *not* opposite and irreconcilable. The problem reduces itself to the discussion of the relation between our perception of things and their real nature. Our perceptions, Kant held, come into relation with the real nature of things through the character of our processes of thought. In other words, our minds work along the lines that nature wills. Our minds are, as it were, attuned to Nature.

This view has implications with biology. Organisms are composed of parts. These are comprehensible only as conditions for the existence of the whole. The very existence of the whole thus implies an end. True, says Kant, nature exhibits to us nothing in the way of purpose. Nevertheless, we can only understand an organism if we regard it *as though* produced under the guidance of thought for the end.

We may note in passing that in coming to this conclusion Kant sets forth the relations of classes of organisms as though they were historically related to each other. He is thus prepared to accept evolution, and he expressly includes the possibility of organisms developing from lower to higher according to mechanical laws (*Kritik der Urtheilskraft*, 'Critique of Judgment', 1790).

The opposition, so familiar to the biologist, between mechanist and teleological views is, Kant thinks, due to the nature of our knowledge, that is of our experience. But our thoughts must be distinguished from our experience. In thought we pass constantly from the mechanist view of the parts to a teleological view of the whole, and back again. Nor do we separate these classes of view unless deflected by some specific doctrine that the parts are really separate. There is, Kant believes, a hidden basic principle of nature which unites the mechanical and teleological. That principle may be none the less real because our reason is unable to formulate it. In practice, in our use of the language of biology, we all accept such a principle, the most convinced mechanist no less than the teleologist.

The inevitable use of the language of teleology by naturalists of mechanist views might be illustrated from innumerable works from Kant's day to our own. A single passage may suffice. It is from a work of 1923 by the foremost exponent of the cell doctrine of the last generation. Professing himself a determinist and a mechanist, E. B. Wilson (1856–1939) yet wrote:

'The cytologist is struck by the *extraordinary pains that nature seems to take to ensure* the perpetuation and *accurate distribution* of the components of the system in cell division. . . . *Nothing is more impressive than the demonstration of this* offered by the nucleus of the cell; but *its obvious meaning is often disregarded or treated with a blind scepticism which pretends that no meaning exists.* To our limited intelligence it would seem a simple task to divide a nucleus into equal parts. . . . *The cell manifestly entertains a very different opinion.* Nothing could be more unlike our expectation than the astonishing sight that is step by step unfolded to our view by the actual *performance.*'

(Italics inserted. From E. B. Wilson, *Physical Basis of Life*, 1923.)

So the modern naturalist returned to the antithesis which Kant was seeking to resolve. A new system of philosophy had intervened. An imperfect fusion of the Kantian scheme with atomistic materialism was, in fact, the working philosophy of most nineteenth-century biologists.

Kant undoubtedly exerted great influence on the biological thinkers of his day. His hand is to be traced especially in the writings of the 'Naturphilosophen'. Of these Goethe and Oken are the most important.

From the seventeenth century onward, schemes of classification of animals were based upon their anatomy. Implicit in such attempts was the conception of some uniformity of anatomical plan. But it is remarkable how far a science may advance before coherent expression is given to its principles. The German poet and philosopher Johann Wolfgang von Goethe was perhaps the first since Aristotle to point out explicitly that the structures of animals exhibit uniformity of anatomical plan. He sought general expressions for the constant factors in their anatomical composition. His generalizations were often vague or extravagant, but in directing the search for them he rendered many important services. Not the least of these was that he persuaded biologists to turn their eyes away from the anatomy of man as the type to which all other creatures were to be referred.

Searching for correspondences in the structures of different creatures, Goethe had his first success in 1784. He noted that the upper jawbone in man is formed of one piece only on each side, whereas in other animals it is of two. He inferred that in early life the human upper jawbone must consist of two pieces on each side, and he marked out the division between the two. Later he was able to demonstrate the truth of his view (1822).[1]

---

[1] Vicq d'Azyr had come to similar conclusions in 1780, and Fallopio in 1561. Goethe and his circle, however, did not know this.

In 1790 Goethe completed an essay on plant meta-morphosis which exhibits the influence of Kant. He set forth three doctrines of great importance.

(*a*) The genera of a larger group, such as a Family or Order, present something in the nature of variants on a common plan. They all express the same *idea*. The *idea* of Goethe became the *type* of A. P. de Candolle (1827, see pp. 197–8) and later writers, and is not far from Plato's.

(*b*) He amplified the view, hinted by Jung (p. 181) and expounded by C. F. Wolff (pp. 469–70), that the various parts of the flower are but modifications of leaves. He stumbled in details and his scheme was needlessly complex, but he reached an important conception (Figs. 151–3, p. 391).

(*c*) The so-called *cotyledons* of flowering plants, which give their name to the great groups, *Monocotyledons* and *Dicotyledons* (p. 184), are nothing but the first leaves borne by the infant shoot (Fig. 152, p. 391).

Goethe's *Preliminary Sketch to a general introduction to comparative anatomy* (1795) and his *Formation and transformation of living things* (1807) set forth several further important biological ideas.

(*d*) Extending his view of the leaf origin of the parts of flowers, Goethe considered that every living being is a complex of independent elements, each referable to one type. Thus not only is there a primordial animal and a primordial plant, representing the animal 'idea' and the plant 'idea', but also each of the parts of each organism represents one primordial part. This position is now indefensible but can be illustrated by the vertebrae. These, fundamentally of the same origin and structure, perform very different functions, and have different forms in different parts of the backbone. So it is, Goethe believed, with other organs. There are in fact animal groups, e.g. the Annelids or the Crustacea, in which he might have brought out the point better, but he did not know enough of their structure.

(*e*) Applying this view to the skull, Goethe expounded the famous *vertebral theory*, more explicitly set forth by Oken. Goethe maintained that the structure of the skull can be explained as a fused series of bone groups, comparable to the series of vertebrae.

(*f*) He enunciated a principle that has been termed the *law of balance*, according to which 'no part can be added without something being taken away from another part, and *vice versa*'. This law played a considerable part in the biological thought of the succeeding generation.

(*g*) He foresaw the value of embryology for the interpretation of adult structure. Thus he perceived that certain bones might be shown to be of compound origin by tracing them back to the embryo.

(*h*) He invented the useful word *morphology* (Greek *morphē*, form). The word describes the science which concerns the structure of living things, the relation of their structures to other living things, the way that these structures arise, and the factors that go to their production.

The Swabian, Lorenz Oken (1779–1851), came very early under the influence of the philosophy of Kant. He developed what he regarded as the ideas of his master into an extreme form, and he became the typical *Naturphilosoph*. This did not prevent him from doing some good routine biological work. Despite Oken's great learning and lofty character, there is hardly a biologist of the past with whom the man of science of to-day would feel so out of sympathy. On this account it is worth devoting a little attention to him. Through him, and through our differences from him, we may learn something of the nature of our own scientific method.

Before we pass to his forgotten methods we may glance

at his more positive achievements. Oken gave an interesting forecast of the modern cell doctrine which included a remarkable appreciation of the true nature of the protozoa (1805, p. 332). He made definite advances in embryological science. He founded a biological journal (*Isis*, 1816–48) which for thirty years published articles of unquestioned value. He instituted the type of annual meeting of men of science (1821) which has since had innumerable imitators, among them being the British Association for the Advancement of Science (first meeting: York, 1831).

Oken's effort to construct a biology that should reflect the actions of the mind appears in a work, the title of which may be rendered *Foundations of Naturphilosophie, the theory of Linnaeus and the classification of animals based thereon* (1802). For Oken, man is the summit and crown of nature. Thus the whole animal kingdom is the representation of his several activities and organs—naught else, in fact, than man disintegrated into the five senses through which, alone, his mind can learn of nature. On this basis there must be five classes of animals, and not more than five. These are virtually representatives of the five senses of man. They are:

(a) *Dermatozoa*, in which the sense of feeling leads. Invertebrates.
(b) *Glossozoa*, the lowest grade in which a tongue, the organ of taste, is developed. Fishes.
(c) *Rhinozoa*, in which the organ of smell is separated and the nostrils take in air. Reptiles.
(d) *Otozoa*, in which the organ of hearing begins to be independent and to open externally. Birds.
(e) *Ophthalmozoa* which have all the organs perfect with the sense of sight leading. Mammals.

Oken's personality and the fashion of the day gave him great influence. His doctrines are now all forgotten. One of them, the vertebral theory of the skull, needs discussion.

It is an evident fact that the spinal column contains a number of repeated units, the vertebrae. To each corresponds a group of nerves and a group of muscles, which show similar repetition. Vertebra, nerve, and muscle are thus, as we say, 'segmentally arranged'. This fact is even more evident in the embryo than in the adult; and in the lower forms, such as fish, than in the higher, such as mammals. Science has found no adequate solution to the

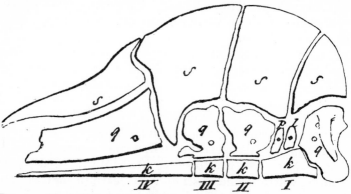

FIG. 104. Oken's diagram of skull of sheep (1819). Bones separated to suggest its origin as from four 'vetebra'. In each 'vertebra' the bones *k*, *q*, and *s* are supposed to be homologous with the similarly lettered bones in the other 'vetebra'.

puzzle of how in most groups of animals many organs come to be segmentally arranged.

For Oken the idea of segmentation provided a plan on which the vertebrate body was built. All parts were segmented. He relates how, resting one day on a walk, he sat contemplating the skull of a deer. It flashed across his mind that it was but a group of fused vertebrae. The skull, like the rest of the body, was segmented. Thus arose his *vertebral theory of the skull* (1807, Fig. 104).

Oken pressed the matter further. Some of the segments of the body of vertebrate animals present appendages—to wit, the fore and hind legs. Doubtless in the complete type of vertebrate all segments should present limbs as they do in some jointed animals, notably Crustacea, Myriapoda, &c. The vertebrate type was, in fact a regular centipede! Were there remains of such limbs in the head? Oken found them in the jaw. So also the mouth cavity represented a series of segments of the intestines, the brain represented a series of segments of the spinal cord, and so on.

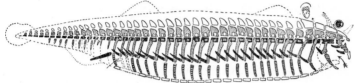

FIG. 105. The 'archetype' of the vertebrate skeleton according to Owen (1846). In the view of Oken, Owen, and others the vertebrate skeleton could be analysed into a series of completely homologous and very similar segments. Of these, several of the anterior elements were supposed to be fused to form the skull. Each segment is equipped with a series of rib-like structures with appendages. From the first two of these segments the jaws were thought to be formed. Other segments developed as limbs.

This theory carried away Goethe, who had already an inkling of it. It misled von Baer, and some of those who followed him, into a blind alley of speculation. Even Johannes Müller (pp. 393–5) adhered to it, and some of his best work was done in the false light that it yielded. The conception of the segmented 'archetype' was elaborated and given colour by several other comparative anatomists, of whom Owen was one (Fig. 105).

§ 5. *The Eclipse of Naturphilosophie*

The vertebral theory was finally refuted by the embryologists. Here we may note the decisive discovery of Martin Heinrich Rathke (1793–1860), the successor of von Baer (pp. 471–6) at Königsberg. Rathke's best-known discovery was of structures homologous with gill slits in bird and mammalian embryos (1825, Fig. 162). At a certain stage of development these creatures present a series of structures which in position and anatomical relations are identical with similar structures in fish embryos. In the latter they continue to develop into gill slits and their associated organs. Malpighi had described arteries arching round the gullet of the chick embryo in the position of these gill slits (p. 152). These arches in the embryo were now explained by the presence of other structures associated with gill slits. They correspond to the vessels that supply the gills in fish. But the skeleton of the gill slits in the embryo was found to correspond to certain parts attached to and ultimately forming part of the jaw, the organ of hearing and the organ of voice in the adult mammal and bird. Thus it became impossible to believe that the parts of the jaw, &c., had originally been of the nature of limbs. So the vertebral theory of the skull became discredited.

The theory was finally dismissed by T. H. Huxley in a classical paper published less than a month before the first communication on the origin of species was read by Wallace and Darwin (1858). Huxley declared his belief that the 'study of the gradations presented by a series of living things may have the utmost value in suggesting *homologies* (p. 226), but the study of development alone can demonstrate them.' Thus he placed embryology in the position of the arbiter of the structural relationships of animals. It was prophetic of the dominant position which it came to occupy in evolutionary comparative anatomy.

The vertebral theory of the skull was an extreme mani-

Description of Fig. 107.

FIG. 107. Drawings by Leonardo to show the correspondence between the horse's *stifle-joint* and the human knee, on the one hand, and the horse's *hock* and the human heel, on the other. Leonardo also attempts to bring out the homology of the muscles which link limb to body in the case of horse and man. He has represented the muscles diagrammatically as strap-like bands, to assist in the interpretation of their action.

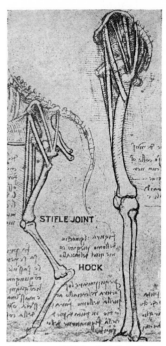

FIGS. 106 and 107. Figures illustrating the nature of biological *homology* and *analogy*.

FIG. 106. Structure of wing in the three classes, Reptiles, Mammals and Birds. In each case the fore-limb forms the main part of the apparatus of flight. The general plan of the limb is found in all. In this sense we trace *homology* in various parts. The other details are, however, so different in the three cases that it is evident that they have arisen quite independently and that the wing in the three classes is an *analogous* and not an *homologous* structure.

In the fossil reptile pterodactyls a membrane stretches between fore- and hind-limb. The main support is an enormously developed fifth or 'little' finger. There are three free fingers. The thumb is absent.

In the mammalian bats the membrane is supported by both limbs, but is largely formed by an enormous web between digits 2, 3, 4, and 5. The thumb alone is free.

In the wing of birds the fore-limb alone is involved. The only digits remaining are 1, 2, and 3, and all are greatly reduced except 2. In the fossil bird *Archaeopteryx* these three digits are still free and provided with claws (Fig. 108). The expanse of the wing in the bird is formed by feathers and not by a membrane. The hind-limb is free and is not involved in the flying apparatus.

FIG. 107. *For description see opposite page.*

festation of the thought of Naturphilosophen. These men were interested in 'types', 'schemata', 'ideal forms'. They linked them with their conception of the *purpose* which lies in living things. The later Darwinians, in dismissing this doctrine—this 'religion'—said that it failed to distinguish between analogy and homology.

The passage just quoted from Huxley is of interest as an early use of the important biological term *homology* (Greek 'agreement'). It was first employed in biology to indicate the relation of an organ to the general type. Organs were said to be 'homologous' with each other when they had the same relation to that type. Later, with the general advent of the doctrine of descent, *homology* became opposed to *analogy*. Homology became a morphological conception based on descent. Analogy is a physiological relationship, without any implication as regards relationship by descent. Thus the human knee, as shown in the drawing of Leonardo (Fig. 107, p. 225) is *analogous* with the hock of the horse but *homologous* with the stifle-joint. The distinction was certainly not adequately recognized by the Naturphilosophen.

Naturphilosophie was destroyed by the evolutionary view and is now placed among the lumber of forgotten theories, but it is well to remember that the method had its victories. The paths of discovery are as diverse as the human mind. But the Naturphilosophen went to fantastic excesses. If Goethe succeeded, it is because he was a man of genius and was more interested in Nature than in his own method. Others were more interested in their method than in Nature. Such men, Oken among them, developed enormous and bizarre systems which remain of interest both for historical reasons and as object lessons.

Naturphilosophie infected other countries besides Germany. Étienne Geoffrey St. Hilaire (1772–1844) in France and Richard Owen (pp. 237–9) in England carried deep marks of its influence. Yet both were comparative anatomists of the very highest order.

It is the prime task of morphology to analyse the bodies of creatures into organs and to compare them with those of other creatures. But morphology is a poor thing unless it can regard the creature as a whole, and no creature is a whole unless alive. This was Goethe's standpoint when he expresses his 'yearning to apprehend living forms as such, to grasp the connexion of their external visible parts, to interpret them as indications of the inner activity, and so, in a certain sense, to master the whole conceptually'. For him the science of morphology should separate itself neither from physiology nor from an understanding of the organism as a vitally active work of art. 'If the creative spirit brings creatures into being and shapes their evolution according to a general plan, should it not be possible to represent this plan, if not to the sense, at least to the mind? This was the view toward which A. P. de Candolle was working (*Organographie vegetale,* 1827, pp. 197–8). This was also the view of Cuvier.

§ 6. *Cuvier* (1769–1832) *and the Principle of Correlation of Parts*

Georges Cuvier has been called the 'dictator of biology'. Few men of science have attained to so influential a position. For years he was the admitted leader of science. His own investigations determined very largely those of his contemporaries. His influence was stimulating to research, though it cannot be said that he invariably exerted his power with the greatest wisdom. He held strong opinions, and tended to suppress views opposed to his own, notably those of the evolutionist Lamarck (pp. 296–2). Cuvier himself believed firmly in the fixity of species.

Cuvier was the son of a Swiss Protestant officer in the French army. He was born near Belfort, then in the Duchy of Württemberg. He studied at Stuttgart, and in 1795 became assistant at the Musée d'Histoire Naturelle at Paris. He rapidly attained distinction by his anatomical descriptions of a variety of animals. In 1800 he published a work on fossil elephants which he compared with living species. The study of fossil forms soon became a separate science, *palaeontology*. Cuvier had a leading part in this development. His *Récherches sur les ossemens fossiles* (1812) is a classic.

The Emperor Napoleon selected Cuvier to direct the reform of education in France. This detached him from his special studies to which, however, he soon returned. On several occasions he was again withdrawn for administrative functions, but these never permanently held him. After the downfall of Napoleon he obtained the favour of the restored dynasty.

With encyclopedic knowledge and boundless energy, Cuvier formed vast scientific schemes. He had a capacity for winning the aid of the ruling powers, and he brought many of his projects to fruition. With the gift of inspiring others, and of causing them to work for him, he never lacked for skilled assistance. He united general culture to comprehensive scientific outlook. Thus he perceived the interest and importance of the history of science, and lectured and wrote well upon that subject. His historical writings can still be consulted with profit.

Cuvier was a 'morphologist' in Goethe's sense. He was, in fact, under the influence of the great German, and his view of organisms was based on their activities as living things. The main conception that guided his work was the *principle of correlation of parts*. The nature of this principle must be discussed.

Organs do not exist or function in nature as separate entities but as parts of organic living wholes. In these living wholes, certain relations are observed which are fundamental to their mode of life. Thus feathers are always found in birds and never in other creatures. The presence of feathers is related to a certain formation of the forelimb with reference to its use as wing. This, in its turn, is related to certain formations of the collar-bone and breast-bone, with reference to the function of flight; these,

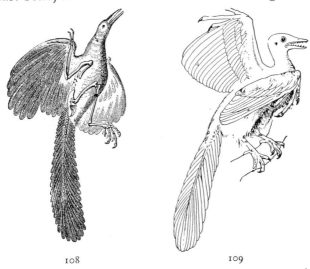

108   109

FIG. 108. Owen's first restoration of *Archaeopteryx lithographica*, a Jurassic fossil bird discovered in 1861 in lithographic slate in Bavaria.

FIG. 109. Modern restoration of *Archaeopteryx*. The animal, though covered with feathers, has yet many reptilian features, e.g. teeth, claws on fore-limb, and long, free, jointed tail.

again, to the form and movement of the chest; these, again, to the function of breathing. So the 'principle of correlation' might be followed through the whole being of the bird, down to its minutest parts. Yes, and even to its psychology.

It is with structure that, for the moment, we are concerned. Given a feather, it is possible to infer that its owner had a particular form of collar-bone. Again, given a particular form of collar-bone, it is possible to infer a feather. The existence of the extraordinary fossil bird *Archaeopteryx* was, in fact, revealed by the discovery (1861) of the impression of a feather (Fig. 108). If enough be known of the comparative morphology of the bird group, it is possible by the use of this principle to make most sweeping inferences.

It was not reserved for Cuvier to discover the principle of correlation. In an elementary form it is obvious, and it is known to us all. If any one were to find a severed hand, he would be confident that it had once been attached to a human body and not to that of an animal. The race, sex, occupation, state of health, &c., of the owner of the hand might easily be inferred. The 'principle of correlation' is the theme of most detective stories. Anatomists before Cuvier and back to Aristotle had, to some extent, been able to act upon the principle of correlation with reference to animal bodies. As with many other scientific principles, it was applied long before it was adequately formulated.

An achievement of Cuvier was that he enunciated this principle clearly, and made it the guiding element in his work. He refined and extended its application far beyond all previous knowledge. As a result of his wide and deep studies, the principle could often be brought to bear upon the merest fragment of an organized body.

The principle of correlation was especially of value to Cuvier in his studies of fossils, as these are usually fragmentary. From fossil elephants he passed to the fossil bones of other mammals and of reptiles. He discerned that these mostly belonged to species no longer existing. At the same time he was constantly working at the skeletons of living forms. Thus he came to be in a good position to elucidate the relationship between the living and the extinct forms. By practice in restoring the missing parts of fossil skeletons, Cuvier became able to recognize the species from a few bones, or even from a fragment of a bone. In this he became very expert, and founded a method of dealing with extinct forms based on the anatomy of extant species that has since proved of the greatest value. The science of palaeontology owes to no one so great a debt as to Cuvier.

Inferences correctly made by Cuvier and his followers, from small portions of fossil bones and subsequently confirmed by discoveries of complete skeletons, excited great astonishment at the time, and are still sometimes cited with wonder. They were, however, the natural outcome of such intensive study of the physiology and anatomy of animals as had long been applied to the body of man. Further, the principle of correlation formed the basis of the whole of Cuvier's classificatory system, and of the beautiful and exact view that he gave of the animal kingdom.

§ 7. '*Le Règne Animal*' (1817)

Cuvier's most comprehensive effort, and that by which he is best known, is *Le Règne animal distribué d'après son organisation pour servir de base à l'histoire naturelle des animaux, et d'introduction à l'anatomie comparée* (first edition, 1817. Many subsequent editions and translations). This formidable work embodies twenty-five years of research on living and fossil forms. It was the most comprehensive biological work since Linnaeus. Knowledge had accumulated in the meantime, and Cuvier had investigated more forms than any before him. The work describes a species from almost every genus then recognized, and is illustrated by hundreds of beautiful plates.

Cuvier's arrangement of the animal kingdom, that is to say his 'system of classification', is of interest. In contrast to Lamarck (p. 296) he will have nothing to do with a 'scala naturae'. He divides animals into four great *em-*

branchements, each of which is built on its own peculiar and definite plan.

I.   *VERTEBRATA*, animals with a backbone.
II. *MOLLUSCA*, such as slugs, oysters, snails, &c.
III. *ARTICULATA* or jointed animals, such as insects, spiders, and lobsters.
IV. *RADIATA*, a group containing all remaining animals.

The Radiata is a very miscellaneous collection, but the other three are all 'natural' groups.

In placing animals in these groups, Cuvier was guided by an analysis of two main sets of functions, which he regards as fundamental, together with the organs subservient to them. Firstly, the heart and circulation form a kind of centre for what he calls the *vegetative functions*, to which the breathing apparatus is attached. Secondly, the brain and spinal cord preside over the *animal functions* and are attached to and served by the muscular system. The vegetative and animal activities are reminiscent of the *vegetative soul* and the *animal soul* of Aristotle (p. 38). Indeed, the whole scheme is infused with the thought of that naturalist.

We turn now to the structures and functions manifested by the four *embranchements*. Vertebrata, Mollusca, and Articulata are bilaterally symmetrical. Radiata are radially symmetrical. Vertebrata have a heart and blood-vessels and a continuous brain and spinal cord. Their skeleton is internal. Its basis is an axis—skull and vertebral column—and appendages—fore and hind limbs. Mollusca have a heart and blood-vessels. Their nervous system consists of discontinuous separate masses. They have no internal skeletons, and their muscles are attached to the external parts, namely the skin and shell. Articulata exhibit in the vegetative parts a functional transition from a system of blood-vessels to a tracheal system (Fig. 89). Their nervous system consists of two long cords running along

the lower part of the body. They have a hard external skeleton to which muscles are attached. Their limbs are jointed.

So far, the position of Cuvier is not greatly different
from that of a biologist of our own time. His treatment of
his 'Radiata', however, separates him widely from us.

The Radiata of Cuvier are, according to our standards,
a random mixture of types. He holds that the Radiata
have ill-defined nervous and muscular systems, that they
have no vascular system, and that their bodies tend to
approach plants in homogeneity. The lower forms are a
homogeneous pulp devoid of organs. There is nothing in
all this to which the modern biologist will subscribe.

The existence of this group, Radiata, in Cuvier's scheme,
and the qualities ascribed to them, give an index of the
progress of biology since his time. The Radiata of Cuvier
are now split up into a dozen or more Phyla and include
the vastly important sub-kingdom of Protozoa or unicellular animals. In estimating his work, we must remember
that there was still no knowledge of the part that the cell
plays in the animal economy, and that he was devoid of the
resolving power of the modern 'achromatic' microscope
(p. 515). Further, despite Cuvier's faith in function as a
guide to form, the technical study of physiology was still
too little advanced to be of use to him, and comparative
physiology as a separate science was non-existent. In
assigning to Cuvier his position as the supreme comparative anatomist, we are therefore justified in passing
over his Radiata and in concentrating on his other three
embranchements.

The classificatory system of Cuvier contains Classes, of
which the titles are still familiar. The scheme may be
drawn up thus:

EMBRANCHEMENT I. VERTEBRATA

Class 1. Mammalia, identical with the modern Class of the same
    name.

Class 2. AVES or birds, identical with the modern Class of the same name.

Class 3. REPTILIA, including, besides reptiles, such forms as frogs and newts, now placed in a separate Class as Amphibia.

Class 4. PISCES, identical with the modern Class of Fishes, from which certain forms, such as lampreys, must now be excluded.

EMBRANCHEMENT II. MOLLUSCA

Class 1. CEPHALOPODA, including forms like the Octopus, and almost identical with the modern Class of the same name.

Class 2. PTEROPODA, a group of free swimming forms, roughly corresponding to the modern Class of the same name.

Class 3. GASTEROPODA, including the ordinary land and water molluscs, such as slugs and snails, roughly corresponding to the modern Class of the same name.

Class 4. ACEPHALA, molluscs with two shells, such as the mussels and oysters. This Class corresponds to the so-called 'bi-valve' molluscs.

Class 5. CIRRIPEDIA or barnacles. These are not in fact molluscs at all, but a group of degenerate Crustacea as was shown by Vaughan Thompson (pp. 495).

Class 6. BRACHIOPODA, a very ancient well-defined group, not now regarded as related to molluscs.

EMBRANCHEMENT III. ARTICULATA

Class 1. CRUSTACEA, much as in modern classifications, but omitting the Cirripedia and certain parasitic forms.

Class 2. ARACHNIDES or spiders, &c., much as in modern classifications.

Class 3. INSECTA, much as in modern classifications.

Class 4. ANNELIDES, a miscellaneous and artificial group, many unrelated to each other and none closely related to the Crustacea, Spiders, or Insects. The group was broken up later by Lamarck.

This classification was a workable scheme, and much superior to anything that preceded it. By its use a naturalist could make some general survey of variations in structure of the animal kingdom as a whole.

Apart from the important generalities summed up in this classification and apart from his great work in the foundation of Palaeontology as a science, Cuvier made two very important contributions to comparative studies.

First was his exploration of the anatomy of the Molluscs. He dissected a great many of these forms. He was thus able to give a good general account of the group and to place its internal classification on a satisfactory basis. No other group except the vertebrata had been so intensively investigated (*Mémoires pour servir à l'histoire des Mollusques*, 1817).

Second was his systematic treatment of the vast Class of Fishes. These had, in fact, been among the first animals to be scientifically treated, as by Belon (p. 90), Rondelet (p. 93), Gesner (p. 94), Willughby (p. 183), and Artedi (p. 187–8). Their anatomy had, however, been neglected. Cuvier placed their affinities, as revealed by their structure, on an entirely new basis. Moreover, he united in his scheme fossil with living forms (*Histoire naturelle des poissons* 1828–31).

§ 8. *The Doctrine of Catastrophes*

The effect on the mind of Cuvier of the discovery of a large number of fossil forms may seem strange.

Cuvier realized that the evidence of Geology showed that there had been a succession of animal populations. He perceived that vast numbers of species, many no longer existing, had appeared upon the earth at different periods. But following Linnaeus, he was a firm believer in the fixity and unalterability of species, though his colleague and early friend, Lamarck, with whom he had quarrelled, was engaged in putting forward the opposite view (pp. 296–300).

Cuvier had, however, to account for the extinction of some forms of life and what seemed to be the creation or at least the appearance of new forms. His explanation of these remarkable facts was that the earth had been the scene of a series of great *catastrophes*. He believed that of the last of these catastrophes we have an historic record. It is the flood recorded in the Book of Genesis! He expressly denied the existence of fossil man.

'If there be one thing certain in Geology', he wrote, 'it is that the surface of our globe has been subject of a great and sudden catastrophe of which the date cannot go back beyond five or six thousand years; that this catastrophe has overwhelmed the countries previously inhabited by men and by those species of animals with which we are to-day familiar; that it caused the bed of the previous marine area to dry up and thus to form the land areas now inhabited; that it is since this catastrophe that such few beings as escaped have spread and propagated their kind on the newly uncovered lands; that these countries laid bare by the last catastrophe had been inhabited previously by terrestrial animals if not by man and that therefore an earlier catastrophe had engulphed them beneath its waves. Moreover, to judge by the different orders of animals of which remains have been revealed, there were several of these marine irruptions.'

It will be seen that Cuvier does not commit himself to the doctrine of a special creation following each catastrophe. What he suggests is that the earth was repeopled from the remnant left. This does not explain the appearance of new species in geological time. He believed, however, that the species which appeared as new came from parts of the world still inadequately explored by geologists.

His followers carried the matter farther and elevated his teaching into a doctrine of successive creations. This came to assume fantastic forms even in the hands of serious scientific exponents. Thus Alcide d'Orbigny (1802–57) expounded the science of palaeontology on the basis of twenty-seven successive creations (1849). There were many variations on this theme which need not detain us, despite the prodigious literature to which it gave rise.

§ 9. *Owen* (1804–92) *and Palaeontology*

The personality of Cuvier lit up a zeal for comparative anatomy and palaeontology which lasted throughout the nineteenth century. Of those inspired by this great movement, Owen was perhaps the most typical. He is also interesting as he was influenced by Naturphilosophie on the one hand and was an obstinate opponent of Darwinian evolution on the other.

Richard Owen, after a few years as surgeon's apprentice, went to Edinburgh University and thence to London. In 1827 he became an assistant at the Hunterian Museum. From Clift (p. 214), whom he succeeded, Owen imbided a reverence for the work of Hunter. In 1830 when Cuvier visited London, Owen made his acquaintance and he went to study in Paris. He began at once the publication of works on comparative anatomy and palaeontology. A continuous stream of these proceeded from his pen throughout his active life.

His monumental *Catalogue of the physiological series of comparative anatomy contained in the museum of the Royal College* (5 vols., 1833–8) is still of great value. To identify the species from which Hunter's specimens had been derived, Owen dissected a large number of animals. The dissections of the rarer of these he carefully recorded, and in many cases his accounts of these creatures are still consulted by naturalists.

Owen next embarked on an immense investigation of the teeth of mammals (*Odontography*, 1840–5). Teeth, being the hardest bones in the body, are most often found fossilized. Thus his investigations led him into palaeontology, of which he soon became one of the admitted masters. He published many monographs of extinct forms. Perhaps his best-known are those of the giant bird, the recent but extinct Dinornis of New Zealand (1846), and the much more ancient giant walking sloth,

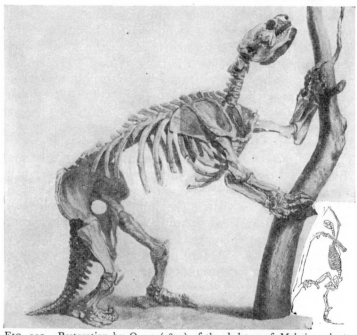

FIG. 110. Restoration by Owen (1842) of the skeleton of *Mylodon robustus*. This massive creature, as big as a Rhinoceros, was an inhabitant of South America. There it occupied an area which included that of its diminutive arboreal ally, the slender modern sloth *Bradypus tridactylus*. The skeleton of the sloth is represented inset and drawn to scale. Both in size and build it is a great contrast to that of the *Mylodon*. (Compare Fig. 111.)

the fossil Mylodon of South America (1842, Fig. 111).

In 1856 Owen became director of the Natural History department of the British Museum. The immense wealth of material there gave him unrivalled opportunities, and his activity and industry rose to the occasion. His great work on the *Anatomy and Physiology of the Vertebrates*

(1866–8) was based entirely on personal observation and was the most important of its kind since Cuvier. The system of classification adopted by Owen has not won favour, but as a record of facts the book is still valued.

When Owen first became Director of the Natural History Department of the British Museum, that department was part of the main museum building and was greatly hampered for want of space. As a result of much agitation and representation in official quarters, Owen at last succeeded in persuading the authorities to move it to South Kensington.

In leaving the subject of comparative method we note that since the general acceptance of evolutionary theory, comparative studies have been almost entirely directed by belief as to genetic relationships (chap. viii). The alliance of comparative studies with evolutionary doctrine has had the effect of focussing attention on structure as distinct from function. Comparative physiology almost ceased to be studied in the later nineteenth century and is only now reviving. Comparative anatomy in its turn became largely a study of developmental stages, and embryology became the comparative study *par excellence* (chap. xiii).

FIG. 111. Skeleton of *Megatherium* described by Cuvier as 'an animal of the sloth family, but as big as a rhinoceros'. The first specimen was brought to Madrid from the Argentine in 1789. From Cuvier's *Recherches sur les ossemens fossiles*, 1812.

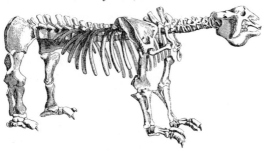

VII
DISTRIBUTION IN SPACE AND TIME

§ 1. *Early Biological Exploration. Joseph Banks* (1743–1820)
and Robert Brown (1773–1858)

IN the eighteenth century the practice was begun of
carrying naturalists on voyages of exploration, with
equipment for observation and collection. One of the
earlier expeditions thus provided sailed the Pacific under
command of James Cook (1728–79). Joseph Banks accompanied him. He possessed great wealth and while a
student at Oxford had paid for a lecturer on botany, the
professor of the subject having given only one lecture in
thirty-five years! Banks devoted himself to Natural History. In 1766 he had accompanied an Admiralty vessel
to Newfoundland on business concerning the fisheries,
and had thus gained his first glimpse of wild life. Through
numerous botanical excursions he had become a com-
petent naturalist. For the voyage with Cook, Banks provided equipment for biological work. He also engaged
Daniel Solander (1736–82), a pupil of Linnaeus, as
naturalist together with four artists.

The *Endeavour*, Lieutenant James Cook, 330 tons,
sailed from Plymouth in August 1768 and was away three
years. She crossed the Atlantic, doubled Cape Horn,
turned north-west, and was at Tahiti to observe the transit
of Venus in June 1769. She now passed south and spent
six months on the coast of New Zealand. Cook was the
first to circumnavigate its islands. Banks and his staff
were very active here, and great additions were made to
the knowledge of plants, birds, and fish.

Cook next turned west, and reaching Australia explored
its eastern seaboard. Cook gave its name to *New South*

Wales. Botany Bay was so called as being a happy hunting-ground for Banks and Solander. By navigating *Endeavour Strait* between Australia and New Guinea, Cook established the separateness of these two land masses.

Many new forms of life were encountered. Of one Cook says that

'In form it is most like the jerboa [but] as big as a sheep. The head, neck and shoulders are very small in proportion to the other parts. The tail is nearly as long as the body, thick near the rump, and tapering towards the end. Its progress is by successive hops, of a great length, in an erect posture. The forelegs are kept bent close to the breast. The head and ears bear a slight resemblance to those of a hare. This animal is called by the natives *kangaroo.*' [*Abbreviated.*]

The speed of the kangaroo was found to outstrip easily that of the ship's greyhound. Many other curious animals were seen, but the best work was done in collecting plants. The herbarium put together by Banks and Solander is at the Natural History Museum at South Kensington, and has formed the nucleus of a great collection there. The map, showing the track of the *Endeavour* marked in by Cook's own hand, is in the British Museum at Bloomsbury. The journals both of Cook and Banks still exist.

The expedition returned in 1771. Cook started his second voyage—again with naturalists on board—in the following year. It established a record in that, despite the hardships of four years, out of 118 men only four died. Of these, three were killed in accidents and one succumbed to a disease contracted before he left England. Cook's account of his methods of preserving health on board ship is one of the most important pronouncements on the subject.

Banks takes a distinguished place as a patron of science. Among those for whom he made a scientific career possible were the two Austrian artists, the brothers Franz and Ferdinand Bauer—among the best of all botanical draughtsmen—and the botanist Robert Brown.

Brown began early to exhibit the industry, thoroughness, and power of generalization that afterwards distinguished him. As a young regimental medical officer, he occupied himself with collecting plants. While stationed in Ireland he met Sir Joseph Banks, then President of the Royal Society. Banks had collected a very fine biological library and had greatly extended his herbarium. These were placed at Brown's disposal. Through Banks, Brown was appointed as naturalist to a new expedition that sailed in July 1801 under Captain Matthew Flinders (1774–1814), recently returned from exploring Australia and Tasmania. Ferdinand Bauer (1760–1826) went as botanical draughtsman, and William Westall (1781–1850), afterwards well known as an artist, accompanied the expedition to paint landscapes. Their ship, the *Investigator*, 334 tons, was fitted for scientific purposes. They arrived in December 1801 at their first objective, King George's Sound, in Western Australia.

The whole south coast of Australia was systematically explored. Many of the geographical names in that region are those given by Flinders. *Cape Catastrophe*, west of Adelaide, records the loss of a cutter and crew. Close by is the *Sir Joseph Banks Group of Islands*. A little farther on, covering the mouth of the great bay of Adelaide is *Kangaroo Island*. Farther east they met the French exploring vessel, the *Géographe*, and named the place *Encounter Bay*.

In May 1802 Flinders reached Port Jackson, where he met his supply ship and the *Géographe* again. The crews of all three vessels—less scientifically commanded than Cook's—were suffering from scurvy owing to want of fresh food. The *Investigator* was now unseaworthy. Flinders returned to England in another ship, and being wrecked, lost Brown's duplicate specimens and living plants. Brown and Bauer remained behind to explore the botany of the coasts of Australia and Tasmania. The two reached England in 1805, after four years' absence.

Brown had collected industriously throughout his travels, and Bauer had worked no less well. There was

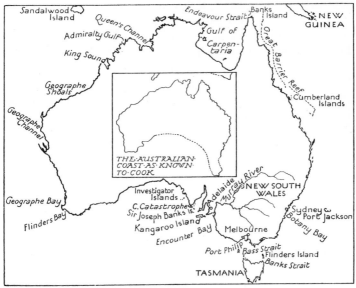

FIG. 112. Australia with Cook's outline map inset. The names inserted were given as a result of the voyages of Cook and Flinders.

a collection of 4,000 species of dried plants, many new to science. During the homeward voyage Brown busied himself with the close study of these, and made many important observations in botanical anatomy and physiology. Soon after his return he became Librarian to the Linnean Society and in 1810 to Sir Joseph Banks. That public-spirited and far-sighted patron of science died in 1820. By his will he endowed Franz Bauer to enable him to continue his work, and left to Brown his house in London together with his library and collections for life. In 1827 Brown permitted the books and herbarium to be stored in the British Museum, where they now form the nucleus of some very important collections.

The rest of Brown's life was passed in the uninterrupted production of a long series of important and original botanical works. He possessed great penetration and pertinacity, combined with unusual powers of generalization. He had a whimsical manner of publishing his discoveries. These he never announced distinctly but buried in memoirs, the titles of which do not suggest the nature of their content. Brown was, moreover, curiously secretive. Darwin in his autobiography writes that

'I saw a good deal of Robert Brown *facile princeps botanicorum* as he was called by Humboldt (p. 269). He seemed to me to be chiefly remarkable for the minuteness of his observations and their perfect accuracy. His knowledge was extraordinary great, and much died with him, owing to his excessive fear of ever making a mistake. He poured out his knowledge to me in the most unreserved manner, yet was strangely jealous on some points. I called on him two or three times before the voyage of the "Beagle", and on one occasion he asked me to look through a microscope and describe what I saw. This I did, and believe now that it was the marvellous currents of protoplasm in some vegetable cell (Fig. 147, p. 353). I then asked him what I had seen; but he answered me, "That is my little secret!" '

Brown is specially associated with four important topics:
(*a*) The Cell Nucleus (p. 333).
(*b*) The nature of the sexual process in higher plants (p. 516 and Fig. 175).
(*c*) 'Brownian movements' (p. 350).
(*d*) The microscopical examination of fossil plants (p. 278).

In a series of monographs he described the plants collected by himself and by many other travellers. All are models of exact observation combined with wide and deep knowledge. Among the most interesting is his account of the largest flower known. This was first seen in Sumatra by a young collector Joseph Arnold (1782–1818) while travelling with the great colonial administrator, Sir Stamford Raffles (1781–1826). In memory of these men Brown named it *Rafflesia Arnoldi*. The flower of this species first seen by Arnold

'measured a full yard across; the petals being twelve inches from the base to the apex, and it being about a foot from the insertion of the one petal to the opposite one. The nectarium would hold twelve pints, and the weight of this prodigy we calculated to be fifteen pounds.

'There were no leaves or branches; so that it is probable that the stems bearing leaves issue forth at a different period of the year. The soil where this plant grew was very rich, and covered with the excrement of elephants.' [*Abbreviated.*]

Rafflesia, as we now know, is parasitic on the roots of a vine and is devoid of green leaves. Each flower is either male or female. Its mechanism of fertilization is still unknown, but it is suggested that the elephant is the agent!

§ 2. *Pre-evolutionary Geological Theory*

In antiquity and in the middle ages fossils were usually looked on as *lusus Naturae*, 'Nature's little games'. Gesner wrote a book on them (1555). The Dane, Niels Stensen (1648–86), who spent some years in Italy, discussed the formation, displacement, and destruction of stratified rocks in Tuscany (1669). He recognized the organic origin of fossils. Stensen was followed by several Italians. Among them Vallisnieri (pp. 208, 317, 441) figured many marine fossils and recognized the nature of a geological 'fault'. Martin Lister (1638–1712), taking Stensen's view, wrote the first book devoted to fossils which accepted their organic nature (1678), and made suggestions for a geological map. Hooke and Ray had similar views.

The works of all these were known to Buffon (pp. 293–5). In his *Époques de la Nature* (1778) he tried to set forth a history of the earth. There were seven 'epochs'. The first was the 'incandescent' stage. The fifth saw the advent of large pachyderms—rhinoceroses, hippopotamuses, elephants—and their wide distribution even in temperate climes. The sixth saw the separation of the two continents. We live now in a seventh epoch—that of man.

The work of James Hutton (1726–97) has a more modern character. He travelled widely in order to study rocks, and perceived that it is mostly in stratified rocks that fossils occur. He saw clearly that the imposition of successive horizontal layers is inexplicable as a result of a single great flood but suggests rather a quiet orderly deposit over a long period. In his *Theory of the Earth* (1795) he interpreted the strata as having once been the beds of seas, lakes, marshes, &c.

It was soon recognized that rocks often contain fragments from lower layers and that stratified series are often tilted, bent, or broken. Many, encouraged by Cuvier's doctrine of 'catastrophes' (pp. 235–7), ascribed these irregularities to violent upheavals. In this connexion it is interesting to observe that the *Essai sur la géographie minéralogique des environs de Paris* (1811) of Alexandre Brongniart (1770–1847), though written in collaboration with Cuvier, inclines more to the views of Hutton.

William Smith (1769–1839), a civil engineer, obtained an insight into the nature of strata while cutting canals. He produced the first coloured geological map (1815). His *Stratigraphical System of Organised Fossils* (1817) showed that certain layers have each their characteristic series of fossils. Some members of a series are wont to occur also in the layer below, others in the layer above,

others in all three. Therefore changes in the flora and
fauna which these fossils represent could not have been
sudden. He saw, too, that the further back we go, the
less like are the fossils to forms still living.

A third British geologist, Charles Lyell (1797–1875),
finally exorcised the catastrophic demon. He took to the
study of Geology while at Oxford, travelled considerably,
and was influenced both by William Smith and by
Lamarck. He saw that the relative ages of the later
deposits could be determined by the proportion they
yielded of living and of extinct molluscan shells. In his
great *Principles of Geology* (3 vols., 1830–3) he showed
that rocks are now being laid down by seas and rivers
and are still being broken up by glaciers, rain, sand-
storms, and the like: that, in fact, geologically ancient
conditions were in essence similar to those of our time.
Few books have exercised more influence on the course
of biological thought.

We are struck by the overwhelming share of British
investigators in the early development of geology as a
science. The very names of the formations suffice to
establish this fact. Lyell is responsible for *Devonian* (from
its predominance in Devonshire), *Carboniferous* (or 'coal-
bearing'), *Pliocene* (Greek 'more recent'), *Miocene* ('less
recent'), and *Eocene* ('dawn of recent'). Sedgwick (p. 245),
the Cambridge geologist with whom Darwin went on
geological excursions, invented *Cambrian* (Cambria =
Wales), *Palaeozoic* (Greek 'ancient life'), and *Cainozoic*
('new life'). Between the last two formations, John Phil-
lips of Oxford (1800–74) interpolated *Mesozoic* ('intermediate life'). Other British contemporaries are responsible for *Ordovician* and *Silurian* (the Ordovices and
Silures are British tribes mentioned by Caesar), *Permian*
(from the province of Perm in East Russia), and *Cretaceous*
(Latin 'chalky'). On the other hand *Triassic* (Latin *Trias*,

'the number three') and *Jurassic* (from the Jura mountains) were titles given by German geologists at the beginning of the nineteenth century. The term *Tertiary* is older
and was used by eighteenth-century Italian writers. The
tertiary formations were held to be the third of a series of
which the *Secondary* corresponded roughly to the *Mesozoic*
and *Palaeozoic*, and the *Primary* to the non-fossil-bearing
rocks. The word *Geology* itself was introduced (1779) by
H. B. de Saussure (1740–99) of Geneva, founder of
modern mountaineering.

§ 3. *Darwin* (1809–82), *the 'Beagle'* (1831–5), *and Island
    Life*

The name of Charles Darwin is closely associated with
that view of the succession of living things summed up in
the words *Organic Evolution*. Through this doctrine he
has profoundly influenced every branch of biological inquiry. He has also had a large effect on philosophical,
political, religious, and ethical thought. To appreciate his
distinction, it should be remembered that, had he never
written on Organic Evolution, he would still stand in the
front rank among naturalists.

Charles Darwin was the son and grandson of medical
men. His grandfather, Erasmus Darwin, himself wrote
on Evolution (pp. 295–6). The family exhibits hereditary
ability to an extraordinary degree. In all its ramifications
it has probably produced more men of intellect than any
other of which we have a clear record.

After contemplating a career in Medicine while at
Edinburgh and then in the Church while at Cambridge,
Darwin took an undistinguished degree. At Cambridge
the course of his life was determined by a friendship with
the professor of botany, the Reverend John Henslow
(1796–1861), a fine, upright, inspiring man with an
enthusiasm both for his subject and his pupils. Henslow

was the pioneer of practical elementary teaching in botany. Darwin came under his spell.

In 1831 Henslow, with remarkable insight, pressed young Darwin to take up geology. After a geological tour with the Reverend Adam Sedgwick (1785–1873), the first professor of Geology at Cambridge, Darwin received a letter from Henslow offering him the position of naturalist on the *Beagle*, He accepted with diffidence.

The *Beagle*, 238 tons, set sail under Captain Fitzroy (1805–65) in 1831. Her objective was to extend the survey of South America, and to make observations for determining longitude. Darwin went as a naturalist without salary, at the invitation of the captain. His working place was a narrow space at the end of the chart-room. He always held that the need for order and method thus imposed was among his best pieces of training. His equipment was meagre. Acting on what seemed the whimsical but was really the wise advice of the experienced Robert Brown, he took no compound microscope.

On his return in 1835, Darwin put together some of his scientific results. Some of the most significant appeared in the famous *Journal of Researches* (1839–40).

In his old age Darwin wrote:

'The voyage has been the most important event in my life. I owe to it the real training of my mind. The investigation of the geology of all the places visited was [specially] important. On first examining a new district nothing can appear more hopeless than the chaos of rocks; but by recording the stratification and nature of the rocks and fossils at many points, light begins to dawn and the structure of the whole becomes more intelligible. I had with me the first volume of Lyell's *Principles of Geology*, and the book was of the highest service.' [*Abbreviated.*]

Among the most important observations made by Darwin during his voyage were those on the very peculiar animal and plant inhabitants of various isolated oceanic islands which have never been connected with continental lands.

Of all islands the most fruitful for the development of biological ideas have been the Galapagos, a small volcanic archipelago in the Pacific, situated on the equator, some 500 miles west of the nearest South American coast. They are named from the enormous numbers of tortoises (Spanish *galapago*) that once dwelt there. When discovered early in the sixteenth century, the archipelago was uninhabited. Later it formed a lurking place for buccaneers who were not ideal conservators of local fauna and flora. There were probably fifteen species of giant tortoise, all indigenous, on the islands. Seven species survived into the nineteenth century.

The Galapagos were visited by Darwin in 1835. His botanical collections were subsequently (1849–51) investigated by J. D. Hooker (pp. 255, 559). There is no more remarkable passage in Darwin's *Journal of Researches* than his account of these islands.

'Of the flowering plants', says Darwin, 'there are 185 species, 100 confined to this archipelago. It is surprising that more American species have not been introduced, considering that the distance is only 500 miles from the continent; and that drift-wood is often washed on the shores. The peculiarity of the Flora is best shown in certain families. Thus there are 21 species of Compositae, of which 20 are peculiar; these belong to twelve genera, and of these genera no less than 10 are confined to the archipelago! The Flora has an undoubted Western American character [without] affinity with that of the Pacific. [Moreover] a vast majority of all the land animals are aboriginal.

'Why on these small points of land were there aboriginal inhabitants created on American types? The islands of the Cape de Verd group resemble in all their physical conditions, far more closely the Galapagos than these latter resemble the coast of America; yet the inhabitants of the two groups are totally unlike, the Cape de Verd

Islands bearing the impress of Africa, as the Galapagos are stamped with America.

'[Moreover] the different islands are to a considerable extent inhabited by a different set of beings. The aboriginal plants of different islands [are] wonderfully different.'

Name of Island	Total No. of Species	No. found in other parts of the world	No. confined to the Galapagos	No. confined to one island	No. confined to the Galapagos but found on more than one island
James	71	33	38	30	8
Albemarle	46	18	26	22	4
Chatham	32	16	16	12	4
Charles	68	29	21	21	8

'In James Island, of the thirty-eight Galapageian plants found in no other part of the world, thirty are exclusively confined to this one island. In Albemarle Island, of the twenty-six aboriginal Galapageian plants, twenty-two are confined to this one island, that is, only four are at present known to grow in the other islands of the archipelago; and so on. In like manner the different islands have their different species of the [world-wide] genus of tortoise, and of the widely distributed American genus of the mocking thrush, as well as of two of the Galapageian sub-groups of finches.' [*Greatly abbreviated from fourth edition.*][1]

Since Darwin's time, several examinations have been made of the fauna and flora of the Galapagos. His general results have not been greatly modified.

A famous case of a peculiar form native to an oceanic island is the Dodo of Mauritius. When that uninhabited volcanic island was discovered in 1505, it was stocked with large, unwieldy, flightless birds. These the Portuguese

[1] The section on the Galapagos islands was considerably altered by Darwin from edition to edition of the *Journal*. In the first three editions most of the details here given are not to be found. But the impression carried even by the first edition (1839–40) is the same, and he remarks on 'the entire novelty of the act that islands in sight of each other should be characterized by peculiar faunas'.

called *Doudo*, which in their language means 'simpleton'. Later the Dutch brought live specimens to Europe. Several artists depicted it. The Dodo, which was a helpless and harmless as it was unpalatable, was the prey of every lout that could wield a stick. Towards the end of the seventeenth century it became extinct. It was a very aberrant member of the pigeon family.

An allied form was the 'Solitaire' of the island of Rodriguez, not very distant from Mauritius. It had bony knobs on its flightless wings, which it used as clubs in self-defence. Its methods were ineffective, for it became extinct about 1761.

The little island of Réunion, also near Mauritius, lost another dodo-like bird which survived until toward the end of the seventeenth century.

From Mauritius itself have disappeared at least two species of parrot, a dove, and two species of coot. A number of other birds peculiar to each island have been lost from both Réunion and Rodriguez, some surviving until the nineteenth century.

This account of the lost birds of a single group of islands illustrates both the peculiarity and the vulnerability of island life. The numbers of the losses are the more impressive when we recall that island fauna is very poor in species. The flora has suffered no less severely. Nor is man the only destroyer. Dogs, cats, rats, introduced by him, have exterminated many peculiar animal forms, while hogs, rabbits, and above all goats have completely wiped out many island species of plants.

A tragic example is the biologically important island of St. Helena, one of the most isolated of all terrestrial spots. It is 15° south of the equator, 1,100 miles from the coast of Africa, and 1,800 from South America. Discovered in 1502, it has been inhabited since 1513. Darwin called there in 1836.

When first discovered St. Helena was densely covered with forest, now almost utterly destroyed. Its rich soil could only be retained on the steep volcanic slopes so long

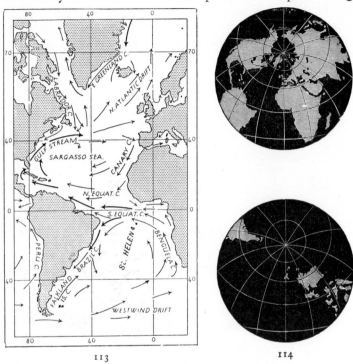

113   114

FIG. 113. The Atlantic currents.

FIG. 114. The land and water hemispheres, from Huxley's *Physiography* of 1877, when the Antarctic continent was unknown.

as they were protected by the vegetation. This was destroyed by imported goats, who were soon aided by the reckless waste of man. Extirpation of the highly peculiar vegetation caused also the destruction of most of the animal species which once lived on the island.

St. Helena has never had indigenous vertebrate fauna

of any kind on land, though there are many peculiar species of shore-haunting fish. No indigenous fresh-water animals or plants are known.

The flora—what is left of it –is remarkable. The *Challenger* reports recognized 65 species of certainly indigenous plants; 24 probably indigenous; 5 doubtfully indigenous (total 94). There are 38 flowering plants, all

Fig. 115. Her Majesty's Ship *Challenger*, 1872–6.

save one being peculiar to the island. Conspicuous among them are Compositae of tree-like proportions. There are 27 ferns, of which 12 are peculiar. Hooker considered that the peculiar species 'cannot be regarded as close allies of any other plants. Seventeen belong to peculiar genera, and of the others all differ so markedly from their congeners that not one comes under the category of an insular form of a continental species.'

The best known land animals of St. Helena are the

beetles. Of the 129 indigenous species all save one are peculiar. This degree of individuality is probably unique. Of these beetles the great majority are wood-borers, as might be expected in an island once forest-clad.

Of the living forms of St. Helena, the plants are nearly allied to South American species, while the beetles have affinities in descending order of frequency with South African, Madeiran, European, and Madagascan forms. These affinities are illuminated by the currents around the island (Fig. 113). Recent research has shown that viable seeds can be carried vast distances by ocean currents. Such seeds have been seen to be washed ashore and to germinate at St. Helena. Nevertheless, most of the indigenous population of St. Helena is descended from forms that probably reached the island in Miocene times or earlier.

Oceanic islands, of which Galapagos and St. Helena are examples, have for the last two centuries drawn the attention of naturalists. Their extraordinary wealth of peculiar forms and their difference from their neighbours —both continental and insular—are among the most striking phenomena in the distribution of living things. Such facts set Darwin thinking of the origin of species. They, more perhaps than any other, suggested to him his solution of the problem.

§ 4. *Oceanic Exploration from the 'Beagle' to the 'Challenger'*

Between 1839 and 1843 the *Erebus* and *Terror* explored the Antarctic under the command of Sir James Ross (1800–62). As naturalist there accompanied him Joseph Dalton Hooker (1817–1911), afterwards director, in succession to his father, of the Royal Botanic Gardens at Kew (1865–85). Hooker took with him Charles Darwin's recently published *Journal*, a gift from Lyell.

Hooker was an industrious collector and skilled systematist. None of his numerous writings is of more weight than the series which appeared in parts from 1844 to 1860 on the flora encountered by the *Erebus* and *Terror*. It includes accounts of the plants of the Antarctic islands as well as those of Tasmania and New Zealand. It may be said to lay the foundation of the systematic study of plant geography. Its last volume is also interesting as the first important botanical work written by an adherent of the doctrine of organic evolution.

The expedition of the *Erebus* and *Terror* was important for the great depths sounded by Ross. Animal life was proved to be abundant as deep as 2,400 feet. Hooker also showed the importance in the economy of marine life of the minute plants known as diatoms (p. 262).

The great wealth of marine life was exhibited by many workers in the 'fifties. The Scandinavians, Anders Retzius (1796–1860) and Michael Sars (1805–69) were specially concerned with living things, notably crustacea and edimoderms up to 2,700 feet. In 1851 Mary Somerville (1780–1872) wrote that she believed "no part of the ocean depth was uninhabited." Nevertheless, it was surprising when G. C. Wallich (1815–99) from H.M.S. *Bulldog* reported in 1861 echimoderms and protozoa from 7,500 feet in the N. Atlantic. Hopes were aroused that the depths might contain forms long extinct elsewhere.

A new outlook on oceanic exploration was introduced by the American naval officer Matthew Fontaine Maury (1806–73). He produced a valuable work on navigation (1836). In 1839 an accident rendered him permanently lame, and he began to occupy himself in extracting from logs of ships the observations of winds, currents, temperatures, and so forth.

The charts that Maury thus drew up led to such a shortening of passages that an international conference was called in 1853 to consider further organization of such observations. His *Physical Geography of the Sea* (1855) was an influential work. Largely as a result of his activities, Government meteorological offices were established by Great Britain and Germany.

All the western maritime nations were now actively interested in the physical geography of the sea. Moreover, numerous and striking advances in the knowledge of marine plants and animals had drawn special attention to deep-sea forms. In 1866 a transatlantic telegraph cable was projected. It was necessary to obtain knowledge of the depth and the character of the floor of the Atlantic Ocean. Improvements had been made in sounding apparatus. Instruments for taking the temperature of the sea-water at great depths and for obtaining samples of the deep water had been invented. Scientific cruises by the Admiralty vessels *Lightning* (1868) and *Porcupine* (1870) had brought back novel biological data from the North Atlantic and from the Mediterranean.

§ 5. *The 'Challenger' Expedition* (1872–6) *and the Rise of Oceanography*

The British Admiralty now determined on oceanic exploration on a hitherto unheard-of scale. The corvette *Challenger*, Captain George S. Nares (1831–1915), 2,300 tons, was commissioned. She carried both sail and steam and was fitted with sounding, dredging, and other apparatus for examining deep water. She had biological, chemical, and other laboratories, and carried a staff of six naturalists under Charles Wyville Thomson (1830–82).

Leaving Portsmouth in December 1872, she crossed and re-crossed the Atlantic. Soundings, dredgings, temperatures, samples of sea-water were constantly taken. Biological material, especially from great depths, was continuously collected. The Atlantic islands, Madeira, Canaries, Azores, Bermudas, Cape Verde, as well as

Robinson Crusoe's Island (Juan Fernandez), were visited and their plant and animal forms collected. A special survey was made of Tristan Da Cunha, a small volcanic peak between the Cape of Good Hope and Cape Horn.

After a call at the Cape, the *Challenger* visited Kerguelen Island, a mountainous mass of desolate land midway between the Cape and Australia. Abundant fossil remains of trees were there found. Much attention was paid to the vegetation, which is of great antiquity and allied to that of America rather than of Africa, as Hooker had already observed. Many species of shore and freshwater algae were found to be peculiar to the island. Proceeding southward, the *Challenger* was the first steamship to cross the Antarctic circle.

Extensive researches were then made in the Pacific, the route leading by Melbourne, New Zealand, Fiji, Torres Strait, the Banda Sea, and the China Sea to Hong Kong, northward to Yokohama, and across the Pacific by Honolulu and Tahiti, making South America at Valparaiso. Following the coast, the *Challenger* went through the Strait of Magellan and reached Sheerness in May 1876. She had travelled 69,000 nautical miles and had taken 372 deep-sea soundings.

The vast collections of the *Challenger* were now investigated by a whole army of naturalists under John Murray (1841–1914). The results were issued by the British Government in fifty thick folio volumes. These provide the best-worked-out account of a biological expedition and form the solid bases of a science of oceanography.

The work made it evident that, for any understanding of the life of our planet as a whole, an exact knowledge of the physical conditions of the sea is essential. Oceanography has since developed in a manner which demonstrates the interdependence of the biological and the physical sciences. A study which involves more than two-thirds of the earth's surface, and implicates the whole past and future history of the other third, is of primary importance to our conception of life as a whole.

The departure of the *Challenger* was soon followed by that of the United States Government steamer *Tuscarora*, whose scientific staff investigated the floor of the Pacific. Other American and Norwegian expeditions followed in rapid succession. Alexander Agassiz (1835–1910), son of Louis Agassiz (1807–73, pp. 309, 489), was especially prominent in this work. Trained as an engineer, he was able greatly to improve the apparatus of oceanic investigation. Among his most remarkable results was his demonstration that the deep-water animals of the Caribbean Sea are more nearly related to those of the Pacific depths than they are to those of the Atlantic. He concluded that the Caribbean was once a bay of the Pacific and that, since Cretaceous times, it has been cut off from the Pacific by the uprise of the Isthmus of Panama.

The relation of biological to physiographical knowledge is illustrated by another discovery that followed on the technical improvements of Agassiz. The *Challenger* naturalists had considered the results of certain dredgings in the Atlantic north of Britain. They noted, as did the investigators of two other survey ships, a great change in the bottom-living fauna along a line between the northwest of Scotland and the Faroe Islands. This area was therefore systematically sounded. A long narrow ridge separating the Arctic from the Atlantic waters was thus revealed. Entirely different physical conditions were found to prevail in the deeper parts of the ocean on the two opposite sides of the ridge (Fig. 116). This very important geographical feature was named the Wyville Thomson Ridge (Fig. 117). Its biological and physical exploration has stimulated similar work in other parts of the world leading to a mapping of the ocean floor.

Much new knowledge of marine biology has resulted from the investigations carried out by Prince Albert of

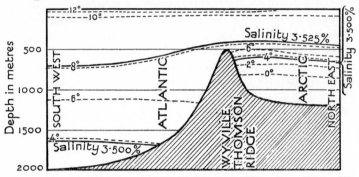

FIG. 116. Section through the Wyville Thomson Ridge from NE to SW. It will be seen that the surface conditions on the two sides of the ridge are very similar but that the deep water conditions are very different.

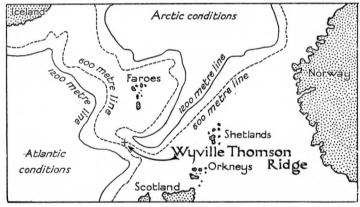

FIG. 117. Showing position of Wyville Thomson Ridge.

Monaco (1848–1922) in a series of specially equipped yachts. Before the twentieth century had dawned, other nations had joined in the work. American, Russian, Belgian, German, Austrian, Italian, Dutch, and Danish

expeditions went forth. In 1902 there met at Copenhagen an International Council for marine exploration.
The result was a fairly complete biological survey of the
North Sea. Other areas were yielding information.
Notably, the detailed investigation of the Pacific Ocean
was soon taken in hand by the Pan-Pacific Conference,
and the biology of the Antarctic began to be explored.

When the bottom, the depths, and the intermediate
depths of the ocean have been explored, we shall attain
to a general view of life in the sea as a whole. This
reached, it will be possible to create a biological science of
the sea which will bear analogies to the sciences both of
physiology and of economics. There is in the sea a complex balance of life as a whole, comparable to that complex
balance of metabolism which goes to making up the life
of the individual on the one hand and of communities of
individuals on the other.

§ 6. *Distribution of Life in the Sea*

More than two-thirds of the earth's surface is covered
by sea (Fig. 114). The oceans in relation to their area are
as shallow as a sheet of water of one hundred yards
diameter and an inch deep. Yet the greatest ocean depth
is over seven miles. If a globe of 40 feet diameter represent the earth (that is, 1 foot to 200 miles) the highest
mountain or the deepest sea would be an elevation or
depression of one-third of an inch.

There is life in the open sea at every depth, but a
great concentration near the surface and at the bottom.
The conditions at the two levels differ greatly.

The surface teems with vegetable life. There is abundance of the larger 'sea-weeds', both around the coasts
and in parts of the open sea. Sea-weeds, however, are
insignificant, both in bulk and bionomic importance, as
compared to the vast flora of microscopic forms. These

are distributed universally over the surface of the ocean and below it for a few hundred feet.

Of surface-living plants, first in importance are the minute unicellular diatoms (Greek 'cut in two'). These are named from the two siliceous valves—the so-called 'skeleton'—in which each dwells (Fig. 121). The group is very rich in species. Diatoms swarm in vast numbers

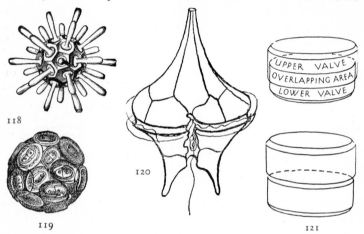

Plankton. Common unicellular plant forms.

Fig. 118. *Rhabdosphaera* × 1,400. Fig. 119. *Coccosphaera* × 1,100.
Fig. 120. *Peridinium*, from Plymouth Sound × 300.

Fig. 121. Diagrams of 'skeleton' of *Coscinodiscus*, a genus of box-like diatoms of world-wide distribution.

in temperate and colder seas. At death they sink and are consumed by the inhabitants of greater depths, while their insoluble and undigested skeletons form the 'diatom ooze' that covers a large part of the ocean floor.

The Dane, Otto Frederik Müller (1730–84), was the first to describe diatoms (1773). He was also the inventor of the naturalist's dredge. The Swede, Carl Adolf Agardh (1785–1859), afterwards a bishop, who was distinguished both as mathematician and botanist, gave the first general account of diatoms. He described forty-nine species in his *Systema algarum* (1824).

By reason of their beauty the diatoms became a classical subject of microscopic research. When the *Challenger* sailed more than 4,000 species were already known. The *Challenger* naturalists found that diatoms so abound toward the polar regions as to tint the water. In warmer waters they found that other lowly vegetable forms, notably the Peridinians, take their place.

The Peridinians have two long whip-like processes which, during life, are constantly lashing in characteristic grooves (Fig. 120). Their cell-wall is of organic matter and is decomposed after death. Many Peridinians are luminous. Related to them is the well-known *Noctiluca miliaris*, a common cause of phosphorescence in warm waters. Otto Frederik Müller was also the first to describe a Peridinian, namely the immensely numerous *Ceratium tripos*.

The vast number of individuals and the variety of species of the oceanic microscopic flora was not realized until about 1860. The survey for the first Atlantic cable then gave opportunities for oceanic research. The *Challenger* naturalists threw much further light on the subject, and brought to notice the extreme importance of oceanic plant-life. Notably, they showed that the curious and supposedly 'crystalline' oceanic 'Coccoliths' (Figs. 118–19) that had given rise to much speculation were of plant origin. They are the calcareous products of the microscopic oceanic plants, since named Coccolithophoridae. Their skeletons, like those of the diatom, persist after death, and form ooze at the bottom at moderate depths. At greater depths they are dissolved. The distribution of these Coccolithophoridae and of other forms can in part be gathered from a map of the sea bottom (Fig. 122).

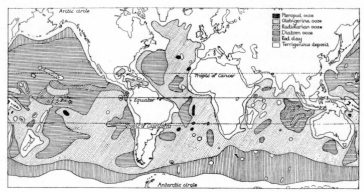

FIG. 122. The sea-bottom, as known to about 1900, corresponded to six general types.

(1) A zone of *Terrigenous Deposit* surrounds the land masses and is a result of their attrition.

(2) The main floor of the Ocean is *Globigerina Ooze*. This was first described in 1853 simultaneously by American and German observers. It forms the floor of most of the Atlantic and Indian and of about one-third of the Pacific Ocean. It consists mainly of the minute calcareous skeletons of *Globigerina* and allied genera of Foraminifera, a group of Protozoa.

(3) *Pteropod Ooze* is a variety of (2), in which abound the spindle-shaped shells of Pteropod Mollusca. These are distintegrated at depths greater than about 3,000 metres. It is chiefly an Atlantic formation.

(4) *Red Clay* was discovered and named by Wyville Thomson on the *Challenger* in 1873. Its origin is mainly the decomposition of volcanic minerals, but it is essentially a residue after other substances have been dissolved out. It is not found at less than about 4,300 metres and chiefly in the Pacific. It contains numerous nodules of manganese dioxide, often formed round indestructible animal remains, as teeth of sharks and ear-bones of whales. These are sometimes of extinct species, and even of forms so ancient that they occur as fossils in Tertiary beds. Common also are globules of meteoric iron. Such remains must fall uniformly on all parts of the ocean floor and yet are very rare, except in Red Clay. It is thus inferred that this deposit accumulates incomparably more slowly than the other types. Red Clay contains very little calcium carbonate, which is dissolved at great depths.

(5) *Radiolarian Ooze*. It is an outlier of Red Clay and is a variety of it containing a high proportion of the siliceous shells of Radiolaria, a group of pelagic Protozoa. It is almost confined to the Pacific.

(6) *Diatom Ooze* was first recognized by Murray on the *Challenger* as the characteristic deposit in the neighbourhood of the Arctic Circle. It is formed chiefly of the siliceous skeletons of the unicellular plants known as Diatoms. There is a complete belt of it in the southern hemisphere. In the northern hemisphere it occurs only in isolated areas, since land masses there intervene in its characteristic latitudes (between about 45° and about 60°).

Oceanic plants were studied on the *Challenger* in conjunction with the floating fauna with which they dwell. The name *plankton* (Greek 'drifting') was invented for the whole community by Victor Hensen (1835–1924) of Kiel. The study of plankton has become of great importance. Hensen, primarily a physiologist, began it while considering the production of nutritive substances under different meteorological conditions. He thus laid the

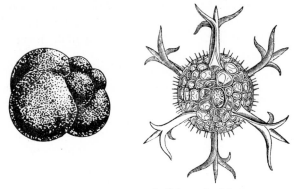

Plankton. Common unicellular animal forms.

FIG. 123.   Shell of *Globigerina*, the spines of which have been dissolved in their fall to the bottom.

FIG. 124.   *Hexanastra quadricuspis*, new genus and species of Radiolarian discovered by *Challenger* naturalists.

foundations of the systematic study of the economics of the life of the ocean, *oceanic bionomics*, as we may call it. The subject is fundamental for our conception of the course of life as a whole upon this planet.

An interesting relationship between animal and plant-life was revealed by the *Challenger* in the Sargasso Sea. Great masses of 'gulf-weed' there abound. The bright yellow of this plant contrasts with the deep blue of the water. For concealment, the shrimps, crabs, and other creatures that swarm in the weed are also yellow. In general, however, animals that live on the ocean surface are colourless.

Among the fauna as among the flora of the ocean the microscopic forms are of far greater economic importance than the more impressive larger creatures. Of very great significance for the life of the sea are the Foraminifera, a group of minute unicellular animals with calcareous shells. Species of one genus, *Globigerina*, float everywhere on the surface. The dead shells make up the vast mass of *Globigerina ooze* which is the usual deposit of the Atlantic bottom. Chalk deposits have been formed, in past ages, from oozes of this type (Fig. 123).

Nearly all the major invertebrate Phyla contribute to the fauna of the open ocean. There are innumerable oceanic crustaceans, many very minute. Jelly-fish and molluscs abound. The shells of a group of the latter, the *Pteropods*, form an important constituent of some oozes. There are even a few oceanic representatives of such essentially land-forms as insects and such essentially shore-forms as anemones. One insect was often taken by the *Challenger* in the open ocean. This was a bug (*Halobates*, family Hydrometridae), with a small round wingless body and long legs. It lives on the juices of jelly-fish. Oceanic anemones, instead of clinging to rocks, have air chambers as floats, and like many oceanic species they form complex floating colonies.

All the great vertebrate groups, except the Amphibians, have oceanic representatives.

Of mammals, the whales and porpoise breed at sea. Their remote ancestors used to come ashore to breed, as do still the seals.

Of birds, the petrels are the most oceanic. The largest is the albatross, whose gliding flight has always excited wonder. It lands to lay and incubate eggs.

Of reptiles, the sea-snakes are viviparous, breed at sea,

and seldom come to shore. Some species cross the ocean, but are usually found near shore. Certain turtles are at times seen far out at sea. They, too, usually dwell near the coast and always come ashore to breed.

Fish frequent the surface of the open ocean less than might be supposed; there are probably more species about the coast and near the bottom. There are, however, some species that live at intermediate depths of the open ocean. The sea is a vast place, especially for creatures that move freely in three dimensions, and the sexes of such fish would have especial difficulty in finding one another. The problem has been solved in certain cases by the female carrying a small male attached to her (Fig. 181).

A feature of oceanic surface-life is phosphorescence. The *Challenger* recorded that often the sea was lit up with sheets of a diffuse light, where the water was broken before the breeze. At other times the water was full of luminous specks. This, the commonest form of phosphorescence, is due to a variety of small creatures, notably crustaceans, each of which gives out its flash. Some crustaceans are phosphorescent on their own account; others derive their light from phosphorescent food in their stomachs. The phosphoresce of harbours is of bacterial origin.

The circumstances of life on the ocean floor, as revealed by the *Challenger* and by later expeditions, are entirely different from those at the surface. The pressure at 30,000 feet is about five tons to the square inch as against fifteen pounds at the surface. No sunlight penetrates. Below 1,200 feet all is dark. The temperature in the depths is uniform and not much above freezing. There are no currents and no seasons. Summer and winter, day and night are alike. The temperature is almost the same on the equator and at the poles. Conditions are substantially uniform the world over. There is no vegetable life to build up the bodies of the animals that dwell there. Thus the animals prey only on one another. They draw their ultimate supplies from the dead matter that rains down from above.

A large proportion of the animals of the depths are blind, with their eyes reduced to rudiments. Many of the blind fish and crustacea have prodigiously long and delicate feelers. Other deep-sea animals have enormously large eyes and thus make the best of the light emitted by themselves or by other phosphorescent animals.

The results of the deep-sea dredging have been in certain respects disappointing. Specimens of numerous new genera and species of known families have been brought up. Many are interestingly specialized but few are widely different from familiar forms. No 'missing links', no new classes or new orders have been found. Further exploration has confirmed that the plants and animals that inhabit the open ocean, whether on the surface or at the bottom, are mostly very widespread. An exception must be made for the inhabitants of the most extreme depths the isolation of whose fauna is comparable to that of oceanic islands (Fig. 65). The distribution of oceanic forms is determined by temperature, degrees of saltness, intensity of light, pressure, &c.

The extension of our knowledge of the conditions that prevail in the Ocean and in its superincumbent atmosphere is leading to a new scientific ideal. As the laws of oceanic plant-life come into relation with the corresponding laws governing their animal associates, and both with those of physical conditions, we begin to perceive a most impressive physico-biological integration and may one day attain to a conception of an internal economy of the Ocean.

In the twentieth century our view of life in ocean depths has changed considerably for several reasons: (*a*) An Antarctic Continent has been demonstrated: (*b*) Ocean depths have been shown to be much more contoured than was thought: (*c*) The ocean floor has been found to have deep trenches which have each its own local fauna: (*d*) These 'hadal' forms from over 30,000 feet include corals, sea-anemones, echinoderms, segmented worms, and crustacea of genera known elsewhere: (*e*) It is thus confirmed that the depths have become inhabited at a (geologically) recent period.

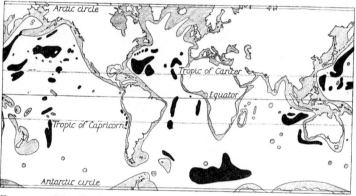

FIG. 125. The distribution of ocean depth. The shaded areas, distributed chiefly around the land masses, are less than 2,000 metres. The large unshaded area is between 2,000 metres and 6,000 metres. The black patches are the true 'Ocean Deeps' from 6,000 metres onward as known to about 1900.

Apart from the open sea there is, round all the land-masses, a relatively shallow 'continental shelf'. This shelf has its own flora and fauna which is of great wealth and variety in all latitudes. That of the North Sea is best known, largely as a result of international effort especially since 1902 (p. 261). There we find two distinct faunas, one in the 'arctic' regions, where the water is usually of a temperature near freezing-point and is very salt, and the other where the water, under the influence of the Gulf Stream, is several degrees warmer and the salinity varies more (Figs. 116 and 117).

Moreover, on the continental shelf, different animals are found at different depths, so that we can conveniently distinguish between littoral, sublittoral, and continental deep-sea zones, with limiting depths of about one hundred, five hundred, and three thousand feet respectively, and each with a characteristic fauna. Finally, within these zones again there are distinct communities of animals, many very diverse species being always found together. These associations are of great importance both economi-

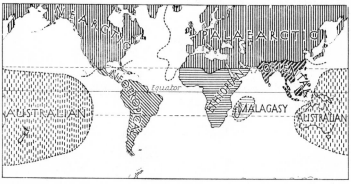

FIG. 126. The main zoogeographical regions. The 'Australian' region includes an immense number of islands, too small to appear on the map.

cally and scientifically. Their discovery is to the credit of the Danish biological station, and their study may well lead to a systematic marine ecology that will take its stand by the side of marine bionomics.

§ 7. *Distribution of Life on Land*

Peculiarities in the distribution of some living forms were remarked by naturalists from the first. In the eighteenth century Buffon (pp. 293–5) drew attention to 'natural barriers' delimiting flora and fauna. The nineteenth century was well advanced before Lyell convinced his readers that present distribution is conditioned by past changes involving the major land-masses. The materials obtained by Darwin on the *Beagle* (published 1839–63), brought out striking facts in the geographical distribution of animals, both living and extinct.

In 1858, the year of the classic contribution of Darwin and Wallace, appeared the pioneer attempt by P. L. Sclater (1829–1913) to divide the world into zoological regions. He discussed the perching birds which lend themselves for the purpose. Their power of flight is small, they are rich in species, and they have been very exactly studied. The Darwin finches of the Galapagos here come to mind.

In the meantime, A. R. Wallace (p. 302) had been at work on the fauna of the Malay peninsula. He was struck both with its resemblances to and with its differences from that of South America where he had also collected. His studies resulted in his *Geographical Distribution of Animals* (1876), still the most important work on the subject.

Wallace based his discussion on mammals. He followed Sclater in dividing the land-surface of the earth into six zoogeographical regions. These he named *Palaearctic, Nearctic, Ethiopian, Oriental, Australian,* and *Neotropical* (Fig. 126).

Wallace's regions have been retained in great part by more modern workers. The most important changes since his time are (*a*) the separation by some writers of the Madagascar (Malagasy) from the Ethiopian region, (*b*) the general recognition that the Palaearctic and Nearctic regions are more nearly allied to each other than to any other region, and their union into a Holarctic region, and (*c*) the subdivision of the 'Australian' or Pacific region (Notogaea). The divisions thus instituted have been grouped as follows:

	⎧ 1. Holarctic	⎰ Nearctic
		⎱ Palaearctic
Arctogaea	⎨ 2. Ethiopian	
	⎪ 3. Malagasy	
	⎩ 4. Oriental	
Neogaea	5. Neotropical	
	⎧ 6. Australian	
Notogaea	⎨ 7. Polynesian	
	⎩ 8. Hawaiian	

Wallace demonstrated many remarkable contrasts. None is more striking than that between Bali and Lombok, near Java. These islands, each about the size of Corsica, are separated by a deep strait which at its narrowest is but 15 miles. Yet, as Wallace remarked, they 'differ far more in their birds and quadrupeds than do England and Japan, the difference being such as to strike even the the most ordinary observer'.

The strait between Bali and Lombok, known as *Wallace's line*, is very deep. It has been generally regarded as delimiting the Oriental from the highly peculiar Australian zoogeographical region. The zoogeographical regions into which the earth's surface can be divided depends upon the particular group of animals chosen. It happens, however, that the divisions of geographical regions based on mammals accords closely with that based on perching birds, and is not vastly different from that based on certain invertebrate groups, e.g. the Spiders. Very different from these, on the other hand, is the division based on such very ancient groups as Reptiles or Molluscs.

Geographical regions are biologically interesting, not so much in themselves, but as revealing or summarizing the history of the various groups from which they are constructed. Thus the distribution in space of living forms is ultimately referable to their distribution in time. The discussion of the one is little profitable without the other. The geographical regions of Wallace are based primarily on mammals and birds which have a fairly similar palaeontological history. If we constructed geographical regions from groups with different histories, such as reptiles or insects, the results would be different.

The general principles that determine plant regions are similar to those of animals, but their application is somewhat different. The subject has been broached mainly in

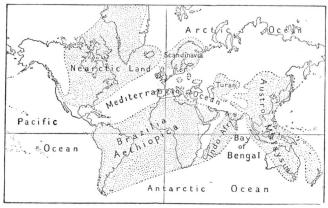

FIG. 127. Hypothetical distribution of land (dotted) in Cretaceous period.

connexion with the flowering plants. These are geologically younger than the groups on which zoogeographical regions are based (Fig. 128). Moreover, temperature and moisture are of overwhelming importance in the life of plants. Even between countries which present but slight differences of climate and are in the same regions of plant geography, certain notable floristic differences may occur. The lawns of the United States are not pied with daisies; the hyacinth, so common in English woods, is not found in those of Germany; our common purple foxglove is not an inhabitant of Switzerland. It is doubtless material needs

—of the nature of which we are largely ignorant—and not physical barriers, that determine such apparently arbitrary distributions.

On the other hand, the means of dispersal of flowering plants are more effective than those of vertebrates or of most other animal groups. The effects of this are sufficiently evident on oceanic islands.

A pioneer plant geographer was the German philosopher and traveller, Alexander von Humboldt (1769–1859). His interests were largely determined by a youthful friendship with a companion of Captain Cook. He was the founder of the science of physical geography and was the first to delineate 'isothermal lines'. Von Humboldt began his *Kosmos* (1845–7) in his seventy-sixth year. This great book did good service in emphasizing the relations between the forms and habits of plants and the character and soil of their habitat. Humboldt's presentation is rendered attractive by his magnificent descriptions of tropical vegetation.

Certain resemblances between the flora of Africa, South America, and Australia impressed Humboldt and other naturalists. In 1847 J. D. Hooker (p. 254 and Fig. 127) suggested in explanation a land connexion between South America and Australia as late as Jurassic times. Various names, forms, and areas have been ascribed to this now fragmented continent (Fig. 127). The known facts are marshalled in the 'theory of continental drift' which regards modern continents as the result of the sundering of one vast land mass.

Attempts to delimit definite plant regions have been less successful than those of the zoogeographers. The subject was approached by the German naturalist, A. H. R. Grisebach (1813–79), of Göttingen, in a series of papers leading up to his *Vegetation der Erde* (1872) and *Pflanzen-Geographie* (1878). These works are detailed lists of floras, based primarily on those of the West Indies. They show each flora to be adapted closely to its climatic and environmental circumstances. Oddly enough, Grisebach rejected the doctrine of descent.

Unsuccessful attempts were next made to divide the earth into floral areas, corresponding to the zoogeographical regions. Later botanists have mostly laid far more stress on climate than on geographical regions. The most successful has been A. F. W. Schimper (1856–1901) of Basel, in his *Pflanzen-Geographie* of 1898.

A simple scheme of plant distribution that covers a very large number of phenomena was set forth by W. T. Thiselton-Dyer (1843–1929), director of the Royal Botanic Gardens at Kew, who had an enormous floristic experience. Thiselton-Dyer (1909) divided the earth's flora into three great primary areas: (*a*) the North Temperate Zone, (*b*) the Tropical Zone, and (*c*) the South Temperate Zone. The northern tropic cuts off (*a*) from (*b*) with considerable accuracy. The southern tropic separates (*b*) from (*c*) with less precision, and, indeed, (*b*) and (*c*) are less distinct from each other than are (*a*) and (*b*). The characteristics of these divisions are as simple as their geographical boundaries.

(*a*) The North Temperate Zone contains most land. It is continuous save for the geologically recent break at the Bering Straits. It is characterized (i) by needle-leaved cone-bearing trees; (ii) by catkin-bearing and other trees, which form an important family (*Amentiferae*) that lose their leaves in winter; and (iii) by a great number of herbaceous plants that die down annually.

(*b*) The Tropical Region occupied areas widely separated by intervening ocean. It includes the major part of Africa, is imperfectly continuous with Southern Asia, and discontinuous with tropical America and with the Eastern Archipelago. It is characterized (i) by gigantic Monocotyledons, notably the palms, by the natural order Musaceae and by the enormous grasses known as 'bamboos'; (ii) by evergreen polypetalous trees and by figs; (iii) by the rarity of herbaceous plants which, when tropical, are mostly parasitic on other plants.

(*c*) The South Temperate Zone occupies very widely separated areas of South Africa, South America, Australia, and New Zealand. It is characterized by the possession of a number of peculiar Natural Orders, which are mostly of shrub-like habit. Many are intolerant of moisture. Individual species are very numerous and often very restricted in area of distribution.

The great supremacy of the tropical region in wealth of forms was well brought out by nineteenth-century botanists. Apart from families of very restricted distribution (17) and those universally dispersed (92) Drude recognized 131 families of flowering plants. Of these 30·5 per cent. are characteristic of the North Temperate, 52·5 per cent. of the Tropical, and 17 per cent. of the South Temperate Zone. If genera and species instead of families were considered, the wealth of the tropics would be found to be far greater. The tropics are the storehouse and perhaps the nursery of species. It seems probable that the other regions are constantly being invaded from the tropics. In general we may say of plants as of animals that their distribution in space must be discussed in connexion with their distribution in time.

§ 8. *Geological Succession*

We have glanced at the general history of palaeontology up to Darwin (pp. 245–8). His optimistic followers expected the geological record to reveal a series of forms becoming progressively 'higher' and 'more differentiated'. Research, they thought, would bring to light a number of 'links' between existing species, genera, families, orders, classes, and even phyla. It has now long been apparent that such 'links' are, in fact, conspicuous by their absence.

Most major groups of organisms go back very far. It is true that ancient forms often present features which suggest relationship to other groups. But anything to which the palaeontologist can point as actual ancestral

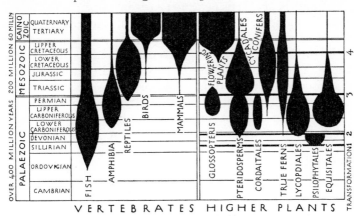

FIG. 128. Diagram of geological succession of higher organisms. The varying width of the black bands is an attempt to represent the relative dominance of the various classes at the different geological periods. (Animals and plants must be considered separately in this connexion.) The time estimates to the left must be taken as the roughest of approximations. The numbers to the extreme right and the black horizontal lines corresponding to them indicate the four great floral transformations (pp. 280–3).

forms are few indeed. Among higher animals one of the best authenticated and most frequently quoted continuous series leads to the modern horse (Fig. 136). If we consider larger groups, the record often supplies helpful suggestions without giving exact information. Thus the relationship of birds and reptiles is fairly clear, as is the line of descent of the survivors of the immense group of mesozoic reptiles. On the other hand, the relationship between amphibia and reptiles is extremely obscure. The mammals, too, go back very far without junction with any other group. For junctions of invertebrate phyla, we gain little help from geology.

Fossil botany became a science in the 'twenties with the application of the binomial system to coal-measure plants. Attempts were made by A. T. Brongniart (1801–76) in 1822 and E. T. Artis (1789–1847) in 1825. Most of the specific names of Artis are still used. Several botanists followed on similar lines, but the work was redirected by William Nicol (1768–1851), remembered for his prisms. In 1851, he ground slices of fossil wood thin enough for microscopic examination. The process revealed the cell walls clearly and was at once adopted by Robert Brown (1851). The subject owed more, however, to W. C. Williamson (1816–95), a disciple of William Smith (p. 246) who began his work in 1858.

Williamson demonstrated that in coal are to be found gigantic woody forms similar to the higher existing flowerless plants, such as horse-tails, ferns, and club-mosses. His results met with neglect until 1882, but from that time the great importance of palaeobotany, and the immense mass of material available in coal and other formations have been recognized. This extension of the knowledge of fossil plants has been due to many workers; to none, however, as much as to D. H. Scott (1854–1934) who was, at first, an associate of Williamson. The demonstration of the relationships of several of the major groups of fossil plants to each other has been largely his work.

We turn now to consider the broader results of the study of palaeobotany. A plankton flora (p. 265) of algae must have existed at a very early date. We can hardly, in the nature of the case, have geological evidence for it. In time these algae came to be attached and rooted as shore-forms and, in the Silurian, reached a high development. The land was next invaded.

FIG. 129. *Lyginodendron Oldhamium* found in the lower Carboniferous. It was the first Pteridosperm (seed-fern) of which the seed was identified, and is now one of the most completely known of all fossil plants. The piecing together of its various fragmentary remains has been the work of nearly a century. The foliage was described in 1829, the male organs not till 1905. A contemporary dragon-fly is shown near by. From D. H. Scott. Inset is the ripe seed studded with glands, from F. W. Oliver.

A feature of the geological history of land-plants is a series of four more or less abrupt changes of the world flora. There is no corresponding faunal change and only the third transformation can at present be adequately explained (Fig. 128).

(a) *The First Transformation* was from a marine to a land flora.

Of land plants we have no convincing evidence until the Upper Silurian. In 1888 Sir J. W. Dawson (1820–99) of Montreal produced a classical restoration from the Lower Devonian of the simple marsh-living *Psilophyton princeps*, since traced back to the Silurian. It was leafless, two feet high, with spiny stems half an inch across. Naked oval spore-cases were borne on special curved stems. *Psilophyton* gives its name to the important group Psilophytales (p. 202). They disappeared with the second transformation. Another typical plant of the Upper Silurian was the rootless and leafless Aberdeenshire *Rhynia major*. An allied genus to this, *Asteroxylon*, bore elementary leaves. *Asteroxylon* and *Rhynia* apparently grew in a soil subject to periodic inundations.

Before the second transformation appear a few 'seed-ferns' or Pteridosperms (p. 202) and a few Lycopodials (giant club-mosses), groups characteristic of the next phase. On the whole, however, the transformation is sharp.

(b) *The second transformation* led to the coal measures of the extremely peculiar flora of which we have considerable knowledge. It included many plants with the general habit of ferns. Among these were a few true ferns very distantly related to forms now living. More characteristic were the 'seed-ferns' or Pteridosperms which, though superficially similar to ferns, were not closely related to them. A characteristic Pteridosperm was *Lyginodendron* (Fig. 129). Lycopodials were large and numerous and have left an insignificant remnant in the modern 'club-mosses'. A characteristic Lycopodial was the well-known *Lepidodendron* (Fig. 130). One was found some years ago in an English coal-mine, measuring 114 feet to its first branches. Another beautiful Lycopodial was *Sigillaria* (Fig. 130).

FIG. 130. Idealized Carboniferous scene.

The commonest plants of the coal-measures are the *Calamites*, a magnificent group of which our living 'horse-tails' (*Equisetum*) with their twenty-eight species preserve a faint memory. Their size may be gleaned from their cones, some of which were a foot long (Fig. 130).

Very conspicuous were the Cordaitales, called so from their leading genus *Cordaites*. Some important beds of coal are entirely made up of their leaves. They were tall, slender trees with large simple leaves. The 'flowers' were catkins, which arose just above the leaves (Fig. 130).

In carboniferous times there were uniform climate conditions over a large part of the world, associated with an equally uniform flora. A severe glaciation followed in Permian times. It affected an enormous area, including part of Australia, India, South Africa, and much of South America, then combined into a great southern continent (Fig. 127). The cold was disastrous to the flora. In the northern hemisphere is lost ground, never afterwards regained. In the southern it was blotted out and replaced by a peculiar and characteristic 'Glossopteria flora'. Its chief representative, *Glossopteris*, flourished all over the southern hemisphere and extended to land that is now near the South Pole. Fossil specimens were obtained on the Beardmore Glacier by Captain Scott's expedition of 1912. They were found with the dead bodies of the explorers.

(c) *The third transformation* appears very suddenly and marks the dawn of mesozoic times. For the earliest or triassic period land conditions were relatively dry. This, following the Permian glaciation, may explain the rapid change.

The Pteridosperms were now replaced by a vast array of Cycad-like forms, of which a remnant still survives. True ferns become numerous but quite different from Palaeozoic forms, since most belong to still existing families. The Cordaites are succeeded by Conifers allied to existing families. The great horse-tails dwindle prodigiously both in numbers and size. The giant Lycopods are seen no more. Their allies, the club-mosses, become both few and modest. A few flowering plants are found.

(d) *The fourth transformation* is to the modern flora, which becomes evident in the Upper Cretaceous times.

Flowering plants are its feature. There is nothing primitive about them even at their first known appearance. In the Upper Cretaceous, the vegetation looks quite modern. Monocotyledons include palms, reeds, lilies. They bear about the same proportion to the Dicotyledons as in the modern flora. Among Dicotyledons are many modern families. From the Middle Cretaceous onward, species are encountered of genera that are still represented in our modern flora.

There are two striking features in the geological succession of plants. First is the disappearance of very highly developed groups, Pteridosperms, Psilophytales, Glossopteris, and the like. Second the immense extension backwards of living groups, flowering plants, conifers, ferns, &c., without any definite sign of junction either with each other or with other groups.

§ 9. *Interrelations of Species*

Only since Darwin has any investigation been made of the relative distribution of allied species. This has been possible through the development of *Ecology* into an independent science. The word *Ecology* was given circulation by Haeckel (1886). It is on the model of *Economy* (Greek *oikos*, house, and *nomos*, law). Haeckel considered that the scope of Ecology was to treat of the reciprocal relations of organisms and the external world. For this discussion plants are specially suitable. The external world with which ecologists have had to deal can be divided into two departments.

(*a*) The physical conditions, light, temperature, moisture, terrain, &c.
(*b*) Other species.

The Ecology of Plants (1895) by the Dane, E. Warming (1841–1924), dealt largely with the former theme. *Plant Geography upon a Physiological Basis* (1898) by the Swiss

A. F. W. Schimper (1856–1901), with the latter. These two important works have set the subject on its way. From the beginning, however, it has been cursed, more than most sciences, by a horde of technical terms equally hideous and obfuscating.

Ecology has elicited some suggestive facts for the solution of the problem of the distribution of allied species. When, in the light of ecological principles, we examine the life-habits of plants in a state of nature, we encounter certain evident phenomena that have often been overlooked. It has always been known that plants are immensely fertile, that a single plant of foxglove, for example produces tens of thousands of seeds. Yet foxgloves do not become more common. In fact the relative proportion of different plants in a given wild area remains substantially the same from season to season. Neither plants nor animals often spread beyond their normal haunts. In undisturbed country a clump of some particular plant will come up in the same spot year after year. So too with animals.

Many plants have special modes of distribution—the air-borne seeds of the dandelion occur to the mind. Yet such means of distribution evidently do not serve them. The surface of the earth does not become covered with dandelions, despite the pessimistic opinion of gardeners.

We tend to think of individual plants as we encounter them in fields and gardens. But fields and gardens are ploughed and harrowed, hoed and weeded and manured. Thus each plant is given its place. Nature has less caring for individuals. Plants in nature exist in so-called 'communities'. These do not usually consist of one species but of groups of species.

In nature, when man has not interfered, every spot where life can maintain itself is fully occupied by such groups and has been so for ages. The frontiers of communities may be altered a little by extraordinarily wet or dry seasons and the like. Despite fecundity, despite facilities for distribution, despite apparent strength, it is an excessively difficult thing for a foreign individual to gain entry into a community—indeed the mind has difficulty in grasping the excessive infrequency of such an event. Each district has its own flora and there is as little hope of that of a mountain, for instance, invading that of a valley at its foot as of the reverse process. Wild nature is extraordinarily stable.

The matter demands illustration from a completely wild district where man has never intervened. Mount Ritigala is a solitary precipitous peak of 2,800 feet in the north of Ceylon. That mountainous island has in general a typical damp, tropical climate. The northern part of the island where Ritigala is situated is more dry and arid. But Ritigala is cloud-clapped and its peak always moist.

The flora of Ritigala peak consists mostly of species of genera which flourish in the dry zone below. Despite the chances of vast ages it has received only 103 species characteristic of the wet zone of the island where the climate is similar to its own. Of these 103 species, 24 have fruits suited to carriage by birds, 49 have light seeds, or spores suited to carriage by wind. Thus some three-fourths of the immigrant species have been brought to this remote spot by known means. Yet the poverty in species of the immigrant population emphasizes the extreme and almost inconceivable slowness with which organisms are diffused in nature and the enormous difficulty for the foreign stock to obtain a foothold in an established community [Investigation of S. C. Willis, 1868–1958].

Where Nature has cleared the ground, the phenomena are utterly different. The classical instance is the island of Krakatoa between Sumatra and Java. In 1883 there was a terrific eruption on the island. Life was completely exterminated. By 1886 some 17 species of flowering

plants had re-established themselves on the island. In
1897 the number had risen to 50, and in 1905 to 137.
Thus, in 22 years more immigrants had established themselves on vacant Krakatoa than had succeeded in geological
ages on occupied Ritigala!

From the end of the nineteenth century botanists found
it worth while to record the actual distribution of individual plants or plant colonies.   Such intensive work,
especially in wild or unsettled areas, has brought some
remarkable facts to notice. If any genus containing a large
number of species be taken, some of these species will be
rare, others less rare, and others common. Now it has often
been found that the rare species have a particularly
narrow distribution as compared with the common.   In
some cases it has been shown that a rare species exists
only in an area of a few square yards.

It was one thought that such extremely local species
were failures, driven to mountain peaks or other unfavourable sites as their last stronghold.   Since the advent of
the view of the origin of species by mutation (chap. xv),
it has seemed likely that very localized species are, more
often, new productions. Formed by mutation in isolated
localities, they are able to perpetuate themselves on an
isolated spot, though many ages must be needed for their
spread. A species that is really dying out would be much
more likely to exist sparsely distributed over a wide area
than densely concentrated in a narrow one.

§ 10. *Migration*

A very remarkable phenomenon is presented by the
wandering of animals into distant localities. In the most
striking cases the process is annual and is associated with
breeding. Since the days of Aristotle migration has been
observed among birds, but examples no less regular and
impressive are known among fishes. Certain mammals,

as seals, perform similar journeys. Less regular migrations, bearing analogies to those less ordered movements
of population encountered in the human species, are presented by many groups, from mammals downwards.

The attention of many learned naturalists has been

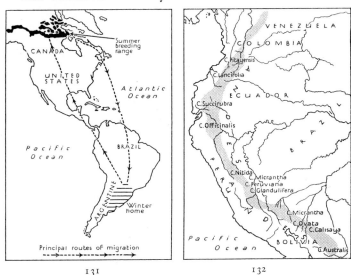

131   132

FIG. 131. To show the route of migration and range of the American Golden
Plover.

FIG. 132. A part of South America to show the distribution of closely allied
species of the Chinchona trees from which Quinine is derived. The shaded area
marks the range of the Andes (Clements R. Markham, 1860). Twelve distinct
species occupy each its own area.

attracted to the migration of birds. Ray (1676), Linnaeus
(1757), Gilbert White (1789), Blumenbach of Göttingen
(1823), and von Baer (1834) treat of it. The subject was
dealt with by Marcel de Serres of Montpellier (1782–
1862). From 1822 onward he attempted to correlate the
geological distribution of animals to their modern geographical distribution.

During the next half-century a mass of data concerning the movement of birds was collected, especially by Scandinavian, Russian, and German naturalists. Among the most important facts elicited was that in many genera of birds, those species that have the widest northerly have also the widest southerly range, and those species which go next farthest north in summer pass next farthest southward in winter.

An extreme case is that of the American Golden Plover. It breeds and nests in the extreme north of Canada within the Arctic circle around Baffin Bay. It spends the winter on the pampas in the south of Argentina (Fig. 131). The route taken by this species is now fairly known, as is that of a large number of other forms.

The conditions that determine migration of birds are partly understood. They are the usual ecological factors—food-supply, light, temperature, moisture, safety from enemies, and the rest. The mode of origin of migration as well as the method by which these birds find their way over vast distances remain a mystery.

A helpful suggestion was put forth by A. R. Wallace in 1878:

'Suppose', he said, 'that in any species of migratory bird, breeding can only be safely accomplished in a given area and that during a great part of the year sufficient food cannot be obtained in that area. Those birds which do not leave the breeding area at the proper season will become extinct, which will also be the fate of those which do not leave the feeding area. If the two areas were, for some remote ancestor, coincident, but geological and climatic changes gradually diverged, we can understand how the habit of migration would become so fixed as to be what we term an instinct. It will probably be found that every gradation still exists, from a complete coincidence to a complete separation of the breeding and subsistence areas.'

This prophecy has been largely fulfilled. But its fulfilment does not bring us much nearer to a knowledge of the mechanism by which the bird finds its way, nor will it explain the numerous cases in which migration takes place over a large part of the earth's circumference. In this connexion many statements have been made which we need not reproduce. The subject is under current discussion.

Migration is known, in some other animal groups in insects, such as locusts, in mammals, such as seals. None is more inexplicable than the seemingly objectless sporadic mass-movements of the lemmings.

Regular migrations in connexion with breeding are found in fish. Thus the salmon breeds in rivers—a safe retreat—but feeds in the sea. Some few fish reverse the process, the most extraordinary and best authenticated case being that of the eel.

The problem of the breeding habits and migratory movements of the eel have been before naturalists since the days of Aristotle. Rondelet (p. 93), Gesner (pp. 94–6), and the other sixteenth-century naturalists discuss them. Francis Bacon (p. 121) refers to them. Isaac Walton (1593–1683) expresses the difficulties in a very quaint passage in his *Compleat Angler* (1653). Naturalists of the eighteenth century described the eel's general habits of migration. They distinguished two migrational movements in the year, in the autumn to the sea, in the spring from the sea. They observed that in the autumn movement the eels were larger than in the spring. It was seen that they fed in the rivers but bred in the sea. The migration itself, part of which is over-land, was repeatedly observed in the nineteenth century. The breeding-place was quite unknown but believed to be on coasts and estuaries.

Meanwhile, a group of small laterally compressed transparent fishes had been observed early in the nine-

teenth century in the Eastern Mediterranean and the
Atlantic. The genus *Leptocephalus* was erected to include
them. In 1896 G. B. Grassi (1854–1925) of Rome—who
afterwards did distinguished work on the parasites of
malaria in man—made an important discovery bearing on
a species of *Leptocephalus*. In the Straits of Messina he
found a number of specimens transitional between *Lepto-*

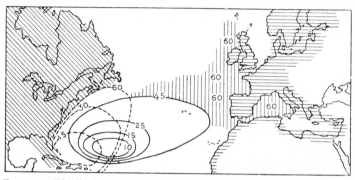

FIG. 133. Distribution of European and American eel. The continuous concentric curves show the limits of occurrence of the larvae of the European eel,
the length being given in millimetres. By the time they have attained the length
of 60 millimetres (vertical shading) they have reached the European and African
continental shelf. As elvers they ascend the rivers (shaded horizontally). Area
of distribution of the larvae of the American eel is indicated by concentric
dotted lines and its adult form by oblique shading.

cephali and young eels. This suggested that the eel bred
in deep water near the coast. In 1904 the Dane, Johannes
Schmidt (1877–   ), showed that the real breeding-
ground of the European eel is a small area on the bed of
the Ocean at a depth of about 300 fathoms to the south-
east of Bermuda. From thence these creatures radiate as
Leptocephali, increasing in size and becoming more eel-
like as they proceed towards their European home. Even
more remarkable is the fact that a breeding-ground not
very far distant produces a very closely allied species which

passes not to European but to American rivers (Fig. 133).

A hypothetical explanation of this extraordinary movement is now forthcoming. It has been suggested that the breeding-ground was once a freshwater lake or river within a transatlantic continent now submerged (Fig. 127), and that the eels have preserved their ancestral breeding-ground and continued to follow their old lines of movement. These instinctive reactions would thus be older than the main geographical features of our world! The eel always bred where he now does, but he still frequents for a part of his life his old haunts —the rivers from which his breeding-ground is now widely separated by continental movement. Further, in the course of ages a differentiation must have taken place between the American and the Old World eel. The breeding-ground of the American eel is now a little further west than that of its cogener. The larvae of both species radiate from their breeding-ground. The American eel is born in warmer waters. Its larval stage is completed in about a year, while that of the Old World eel is not completed until about three years. Both species tend to spread northeastward along with the prevalent currents (Fig. 113), the American form, however, spreads more northward and the European more eastward. Thus, by the time the larva of the Old World eel attains to the elver stage, it has approached the European continental shelf or has entered the Mediterranean. At the end of its first year, however, the American species finds itself on the American continental shelf. The elvers now ascend the rivers of the two continents and pass into the adult stage. The countless millions that do not reach the shore never develop further and perish as larvae (Fig. 133).

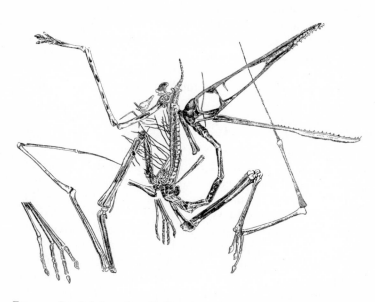

FIG. 134. Fossil skeleton found in Bavaria in 1788. It was regarded as a bird by Blumenbach (1807). Cuvier identified it (1812) as a flying reptile of a previously unknown class, and named it *Pterodactyl*. Compare Fig. 106, p. 225.

VIII
EVOLUTION

§ 1. *Buffon* (1707–88) *and Erasmus Darwin* (1731–1802)

EVEN the most savage and uncultured people have some idea of species. They must naturally distinguish the different animals that they hunt and the various plants that they gather. Some, in fact, exhibit great acuteness in differentiation between species. Their ideas have, however, no formal distinctness. For this we must await the advent of science. Many philosophical writers have included some idea of evolution in their schemes. We shall not consider these but shall concentrate on the concrete idea of evolution of organic species by descent.

We have glanced at the conceptions of species of Aristotle (p. 43), Bauhin (pp. 176–9), Jung (p. 182), and others. By most older writers, species are treated, after the biblical account, as created once for all. Linnaeus held that there are 'as many species as issued in pairs from the hands of the Creator'. Later he admitted that it was in some cases difficult to separate one species from another. Without fully abandoning his original position, he substituted, in effect, the *genus* for the *species* as the original creation, concluding that 'all the species of one genus constituted at first one species'. New species, he thought, had arisen by intercrossing. The first naturalist of modern times clearly to set forth the idea that species are not permanent was Buffon (1707–88).

George Louis Leclerc, Comte de Buffon, scion of a distinguished Burgundian family, early showed a taste for mathematics and the physical sciences. Gradually he moved towards biological problems. He was influenced in this by the writings of Stephen Hales (pp. 363–6). The wealth and social position of Buffon enabled him to devote his time to science. His great industry and natural acuteness were applied in obtaining a broadly based knowledge of biological matters. His attractive literary style rendered very great service in drawing public attention to the value and interest of science. He was not a painstaking investigator, but he made many excellent scientific suggestions.

Buffon's great *Histoire naturelle* sought to embrace all scientific knowledge, and was the first modern attempt of the kind. It appeared in forty-four volumes, and its publication occupied fifty-five years (1749–1804), being completed after his death by an assistant. It was well illustrated. There were several editions and translations.

Buffon's popularity led to his belittlement by later specialists. Nevertheless many of his ideas acted as a ferment in the minds of his scientific successors. His influence can be traced in Erasmus Darwin, in Lamarck, in Geoffrey St. Hilaire, in Goethe, in Cuvier. His more fertile conceptions may be thus summarized:

(*a*) The grand demonstration by Isaac Newton (1642–1727) that the forces that can be investigated on earth are identical with those that move the celestial orbs, deeply impressed Buffon. He, too, would consider Nature as a whole. But, for Buffon, Nature must include living Nature, which Newton had disregarded. Yet Buffon was no mere physiological mechanist, like Borelli, for instance (p. 360), content to explain the more evident activities of the living body. For Buffon all parts and all activities of the world are interrelated.

(*b*) Arising out of (*a*) is his attitude to classificatory systems, especially that of Linnaeus. These he held to be trifling and artificial abstractions, since they fail to present living things as part of the general order of Nature.

(*c*) In the series of living things he paid little attention to the minute differences between species that Linnaeus and the systematists were ever seeking. He looked rather at the common factors, the likenesses. He sought some universal element in living things. The cell doctrine (chap. ix), had it been available, would particularly have rejoiced him. He fastened, however, on the phenomenon of reproduction as a universal accompaniment of life. Spermatozoa were then being brought to light in a constantly increasing range of organisms. Knowing nothing of the cellular phenomena of sex (chap. xiv), he believed that spermatozoa—and comparable bodies which he thought he saw in the ovaries—represented units out of which individuals could be built. The conception has distant affinities with that of the 'monads' of the philosopher Leibniz (1646–1716). It is in some degree anticipatory of the many nineteenth-century theories of the particulate nature of living substance (chap. x).

(*d*) He sought to trace the history of the earth through a series of 'epochs', fossils providing some key to this history (p. 246).

(*e*) Buffon expresses himself variously on the subject of the fixity of species. He was, however, moving ever farther from the position of Linnaeus. He noted that animals have parts to which no special or adequate use can be ascribed. Thus 'the pig is not formed on an original perfect plan, since it is a compound of other animals. It has parts which can never come into action, as lateral toes, the bones of which are perfect, yet useless.' Later he conceived that species alter in type from time to time, but retain marks of their previous types, as the pig retains its disused toes. Then, moving a little farther, he concluded that some species are degenerate forms of others. Thus the ape is a degraded man, the ass a degraded horse, and so on.

These ideas of Buffon were examined by Erasmus Darwin (1731–1802), grandfather of Charles Darwin.

He excelled as a thinker and critic rather than as an observer, but was no mean naturalist. In his *Zoonomia; or the Laws of Organic Life* (1794–6), he sums up in clear though verbose language the general nature of the difficulties among which Buffon was groping. His solution is striking and it is noteworthy that he gathers just those classes of facts which most impressed his grandson.

'When we revolve in our minds,' writes Erasmus Darwin, 'first the great *changes which we see naturally produced* in animals after their birth, as in the production of the butterfly with painted wings from the crawling caterpillar, or of the (air-breathing) frog from the (water-breathing) tadpole; secondly, the great *changes by artificial cultivation*, as in horses which we have exercised for strength and swiftness, or dogs which have been cultivated for strength and courage, as the bulldog, or acuteness of smell, as the spaniel, or swiftness, as the greyhound; thirdly, the great *changes produced by climate*, the sheep of warm climates being covered with hair instead of wool, and the hares and partridges of latitudes which are long buried in snow becoming white during the winter months; fourthly, *the changes produced before birth by crossing or mutilation*; fifthly, *the similarity of structure which obtains in all the warm-blooded animals, including mankind*, from the mouse and bat to the elephant and whale; one is led to conclude that they have alike been produced from a similar living filament.' [*Much abbreviated. Italics inserted.*]

The 'filament' to which he refers is a spermatozoon which he regarded, following Buffon, as a sort of biological unit (p. 295).

Erasmus Darwin thus held that the changes that species undergo in the course of time are due to influences that bear on the individuals from without. These changes he held were passed on to the offspring. This view is now known as the *inheritance of acquired characters*.

§ 2. *Lamarck* (1744–1829) *and his Successors*

The question of the inheritance of acquired characters

was among the most discussed of nineteenth-century problems. Erasmus Darwin was the first to state the thesis. Charles Darwin took such inheritance for granted, Lamarck, a younger contemporary of Erasmus Darwin. made it the key to his evolutionary hypothesis.

Jean Baptiste de Monet Lamarck (1744–1829) of Amiens was educated at a Jesuit College. In 1761, during the Seven Years' War, he entered the army. Failure in health forced him to forego a military career. He turned to medicine, giving special attention to botany, then an important part of the medical curriculum. He became intimate with the famous Genevan philosopher, Jean Jacques Rousseau (1712–78), who influenced him a good deal. Later he worked also with another distinguished Genevan, the botanist, A. P. de Candolle (pp. 197–9).

Lamarck's first biological work was a flora of France (1778). It drew the attention of Buffon, who assisted its author to travel and to publish other botanical works. In 1788 Lamarck took up a botanical post at the 'Jardin du Roi', now the Jardin des Plantes, at Paris. Owing to rearrangement there in 1793 he was forced, at 50, to change over to zoology, to which he devoted himself with enthusiasm. His best-known work is the *Philosophie zoologique* (1809). Its principles are worked out in detail in his *Histoire naturelle des animaux sans vertèbres* (1815–22).

Lamarck was unlucky in his mode of address. His literary style is arid. He was personally eccentric. He married four times, and had to support a very large family on a very small salary. His over-proneness to speculation often made him a laughing stock. Many of his views were extremely fanciful and were held in light esteem by his contemporaries in general. Cuvier, the scientific dictator of the time, who adhered to the view of the fixity of species, formed a low opinion of his abilities. Charles Darwin, among his successors, held him in contempt.

The interest of the theory which by Lamarck is remembered was not fully realized until he had long been dead.

As a systematist, Lamarck certainly made important and lasting contributions. He separated spiders and crustaceans from insects, and defined all these classes. He began to bring order into the class *Vermes* of Linnaeus (p. 192), segregating the truly worm-like forms. He also made some advance in the classification of the echinoderms. He introduced the classification of animals into vertebrates and invertebrates.

With his eye fixed on 'system' Lamarck was convinced that there is a 'natural sequence' for living organisms. If we knew all the species that are or have been, they would, he believed, form a long ladder or scale with comparatively few branches. On this scale each species would differ but little from its immediate neighbours.

There are, in fact, existing species which differ very considerably from their nearest known allies. Lamarck was quite aware of this and he attributed such isolation to gaps in our knowledge. Further research would, he believed, fill these vacant places. The progress of palaeontology seemed to him to offer hopes of such a consummation. Lamarck, therefore, argues that all systems of classification are really artificial, though necessary as summaries of our knowledge.

This idea of the continuity of living things led Lamarck to consider that the animal and the plant series must, at some point, be continuous with each other. He thus emphasized the view that living things should be studied as a whole, an attitude to which he was helped by having himself practised as both botanist and zoologist. For this unified study he invented the term Biology (1802).[1]

---

[1] In the same year the word was used by G. R. Treviranus of Bremen (1776–1837) as the title of a book expounding some of the principles of the Naturphilosophen. The word first appears in English in the modern sense in the *Lectures on Physiology* (edition of 1818) of Sir William Lawrence (1783–1867).

Since Lamarck held that no distinction can ultimately be found to exist between species, it seemed to him intrinsically improbable that they are permanently fixed. He laid much stress on the domesticated animals, which vary greatly from their wild originals. Who, seeing a greyhound, a spaniel, and a bulldog for the first time, would not think of them as different species. Yet all have a common ancestor. These variations have been produced by selective breeding by man. Variations comparable to these in kind, if not in degree, occur also in nature. There must be some agent in nature producing these differences. This agent, according to Lamarck, is the environment. Species, he thought, maintain their constancy of form only so long as their environment remains unchanged.

However strained Lamarck's conclusions in detail, we can now see that he had reached three important and interconnected conceptions.

(*a*) Species vary under changing external influences.

(*b*) There is a fundamental unity underlying the diversity of species.

(*c*) Species are subject to a progressive development.

Lamarck had now to consider the mechanism of this progressive development. What is it? How is variation caused? How do changes of environment give rise to changes of species?

In answering these questions, Lamarck enunciated the 'law of use and disuse' that is inseparably connected with his name. He supposed that changes of environment lead to special demands on certain organs. These being specially exercised become specially developed. Such development, or some degree of it, is transmitted to the offspring. Thus a deer-like animal, finding herbage scanty, took to feeding on leaves of trees. It needed a longer neck to reach the leaves. In the course of generations, the long neck became a more accentuated feature of the creature's anatomy. Thus emerged a beast recognizable as a giraffe. Conversely, useless organs, such as the eyes of animals that live in darkness, being unexercised, gradually became functionless and finally disappeared. The character of a longer neck or of defective eyes was acquired by the individual in the course of its lifetime and transmitted in some degree to its descendants.

The great assumption here is that acquired characters are inherited. Whether they are or are not is still in dispute. In the way suggested by Lamarck, it is certain that they are not. But Lamarck's work has been of value in directing the attention of naturalists to one of the most important problems in the whole range of biological thought.

Lamarck, moreover, brought the question of the constancy of species into the sphere of discussion. His conception of progressive development, or 'Evolution' as it is now called, together with certain ideas on the subject of homologous changes set forth by his colleague, Étienne Geoffroy St. Hilaire (1772–1844), attracted some attention. Unfortunately the details of Geoffroy's scheme were often fantastic to the last degree. Their refutation resulted in all biological speculation falling into disrepute for a while. Much of the discussion on the subject of Evolution during the first half of the nineteenth century was of a vague and fanciful character.

Yet there was a writer whose work bore upon the subject, against whom this charge could certainly not be made. The Rev. T. R. Malthus (1766–1834) was primarily a mathematician and economist, not a biologist. His importance is due to his *Essay on Population* which suggested to Darwin and Wallace simultaneously the conception of the Struggle for Existence and of the Survival of the Fittest.

The *Essay* of Malthus was first published anonymously

in 1798. At that moment political theory was a common subject of discussion, especially in connexion with the French Revolution. Such topics as the 'rights of man', 'natural justice', and the like were in the public mind. The utilitarian philosophers, with Adam Smith (1723–90), Joseph Priestley (1733–1804), and Jeremy Bentham (1748–1832) as their chief spokesmen, formed an advanced and influential school of thought in England. Many believed that a day was dawning when, amidst universal peace, all men would enjoy complete liberty combined with complete equality. Malthus brought out forcefully and systematically the difficulties that must arise, in such a state, from over-population.

In his *Essay* Malthus laid down his famous principle that populations increase in geometrical, but subsistence at best only in arithmetical ratio. He argued that a stage is therefore reached at which increase in population must necessarily be limited by sheer want. Thus he held that 'checks' on population are a necessity in order to reduce vice and misery.

§ 3. *The 'Origin of Species' and the Validity of its Argument*

After the voyage of the *Beagle* (1837) Darwin became recognized as a painstaking naturalist, remarkable for the general breadth of his outlook. He did good work on the barnacles and their allies (1851–4, pp. 519–22), became known for his investigations of mammalian fossil forms, wrote much on geology, and was the author of an admirable book on coral reefs (1842). He was, moreover, an observer of living things, a patient recorder of what he law, a man of reflective mind, and one content to ponder song before giving his views to the world.

'When on board H.M.S. *Beagle*,' he writes, 'I was much struck with the distribution of the organic beings inhabiting South America, and the geological relations of the present to the past inhabitants. These facts seemed to throw light on the origin of species. On my return, it occurred to me that something might be made of this question by accumulating all sorts of facts which could possibly bear on it. My first note-book was opened in 1837. In 1838 I read *Malthus on Population*. Being prepared to appreciate the struggle for existence which goes on everywhere, it struck me that favourable variations would tend to be preserved, unfavourable to be destroyed. The result would be formation of a new species. I had at last a theory by which to work. After five years' work I drew up some short notes; these I enlarged in 1844, into a sketch of the conclusions which seemed to me probable; from that period to the present (1858) I have steadily pursued the same object. Mr. Wallace, who is now studying the natural history of the Malay Archipelago, has arrived at almost exactly the same general conclusions. Sir C. Lyell and Dr. Hooker, who both knew of my work— the latter having read my sketch of 1844—honoured me by thinking it advisable to publish, with Mr. Wallace's excellent memoir, some brief extracts from my manuscripts.' [*Abbreviated.*]

The naturalist and explorer, Alfred Russel Wallace, subsequently made many important contributions bearing on the distribution and variation of living forms. The publication of the joint essays was in July 1858. It was the first public pronouncement in which an effective mechanism was suggested to explain the evolution of organic forms.

Darwin did not initiate the doctrine of Organic Evolution. But, by careful and scientific procedure, he persuaded the scientific world, once and for all, that many diverse organic forms are of common descent, and that species are inconstant and in some cases impossible of definition. Moreover, he directed scientific attention to the occurrence of variation, to its persistence, and to the question of its origin and its fate.

In November 1859 appeared Darwin's great book the *Origin of Species by means of Natural Selection*. Despite its great value and stimulating character and despite the conviction that it carried, its arguments are frequently fallacious. It often confuses two distinct themes. On the one hand there is the question whether living forms have, or have not, an evolutionary origin. On the other hand is the suggestion that Natural Selection is the main factor in Evolution. These themes can be and should be discussed independently. In the *Origin* they are inextricably confused.

Darwin claims that 'complex organs and instincts' have 'been perfected, not by means superior to, though analogous with, human reason, but by the accumulation of innumerable slight variations, each good for the individual'. For this, he says, it is necessary to admit only three propositions.

'That gradations in the perfection of any organ or instinct, either do now exist or could have existed, each good of its kind.

'That all organs and instincts are, in ever so slight a degree, variable.

'And, lastly, that there is a struggle for existence leading to the preservation of each profitable deviation of structure or instinct.'

But this assumes that the 'profitable deviations' are inherited. Thus not three but four propositions are needed.

After discussing the question of the distribution of species in time and space—which provides the most irresistible evidence for organic evolution as an historical process (chap. vii)—he turns to the question of conditions under which a variation is perpetuated.

'Man does not actually produce variability [in domestic animals]; he only exposes beings to new conditions, and then nature acts on the organisation, and causes variability. But man can and does select variations, and thus accumulate them in any desired manner. He thus adapts animals and plants for his own benefit. He can largely influence the character of a breed by selecting, in each successive generation, individual differences so slight as to be quite inappreciable by an uneducated eye. That many of the breeds produced by man have to a large extent the character of natural species, is shown by the doubts whether many are variations or aboriginal species.

'In the preservation of favoured individuals and races, during the constantly recurrent Struggle for Existence, we see the most powerful and ever-acting means of selection. More individuals are born than can survive. A grain in the balance will determine which individual shall live and which shall die—which variety or species shall increase in number, and which shall decrease, or finally become extinct.

'With animals having separated sexes there will in most cases be a struggle between the males for possession of the females. The most vigorous individuals, or those which have most successfully struggled with their conditions of life, will generally leave most progeny. But success will often depend on having special weapons or means of defence, or on the charms of the males; and the slightest advantage will lead to victory.' [*Abbreviated.*]

There are here, as we can now see, a series of fallacies and some erroneous assumptions.

(*a*) All domestic breeds have not been produced by selecting very slight individual differences. On the contrary, some domestic breeds certainly and all domestic breeds possibly have been produced by breeding from individuals which presented very considerable deviations from the normal. The systematic study of individual variations—except in the isolated case of Mendel—did not, however, begin until the last decade of the nineteenth century.

(*b*) That a natural variation should confer an advantage is not enough to secure its perpetuation. The advantage must be effective and moreover it must be transmissible. Now it is difficult to believe that the earlier stages of some developments are effective. A wing, for instance, so little developed as to confer no power of flight, or at least of gliding, would be no advantage.

(c) It is assumed that 'advantages' are inherited. On this matter the modern development of the theory of heredity (chap xv) has entirely changed the whole point of view presented by Darwin and his contemporaries.

(d) It is tacitly assumed that species differ from their nearer relatives in having some special advantages that enable them to adapt themselves to slightly different conditions. In fact, however, allied species are found living in identical areas and under identical conditions. This would hardly be the case if one had the advantage over the other, for then one would flourish and the other decline. There are, in fact, very few characters by which species differ from their fellow-species of the same genus that can be shown to be advantageous.

'As natural selection acts solely by accumulating slight, successive favourable variations, it can produce no great or sudden modification; it can act only by very short and slow steps. Hence the canon of "Natura non facit saltum", which every fresh addition to our knowledge tends to make truer, is on this theory simply intelligible. We can plainly see why nature is prodigal in variety, though niggard in innovation. But why this should be a law of nature if each species has been independently created, no man can explain.'

This passage is of necessity rejected with the modern laws of heredity before us. The aphorism of Linnaeus 'Nature takes no jumps' cannot be maintained. Mutants are, in fact, jumps, *saltus*. The whole Mendelian teaching on heredity is based on *saltus*. There is evidence that Mendelian factors provide at least one and possibly the only effective source of varieties, and certainly the main source of domestic varieties. As to how varieties pass into species we are still in the dark.

Darwin's view of Natural Selection as an effective agent of Evolution is perhaps at its weakest in dealing with the problem of disuse. Here he assumes the inheritance of acquired characters in a form hardly less crude than that of Lamarck.

'Disuse, aided sometimes by natural selection, will often tend to reduce an organ, when it has become useless by changed habits or under changed conditions of life; and we can clearly understand on this view the meaning of rudimentary organs. But disuse and selection will generally act on each creature, when it has come to maturity and has to play its full part in the struggle for existence, and will thus have little power of acting on an organ during early life; hence the organ will not be much reduced or rendered rudimentary at this early stage. The calf, for instance, has inherited teeth, which never cut through the gums of the upper jaw, from an early progenitor having well-developed teeth; and we may believe, that the teeth in the mature animal were reduced, during successive generations, by disuse or by the tongue and palate having been better fitted by natural selection to browse without their aid; whereas in the calf, the teeth have been left untouched by selection or disuse, and on the principle of inheritance at corresponding ages have been inherited from a remote period to the present day.'

Where Darwin concentrates on the theory of descent and forgets his own particular explanation, his work rises to a convincing eloquence.

'If we admit that the geological record is imperfect in an extreme degree, then such facts as the record gives, support the theory of descent with modification. New species have come on the stage slowly and at successive intervals; and the amount of change, after equal intervals of time, is widely different in different groups. The extinction of species and of whole groups of species, which has played so conspicuous a part in the history of the organic world, almost inevitably follows. Neither single species nor groups of species reappear when the chain of ordinary generation has once been broken. The gradual diffusion of dominant forms, with the slow modification of their descendants, causes the forms of life, after long intervals of time, to appear as if they had changed simultaneously throughout the world. The fact of the fossil remains of each formation being in some degree intermediate in character between the fossils in the formations above and below, is simply explained by their intermediate position in the chain of descent. The grand fact that all extinct organic beings either belong to the same system with recent beings, falling either into the same or into intermediate groups, follows from the living and the extinct being the offspring of common parents.

'Looking to geographical distribution, if we admit that there has been during the long course of ages much migration from one part of the world to another, owing to former climatal and geographical changes and to many occasional and unknown means of dispersal, then we can understand, on the theory of descent with modification, most of the great leading facts in Distribution. We can see why there should be so striking a parallelism in the distribution of organic beings throughout space, and in their geological succession throughout time; for in both cases the beings have been connected by the bond of ordinary generation, and the means of modification have been the same. We see the full meaning of the wonderful fact, which must have struck every traveller, namely, that on the same continent, under heat and cold, on mountain and lowland, on deserts and marshes, most of the inhabitants within each great class are plainly related; for they will generally be descendants of the same progenitors and early colonists. On this same principle although two areas may present the same physical conditions of life, we need feel no surprise at their inhabitants being widely different, if they have been for a long period completely separated from each other; for as the relation of organism to organism is the most important of all relations, and as the two areas will have received colonists from some third source or from each other, at various periods and in different proportions, the course of modification in the two areas will inevitably be different.'

We may summarize very briefly the basic criticism of the doctrine of Natural Selection in relation to Evolution. 'Natural Selection means merely that a creature survives.' As to why it survives, or whether its survival be due to one character more than another, and as to how it acquired that character, the theory gives us no information whatever.

Darwin himself compared the action of natural selection

to that of a man building a house from stones of all shapes. The shapes of these stones, he says, would be due to definite causes, but the uses to which the stones were put in building the house would not be explicable by these causes. The conception reveals the general weakness of Darwinism which treats natural selection as an active agent. For when a man builds a house, we have the intervention *of a definite purpose* directed towards a *fixed end* and governed by a *clearly conceived idea*. But these psychic factors have no relation to the causes which produced the stones, and cannot be compared with the action of natural selection.

§ 4. *The Reception of the Doctrine of Evolution*

The *Origin of Species* created a revolution in biology, and indeed in many other departments of thought. The idea that species are not constant was far from new or modern. But here for the first time was a careful and scientific work, by a cautious and painstaking investigator, which set forth a vast amount of evidence on the matter. Moreover, the book suggested a simple and apparently universally acting biological relationship to explain the process of change of form. That relationship, the struggle of living forms leading to natural selection by the survival of the fittest, is certainly far less emphasized by naturalists now than in the years that immediately followed the appearance of Darwin's book. At the time, however, it was an extremely stimulating suggestion.

In 1852, seven years before the publication of the *Origin*, the philosopher, Herbert Spencer (1820–1903), had set forth doctrines of Evolution, in a work where for the first time that word was used to describe the idea of a general process of production of higher from lower forms.[1]

[1] Lyell had used the word some twenty years earlier, in a similar, though less general sense.

The word Evolution was in fact seldom employed by Darwin, but Spencer's application of it rapidly caught on.

Darwinism and *Evolution* thus came to be regarded as synonymous. The doctrines of Darwin, however, only apply and were only meant to apply in the world of life. Nor, even in that department are the two words interchangeable, for Organic Evolution has frequently been conceived quite independently of such 'Darwinian' factors as Natural Selection or Sexual Selection.

The works of Herbert Spencer were very widely read in the nineteenth century, and did much to spread the evolutionary view of life. The phrase 'Survival of the Fittest' was also coined by Spencer, and has caught on as did 'Evolution'.

Among English-speaking biologists Lyell and Hooker, who had long known Darwin's views, at once gave assent. They were joined by the philosopher, Herbert Spencer, by T. H. Huxley, who had previously expressed opinions in favour of constancy of species, and by the American systematic botanist, Asa Gray (1810–88) of Harvard.

The story of the rise of Darwinism has been more often told than any other incident in the history of science. Among its scientific opponents were Owen, the leading comparative anatomist of his day, and Louis Agassiz of Harvard, a very accomplished naturalist. Both were still bemused by Naturphilosophie, as was also von Baer, now in extreme old age. All opposed to evolution the 'idea' or 'type' of Goethe and Cuvier, a metaphysical conception and of its nature insusceptible of demonstration. Owen added the suggestion that with alternation of generations or with a series of metamorphoses, new types might arise by disintegration of the life cycle into its stages. No such disintegration has been demonstrated.

In Germany, then swept by 'liberal' ideas, Darwinism made rapid progress. Fritz Müller and Haeckel were its leading champions. Its ablest scientific opponent was Kölliker. He did not deny the inconstancy of specific forms and he accepted evolution within the limits of certain wider groups. He emphasized what have since been recognized as the three weakest points in the Darwinian position:—

    (*a*) the absence of any experience of the formation of a species;

    (*b*) the absence of any evidence that the unions of different varieties (i.e. incipient species on Darwin's view) are relatively more sterile than unions of the same variety;

    (*c*) the extreme rarity of true intermediate forms between known species, whether living or fossil.

Kölliker and other critics of Darwin pointed out that the 'chance' element in Darwin's scheme was but a veiled teleology. Natural selection had been elevated to the rank of a 'cause', and science has to deal not with causes but with conditions. Darwin was occupying himself with the 'might' and the 'may be' and not with things seen and proved.

In France the reception of Darwinism was on the whole hostile. The influence of Cuvier was still paramount, and positive thinkers such as Bernard gave evolution a cold reception. Its advance was slow, though its ultimate victory fairly complete. The movement led to the revival of interest in Lamarck, and French *transformisme*, as evolution was called, received a strong Lamarckian tinge.

Apart from purely destructive criticism, there were a number of positive suggestions as to the direction and mechanism of evolution. The botanist, Nägeli, was the most productive in this direction. Much of what he wrote on evolution was almost unintelligible, though certain of his views retain their interest. These we may enumerate.

    (*a*) The alteration in the same direction of the conditions of life in allied species (of hawkweed) produce, he found, no convergence of characters. Thus change of environment produced 'no useful variations'.[1]

(*b*) Evolution proceeds along lines controlled by an inner directing force. Under the name *orthogenesis* this idea of determinate evolution was adopted by several naturalists, notably by the American palaeontologist, E. D. Cope (1840–97). The position is, perhaps, tenable. It can only be justified, however, empirically, that is by the discussion of numerous examples.

(*c*) Evolution does not always proceed by minute gradations but may take larger steps. (This had already been pointed out by Huxley and Kölliker.) Nägeli corresponded with Mendel, whose theory is based on the existence of such steps. Unaccountably he failed to realize the value of Mendel's work.

The whole of modern biology has been called 'a commentary on the *Origin of Species*'. In a sense this is true. Biologists are now agreed that living forms correspond to a limited number of common stocks. For more than a generation a main part of biological investigation was directed to the endeavour to trace the history of those stocks. The battle of evolution as a reality was then a stricken field. On the other hand, on the mechanism of evolution there was no new light until the twentieth century. The generation to which the *Origin of Species* was delivered followed Darwin blindly. The last decade of the nineteenth century showed some reaction (chap. xv). This lasted until the rediscovery of Mendel's work ushered in a new era.

§ 5. *Evolutionary History of Living Forms*

As soon as naturalists felt equipped with an ex-

---

[1] Nevertheless evidence has since been produced that the resemblance of some species to each other is secondary and may be the result of similar conditions.

planation of how the diversity of species had arisen, the *relationship* of different groups acquired a new meaning for them. Classificatory schemes became summaries of evolutionary history.

In the great Darwinistic period (1860–1900), naturalists concentrated rather on the structure of animals and plants than on the workings of their bodies or on their

FIG. 135. *Peripatus capensis* from Moseley.

habits. Julian S. Huxley (1887–   ) thus summarizes the recent history of evolutionary theory:—

'The three major types of process operating in evolution are those (1) leading to divergence and variety, (2) to adaptedness and biological improvement, and (3) to stabilization and persistence of type. Darwin, before 1859, realized the universality of (1) and noted examples of (3), the importance and widespread occurrence of which was stressed by T. H. Huxley in 1862. From then to the end of the century most biologists were concerned with establishing (2) the evolutionary interpretation of comparative anatomy, taxonomy and adaptation; the period 1900–30 was devoted chiefly to establishing the mechanism of evolution-natural selection on the basis of particulate variation and inheritance. Only since then has attention been focused on evolutionary process.' [1957, *abbreviated.*]

Thus comparative anatomy rather than comparative experimental physiology became the representative biological study. Even when anatomy was interpreted in embryological terms, the stress was still on structure.

The literature of structural botany and zoology has become bulky almost beyond belief. Out of this vast effort have emerged definite and comparatively simple schemes of classification. The general anatomical relationship within the great plant and animal groups have been deter-

Recent

mined to a degree that commands substantially universal consent. Early stages in the historical development of some of these forms have been reconstructed on sound anatomical and embryological grounds.

Pliocene

Further, for many great groups there have been found actual living or fossil organisms not profoundly unlike the respective ancestral forms which had been pictured. Thus naturalists had inferred the characters of the ancient parent stock of flowering plants, and

Mio-Pliocene

forms were subsequently discovered bearing a remarkable resemblance to this hypothetical ancestral type. Again, *Peripatus*, a curious worm-like creature, regarded as a mollusc when first found a century ago

Miocene

(1826), was proved by anatomical examination fifty years later (1874) to have affinity with the insects. It bears many resemblances to the reconstructed common ancestor of all the insects, of some other groups of creatures with jointed limbs, and of Myriapods (Fig. 135).

Miocene

Amphioxus, a tiny transparent animal, though highly specialized for a peculiar mode of life, reveals many points of similarity to the reconstructed fish-like ancestor of the vertebrates (pp. 481–5). Knowledge of the

Eocene

structure and development of such forms has cleared our ideas as to what we mean by a 'Flowering Plant', an 'Insect', a 'Vertebrate'.

Long before Darwin, study of the geo-

FIG. 136. To illustrate evolution of foot of modern horse by gradual reduction of digits. The horse now walks on its middle toe and finger (Huxley, 1876).

logical record had revealed a long succession of fossil beings differing from, but sometimes linked to, those now living. The acceptance of evolutionary views provided a key to the existence and nature of this series. It was occasionally found possible to continue a geological series right into modern forms, as with the extraordinarily complete series illustrating the evolution of the hoof of the modern horse (Fig. 136). Again, living forms have been found which continue geological series believed to have come to an end. Such was the case with the *Okapi*, discovered in the twentieth century. This creature continues a geological

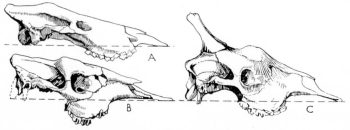

FIG. 137. *A*, Skull of *Helladotherium* from the Miocene of Greece. *B*, Skull of young Okapi. *C*, Skull of Giraffe. (All from Lankester, 1901.)

series which branched off into the giraffe (Fig. 137). One of the most striking of such revelations that geology has to give us is the picture of the vegetation of the coal measures. There, the Club-mosses and Horse-tails, now insignificant families of small and inconspicuous plants, formed the main part of a luxuriant vegetation of giant trees (pp. 280–2, Fig. 130).

§ 6. *Application of the Doctrine of Descent to Man*

There is one species whose origin raised acute controversy. Many ancient and early modern anatomists had drawn attention to the likeness of the anatomy of man to that of the apes. Both Goethe and the American naturalist,

Joseph Leidy (1823–91), interpreted the separate bone in the upper jaw of man (p. 218) as linking him with apes.

Darwin in the *Origin of Species* expressed no opinion as to the relationship of man to lower forms. Several of his supporters, however, notably Huxley and Haeckel, devoted attention to the subject. The formal expression of Darwin's views was reserved till 1871, when he issued the *Descent of Man*. Its opening passage tells that 'Huxley has conclusively shown that in every visible character man differs less from the higher apes than these do from the lower members of the same order of Primates'. This was, however, very different from a demonstration of any intermediate form between man and the higher manlike apes.

A short time before the *Origin* there had been discovered remains of a species exhibiting characters of the supposed ancestral human type. The bones of a manlike creature were unearthed in 1856 in the small ravine of Neanderthal in Rhenish Prussia. The skull was misinterpreted as pathological by Virchow. Huxley ultimately recognized it as human, but the most ape-like yet found. He held that man is 'more nearly allied to the higher apes than the latter are to the lower'. The species to which these bones belong is now entitled *Homo Neanderthalensis*[1] Remains of a considerable number of individuals of this species are known.

Since the discovery of Neanderthal man, a number of other species of fossil man have been discovered. On the other hand, several fossil species of apes approaching somewhat nearer than living forms to the human stem have also been found. The ape-man series is now probably more complete than that of most comparable mammalian groups. The general conclusion is that the junction of man with the apes is farther back than was supposed and

[1] A Neanderthal skull had been found at Gibraltar as early as 1848, but it had not been brought to scientific notice.

that the ape-man group represents a series of rapid species production.

Evolutionary doctrine was soon extended also to man's habits, language, customs, religion, social organization, even his ways of thinking. Thus has arisen the modern science of *Anthropology*. Ultimately the study merges into a consideration of the higher social units. It then discusses the social evolution of man, and develops as *Sociology*. Anthropology and Sociology must lean on the doctrine of descent. They have much to derive from criticism of the mechanism variation and modification.

Extension of evolutionary doctrine to man owes a considerable debt to the French amateur, Jacques Boucher de Perthes (1788–1868). He was a civil servant who devoted his leisure to antiquarian research. As early as 1830 he discovered in the gravels of the river Somme certain flints which he believed bore evidence of very ancient human workmanship. In 1846 he demonstrated the existence of such flints in company with the remains of elephant, rhinoceros, and other tropical or extinct forms. In his *Antiquités celtiques et antédiluviennes* (1847–64) he first established the existence of man from his works in Pleistocene and early Quaternary times. In 1863 he clinched this view by discovering near Abbeville a human jaw associated with worked flints in a Pleistocene deposit.

These conclusions were accepted, though with caution, by Lyell in his *Antiquity of Man* (1863). Since that time the study of the works and arts of stone-age man has developed in parallel with the study of his physical structure. The successions of Palaeolithic cultures, crafts, and art, and their merging into those of the Neolithic are now familiar. They have been equated with geological and geographical changes. Classic works along these lines are *Le Préhistorique* (1882) of Gabriel de Mortillet (1821–98) and the writings of the Abbé Breuil (1877–   ).

§ 7. *Coloration and Mimicry*

A peculiar position in the history of evolution is occupied by the study of the colours of animals. These are often protective in nature. Protective resemblance of a species to its background must have been known to hunters from the beginning. The change of coat of some northern animals to white in winter was known to Theophrastus (fourth century B.C.) and has been familiar through the ages. From Pliny (first century A.D.) and Aelian (*c.* A.D. 200) onward the power of the chameleon to alter its colour according to its surroundings has been frequently remarked. In the seventeenth century Redi (p. 440) drew attention to the peculiarities of the 'stick insects' and like forms. Chameleon and green tree-frog attracted Vallisnieri (p. 208) in the eighteenth century.

In the nineteenth century A. R. Wallace frequently commented on protective coloration and colour-changes as exhibited in Brazil and in the Eastern Archipelago. Protective resemblance to background or to inanimate objects lends itself well to explanation on the basis of Natural Selection. Several modern naturalists have devoted themselves to illustration of this theme. It has been worked out, in great detail, for instance, in connexion with the movement of pigment particles of bottom-living fish, such as plaice. The coloration of these creatures has been shown to adapt itself to that of their background.

Such devices of concealment are manifestly helpful and do not entail any considerable structural modifications. It is remarkable that Darwin passed them over in the first edition of the *Origin*, for they would have made ideal illustrations of this theory. Nor did he deal with the more complex problem of mimicry.

The conception of the coloration of animals was modified by an American artist, Abbott H. Thayer (1849–1921). He pointed out that in most animals the back is dark and the belly relatively light. This distribution of pigment may be explained as an adaptation to conditions of life. If an animal stand in a high light its lower part is in the shade. This is compensated by the lighter colour. On the other hand, the upper part is liable to more brilliant illumination. This is in its turn compensated by the darker colour. Thus the general coloration leads to concealment.

The simple light and shade scheme is often further elaborated. Many animals conspicuous in the museum are well concealed in nature. A classical example is the tiger, whose bright black and yellow banding above the lighter colour below fits in well with his normal surroundings of brilliantly illuminated high grass and reeds. The giraffe and the zebra are similarly coloured in a manner that tends to concealment. In many cases, too, a conspicuous mark or series of conspicuous marks upon the animal's surface distracts attention from the general outline of the creature itself. The 'dazzle pattern' camouflages its bearer. In yet other cases, as in the salamander, the pattern draws attention to the creature. Animals thus conspicuously marked are often distasteful or poisonous. Many brightly coloured insects are in this category.

Coloration has always to be considered in relation to surroundings and cannot be understood without them. The *Challenger* naturalists found that many Crustacea drawn from the depths were of a brilliant scarlet. It was only later realized that this colour could only be perceived in light which includes red rays. In the depths, where these Crustacea dwell, the only light is that emitted by phosphorescent organisms. From this light, however, red rays are absent. There such pigmented forms would not be conspicuous. The explanation of the red pigment must therefore be sought on internal physiological grounds.

Very special interest attaches to the problem of coloration and other forms of special development classed under the term 'mimicry'. By mimicry is meant the deceptive resemblance of one species to another. The subject was brought to notice by Henry Walter Bates (1825–92). This superb self-taught naturalist left England in 1848 with the equally self-taught A. R. Wallace on a collecting expedition to Brazil. Bates remained eleven years. He had with him both Lyell's *Principles* and Darwin's *Journal*. Before the publication of the *Origin* he wrote, concerning certain butterflies, that on their wings 'Nature writes, as on a tablet, the story of the modification of species'. He was vastly impressed with the tropical wealth of species, of which he discovered more than 8,000.

In his classic paper *Contributions to the Insect Fauna of the Amazon Valley* (1861), Bates chose as his text the Heliconidae, an American family of butterflies, with but two genera and about 150 species.

'A most interesting feature of the *Heliconidae*', he says, 'is the mimetic analogies of which many species are the subject. Many inhabit districts tenanted by species which counterfeit them. The *Heliconidae* are the objects imitated, since all have the same family facies, whilst the imitating species are dissimilar to their own nearest allies—perverted, as it were, to produce the resemblance. This resemblance is so close that only after long practice can the true be distinguished from the counterfeit, when on the wing.

'Hundreds of instances of imitative resemblances could be cited. Some show a minute and palpably intentional likeness which is perfectly staggering. It is not difficult to divine the meaning of these analogies. When a Moth wears the appearance of a Wasp, we infer that the imitation is intended to deceive insectivorous animals, which persecute the Moth, but avoid the Wasp. May not the Heliconide dress serve the same purpose to the *Leptalis*? Is it not probable, seeing the excessive abundance of the one species and the fewness of individuals of the other, that the Heliconide is free from the persecution to which the Leptalis is subjected?' [*Slightly abbreviated.*]

This type of resemblance has since become known as 'Batesian mimicry'.

Another early writer on mimicry was the German naturalist, Fritz Müller (1821–97). He also reached his conclusions in Brazil. In 1879 Müller described another type of the phenomenon. In Batesian mimicry the mimicked species is unpalatable or dangerous. Müller showed, however, that there are cases in which both mimicker and mimicked were unpalatable or dangerous.

'If two or more distasteful species be about equally common, resemblance brings them a nearly equal advantage. Each step toward resemblance is preserved by natural selection. They would always match each other numerically, so that finally one would not be able to say which has served as model.'

A large majority of the best and most striking cases of mimetic resemblance have been described from tropical countries. A reason for this is the vastly greater wealth in species of the Tropics as against temperate climes. Thus Bates says of butterflies 'about 700 species are found within an hour's walk of the town [of Para]. The number in the British Isles does not exceed 66. The whole of Europe supports only 321.' The number of Brazilian species known has greatly increased since his day; the European hardly at all.

Further, these tropical species are often very local and are most probably of very recent origin. 'In tropical South America', says Bates, 'a numerous series of gaily-coloured butterflies and moths, of very different families, which occur in abundance in almost every locality, are found all to change their hues and markings together, as if by the touch of an enchanter's wand, at every few hundred yards.' Further, he remarked that 'so close is the accord of some half-dozen species [of widely different genera] in each change, that I have seen them in large collections classed and named respectively as one species' (1879).

There are groups of animals among which almost every member is either imitatively or protectively formed and coloured. Such are the Membracidae, a large family of tropical bug-like creatures containing about forty genera and several hundred species. The bizarre shapes of these insects have attracted observers for over a hundred years. In some Membracidae the adult mimics one species and the larva another. A curious feature of the family is that the mimicking is not usually caused by any modifica-

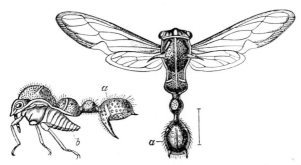

FIG. 138. A Central American Membracid *Heteronotus trinodosus*. Looked at from above the animal looks like an ant. What is apparently the abdomen of the ant, *a*, can be seen from the side to be a part of the huge overgrowth from the thorax. The insect is concealed below this and its true abdomen is seen at *b*.

tion of the bodily shape, but by an outgrowth which has no apparent function save that of imitation (Fig. 138).

Most cases of true mimicry have been adduced among insects. A few are known among serpents and birds. In modern times a degree of scepticism as to the interpretation of mimicry has crept in. Thus Bates reports a case of mimicry of a bird by an insect!

Mimicry in plants is of special interest. There are cases —of which the English bee-orchis is an example—in which a plant seems to imitate an insect. These cases have been a complete mystery in the past, but of late a little light has been shed upon them. In some cases, at least, it has been shown that the insect imitated is the female. The male is deceived and visits the flower as a female of his own species.

It must be remembered that the insects are a group undergoing rapid modification. There are many reasons to believe this, notably the vast number, variety, and frequency of mutations, and the variety and complexity of social systems among insects. It may thus well be that some plants that mimic insects are mimicking extinct species. The 'bee-orchis' in its present state certainly derives no benefit from its mimicry, for it is always self-fertilized.

Despite the enthusiasm with which Darwinian naturalists seized on the phenomena of mimicry, difficulties arise in explaining it on Darwinian lines.

Simple protective coloration can certainly often be regarded as a pure product of the force of Natural Selection. The same cause may be invoked in such a group as the Membracidae, which exhibit as a group a tendency to extensive overgrowth of a part of the body. This overgrowth must take some form. It is an advantage for it to be imitative, and the form of the overgrowth involves the insect in no important physiological change (Fig. 138).

But in the case of an insect the bodily form of which imitates one of a widely different group, e.g. the fly that imitates a bee or ant, or the moth that imitates a hornet, considerable anatomical modifications are involved. No advantage can be claimed for the early stages of such mimicry, and the explanation remains one of the unsolved problems of biology. In general it is true to say that the phenomena of mimicry are little susceptible of explanation by known laws. The early literature of the subject is peculiarly naïve and unscientific.

§ 8. *Parasitism, Saprophytism, Symbiosis*

In the early days of evolutionary theory and in the course of the subsequent development of morphological detail, great interest was raised in certain exceptional modes of living. Apart from the more evident balance of life between plants and animals, stand those organisms that prey on their victims without destroying them, or destroy them but slowly. These are parasites.

The term *parasite* is derived from a Greek word meaning primarily 'mess-mate', and conveyed originally no idea of reproach. It came to imply living at another's expense, and finally assumed the modern sense. Parasitism in its biological usage signifies that an organism depends upon another organism for its existence. Yet all, or almost all, living things are to some extent dependent on other living things. It is thus very difficult to say where parasitism begins, though there is no doubt of the mode of life of the more specialized of parasites.

The dependence characteristic of the parasitic state necessarily demands the service of living hosts. There is a very large group of so-called *saprophytes* (Greek 'putrid growers'). These are organisms that live on organic matter, parasites, as it were, on the dead. In practice the term *parasite* has become confined to organisms which depend for their existence upon physical association with other living organisms or hosts, the relationship being detrimental to the hosts. The work of Darwin undoubtedly focused attention on the subject.

On the zoological side the most prominent early workers in this field were Rudolph Leuckart (1823–98) and Carl Theodore von Siebold (1804–85). Von Siebold was pursuing the subject of parasitism before Darwin, and even more intensively after Darwin's publication *Cirripedia* (pp. 519–22). Leuckart and von Siebold showed that almost every group in the animal kingdom presents parasitic forms. Two of these have become of particular practical importance, the flat-worms and the protozoa.

Many flat worms (*Trematoda* and *Cestoda*) and round-worms (*Nematoda*) are parasitic in man and other animals, and produce disease. They exhibit extreme forms of parasitic degeneration and go through a complicated life-history, often in two or more hosts of widely different species and of different modes of life. Similarly, the Protozoa have parasitic forms, which give rise to a variety of diseases of which Malaria, Amoebic Dysentery, and Syphilis are the best known and commonest. In response to the practical demands made by these conditions, there have arisen in the twentieth century the specialized sciences of 'Helminthology' (Greek *helmins*, 'intestinal worm') and 'Protozoology.' Of the latter science, a basic work is that of the French Army Surgeon, Alphonse Laveran (1845–1922), who described the protozoon parasite of malarial fever in 1881. Patrick Manson (1841–1922) set a lasting example of effective and varied research in this and other tropical fields under conditions of utmost difficulty. Highly important discoveries affecting human health were those of the protozoon *Trypanosoma brucei* causing sleeping sickness, made in 1903 by David Bruce (1855–1931) and D. N. Nabarro (1874–1958) and of the protozoon *Spirochaeta pallida* causing syphilis made in 1905 by Fritz Schaudinn (1871–1906).

In the vegetable kingdom parasitism is as widespread as among animals. Even flowering plants exhibit every degree of parasitism. These vary from the occasional support demanded by weak-stemmed climbers such as woodbine, and the habit of some orchids of growing on the bark or crutch of a tree (*epiphytism*), through the incomplete parasitism of the mistletoe, with green leaves but roots embedded in its host (Figs. 139–40), to the dodder Orobanche, devoid of green colouring matter and dependent on its host for its carbohydrates as well as for its other essential food substances.

It is a well-established biological law that unused organs

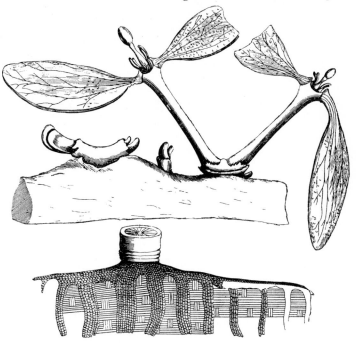

DRAWINGS OF MISTLETOE BY MALPIGHI (1679)

Fig. 139. A plant and two seedlings growing on a branch of an apple-tree.
Fig. 140. Magnified section through an apple branch, showing the way that its tissues are penetrated by the roots of the mistletoe.

degenerate and vanish. The parasite and the saprophyte, once having reached the host or the medium in which their life is to be matured, have no need for organs of sense, of movement or even, in many cases, of digestion. These disappear.

That this disappearance is not related to parasitism as

H.B.—12*

such has now become fairly evident. Deterioration or simplification of the structure of organisms, whether of plants or animals, may take place purely as the result of the character of the medium in which they live and without any degree of parasitism, of symbiosis or of fixation. Thus the cave-living Axolotl, which is a relative of the newt, remains throughout life a blind, unpigmented tadpole. Again, the duckweeds, which dwell entirely in water, have become so extremely simplified that they are not distinguishable as monocotyledons. *Lemna*, for instance, has attained world wide distribution, but has lost all distinctness of shoot and flower, is devoid of most of the structures of a higher plant. Duckweeds very rarely produce flowers, which are always of the simplest type (Fig. 141).

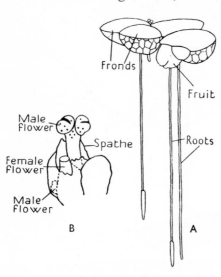

Fig. 141. *A,* Entire plant duckweed *Lemna gibba. B,* Inflorescence of duckweed *Spirodela polyrhiza.* Both from Arber after Hegelmaier (1868–71).

Among the most significant of all plants for the general balance of life is the lowly and perhaps degenerate group of organisms known as *bacteria*.

The study of the lower fungi and especially of the bacteria was initiated as a separate science by Ferdinand Cohn (1828–98) of Breslau, a pupil of Johannes Müller. Cohn described a large number of species of fungi parasitic on plants. Their differentiation and life-history has since become a matter of great economic importance in connexion with agriculture. Probably no department of biology is now more productive of literature than is plant pathology which Cohn initiated.

But not only in disease has the study of fungi and their plant hosts acquired a new importance. During the last two decades of the nineteenth century it became evident that the formation of mould or humus was itself largely the work of fungi. Schleiden had demonstrated that the roots of certain plants are always infected with fungi (1842), and A. B. Frank (1839–1900) showed further (1885) that in some higher plants germination is impossible without the aid of fungus companions. The relationship is of great biological significance. To the fungi which are thus necessary for germination Frank gave the name *Mycorrhiza*.

Experimental evidence has suggested that infection by bacteria of the alimentary canal of new-born higher animals may be necessary to life just as *Mycorrhiza* is essential to seedling plants (p. 384). The matter, however, needs reinvestigation in the light of modern knowledge of vitamins.

The association of two types of living things thus mutually beneficial to each has been termed *symbiosis* (Greek = 'living together'). Such partnerships are of quite common occurrence throughout the animal and vegetable kingdom. There are even vegetable forms, such as lichens, in which, as was first shown by the German botanist H. A. de Bary (1831–88), neither partner can exist without the other (1860). There are animals that have in their tissues numbers of green unicellular plants which thrive on their waste products, while the animal itself feeds upon the compounds elaborated by the plants within it. Such are certain flat-worms which are symbiotic with algea (*Convoluta Rosoffensis*, 1891). There are many other associations of animals and plants of less intimate character. The study of the wider problems of symbiosis comes within the department of Ecology (pp. 283–6).

A particularly interesting aspect of Ecology is that which looks upon the entities that we call organisms as symbiotic colonies of different types of being. Thus, for example, the wood-eating termites or white ants cannot digest the cellulose of the fibres that they consume. This is done for them by certain flagellate protozoa which live in their intestines. The termite consumes the products of activity of these protozoa. Thus the creature called a white ant is really a colony of diverse living entities. Similarly man depends on organisms for certain of the processes which go on in his alimentary canal. The creature that we call a 'man' is a colony of which innumerable bacteria are no less necessary members—though more replaceable—than the sapient being that writes about them. This ecological theme opens out great vistas of development in which organisms can be considered in fundamental relation to their living environment.

The subject of evolution may be pursued into many such paths as these. But we must remember that while evolution illuminates them, they do not illumine evolution. The general position, after forty years of research, may be thus summed up. Evolution, at the end of the nineteenth century, was universally accepted as a general description of the history of organic forms. So far Darwin had completely conquered. He had also conquered on Natural Selection in the sense that, on a mathematical basis, very small advantages, if inherited, must yield species differentiation. The weak point was in the word *if*. But most evolutionary thought has been deflected in the twentieth century to the nuclear expression of Mendelian principles which are answering the *if* (chaps. ix, xiv, xv).

PART III. EMERGENCE OF MAIN THEMES OF MODERN BIOLOGY

IX

CELL AND ORGANISM

§ 1. *Emergence of Cell Doctrine*

THE Latin word *cella* means 'a small room'. It is still so used in such phrases as a 'hermit's cell' or a 'prisoner's cell'. The Latins applied the same word to the cells of a honeycomb. Robert Hooke in his *Micrographia* (1665) likened the structure of cork to honeycomb, describing it as composed of *cellulae*. He was examining the thickened walls of dead cells. Moreover, he perceived that the surface of living plants, when magnified, appears as broken up into similar little divisions. Here he was looking at the walls of living cells, though he failed to recognize their nature (Figs. 142–3).

A little later, Malpighi, examining sections of plants, found them made up of little bodies closely applied together, each surrounded by a definite wall (Figs. 77–9). Such a body he named *utriculus* (diminutive of *uterus*). Leeuwenhoek, too, frequently figured cells in both plant and animal tissues. He formed no clear conception of their real nature.

Grew observed similar structures. He spoke of the structures as *cells* or *bladders*. He noticed that in the younger parts the cells were juicy, had thin walls, and were closely applied. Such parts he called *parenchymatous* and rightly regarded as most actively growing (Figs. 83–4). Grew's word *parenchyma* (Greek 'poured in beside') is an ancient term used by the Alexandrian anatomist, Erasistratus, and still current.

At first the parenchyma seemed crystalline to Grew.

He soon realized that it is more complex. 'Next to the Cuticle (of the bean),' he says, 'we come to the *Parenchyma*. I call it *parenchyma* not that we are so meanly to conceive of it as if it were a meer concreted Juyce. For it is a body very curiously organiz'd' (*Anatomy of Plants*, 1682).

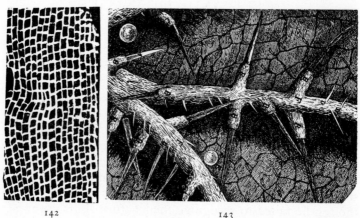

142   143

FIG. 142. Magnified outlines of cell-walls on the cut surface of cork, from Hooke (1665).

FIG. 143. Lower surface of leaf of stinging-nettle, from Hooke (1665). The outlines of the cells are seen. Here and there along the veins *FF* are pointed spikes, each of which is set on a flexible base *BB*. These are the stinging elements. There are also other hair-like spikes *DD* which have not this peculiar structure.

He went farther and reached a correct view of the origin of the vessels. Thus: 'One single *Row* or *File* of *Bladders* (i.e. cells) evenly and perpendicularly piled may sometimes . . . all regularly break one into another and so make one *continued Cavity*.'

After Leeuwenhoek's death (1723) little further progress was made until the nineteenth century. Several observers, however, concluded that fat consists essentially of masses of 'cells', each enormously distended by a drop of oil. Now fat had been regarded as 'non-parenchymatous'. This view, therefore, extended the vague cell idea beyond the parenchymatous structures.

Just at the dawn of the nineteenth century the French investigator, M. F. X. Bichat (1771–1802), perceived that the different bodies and parts may be analysed into certain elements of specific appearance and texture. Of these he distinguished twenty-one. He likened the structure of the body to a woven fabric. His word was *tissu*, an old term for a particular kind of rich cloth.

Bichat, who was a hasty worker and writer, did not use the microscope. Early death prevented him from further developing the knowledge of the tissues. His work was, however, continued by others, and the special study of the minute structure of the tissues came to be called *histology*. This term was introduced by Owen (1844). It is really derived from *tissue*, for 'histology' is formed from a Greek word which means 'something woven'.

During the seventeenth, eighteenth, and early nineteenth centuries knowledge was accumulating of those beings whose bodies, as we now know, consist of but one cell. *Vorticella* had been described in 1677, *Paramecium* in 1702, *Amoeba* in 1755. Several works dealing with the organisms of infusions and with other microscopic plants and animals had appeared. Such organisms were called *Infusoria*.

These 'Infusoria' did not, however, correspond with the group of unicellular creatures to which biologists came later to attach that name. Included in the older 'Infusoria' were bacteria and algae as well as small worms, rotifers, and many other multicullelar forms. This was the case, for instance, in the posthumous *Animalcula infusoria fluviatilia et marina* (Copenhagen, 1786) of Otto Frederik Müller (pp. 260–3). Moreover, so little was yet realized of the true nature of the Foraminifera that *Globigerina*, so vastly abundant in northern seas (Fig. 123), was described in 1826 and later as having affinities with the nautilus! Contemporary accounts ascribed a variety of imaginary organs to the *Infusoria*.

No proper appreciation of the nature of all these forms could be made until the general recognition of the higher animals as cell aggregates. That view had, in fact, been set forth by a naturalist to whose genius biology owes more than is sometimes allowed. Oken (pp. 220–23) in *Die Zeugung* ('= Generation', 1805) enunciated the doctrine with considerable clarity:

'All organic beings', he wrote, 'originate from and consist of vesicles or cells. These, when detached and regarded in their original process of production, are the *infusorial mass or Urschleim* whence all larger organisms fashion themselves or are evolved. Their production is therefore nothing else than a regular agglomeration of *Infusoria*—not, of course, of species already elaborated or perfect, but of *mucous vesicles* which by their union or combination, first form themselves into particular species.'

Three words in this passage need some commentary. *Infusoria*, in the terminology of the day, included a great many organisms that were not unicellular. Oken, however, is clearly using the word as nearly equivalent to independent organisms consisting of but one cell. By *Urschleim* he means something akin to what we mean by 'protoplasm', a word not yet invented. By *mucous vesicles* he means living cells. Oken, in fact, was reaching out to a conception both of protoplasm and of cell.[1]

The passage is an illustration of an interesting feature in the history of science. We often encounter, in works of

---

[1] The recognition of the Infusoria as a special unicellular group was made by Dujardin in 1841. Siebold in 1845 included in his *Infusoria* both *Ciliata* and *Flagellata*. The limits of the group were indicated in 1880 by W. Saville Kent in the title of his *Manual of the Infusoria including a description of all known Flagellate, Ciliate and Tentaculiferous Protozoa* (1880). The first and last class were removed by Butschli in 1887, and the *Infusoria* thus became equivalent to the modern *Ciliata*.

exceptional ability, the enunciation of truths that are beyond the comprehension of their author. The presentation of premature views is naturally liable to be confused. The vocabulary of an age reflects the interests and beliefs of that age. The character of the scientific ideas is reflected in the character of contemporary terminology. With an inevitably changing vocabulary, we must beware of reading into words used in older works meanings that are too modern for such connotation. Yet there are writers with whom such caution may at times be misplaced. Oken can and does use the word *cell* in the modern sense. His works were widely read by his contemporaries, and it is more than probable that he inseminated the minds of the recognized founders of the cell doctrine.

A somewhat similar course can be traced in the knowledge of certain parts of the cell, and notably of the nucleus. The nuclei in the blood-corpuscles of fish had been depicted by Leeuwenhoek and by several observers in the eighteenth century. In 1823, in a drawing illustrating Hunter's specimens, the artist, Franz Bauer (1758–1840), produced figures of these corpuscles. The descriptive legend tells that they 'show the *nucleus*'. This is, perhaps, the first use of the word nucleus in such a connexion.

In 1831, as a result of his Australian experiences, Robert Brown (pp. 242–5) undertook a special study of the Asclepiadaceae. These dicotyledonous plants, many of Australian origin, are allied to our periwinkles. They have, however, a mode of fertilization recalling that of the monocotyledonous Orchids, a group in which Brown was taking special interest. Brown discerned and figured nuclei in the surface layers of cells both in Asclepiadaceae and in Orchids. He realized that the nucleus was a regular feature of plant-cells, and his use of the word normalized it into biological nomenclature.

Between 1833 and 1838 several investigators described the cells of various vegetable and animal tissues. Notably in 1835 Purkinje (pp. 340–1) drew attention to the microscopic structure of the skin of animals, especially in the embryonic state. He pointed out that it was composed of masses of 'cellulae' which he compared to the parenchymatous tissue of plants known since Hooke (p. 330).

At this time there was working in Berlin C. G. Ehrenberg (1795–1876), an old travelling companion of Humboldt (p. 274). Ehrenberg was interested in minute life and had the best microscopes at his disposal. He investigated many microscopic forms and watched organisms, which we now know to be unicellular, digesting food and discharging undigested residue. He was wrong in his interpretation of the nature of these organisms, to which he ascribed many organs which they do not possess, but his finely illustrated writings drew much attention to them. His *Die Infusionsthierchen als vollkommene Organismen* ('Infusorial animalcules as complete organisms,' 1838) was contemporary with the more widely read papers of Schleiden and Schwann.

§ 2. *Schleiden* (1804–81) *and Plant Cells* (1838)

M. J. Schleiden, who began life as a lawyer, was for many years professor of botany at Jena, where Oken's most active years had been passed. Despite his great ability and originality, an impetuous temper and an arrogant character led Schleiden into many errors. He had a bias against the dry systematization into which botany had fallen—the mere naming, describing, and arranging of plants, and the search for new species. To the microscopical analysis of structure and growth, on the other hand, he eagerly applied himself.

Schleiden raised this matter in 1838 in a well-known paper in Müller's *Archiv fur Anatomie und Physiologie* with the title 'On phytogenesis' (Greek, 'plant origin'). Schleiden seized on the idea of the cell as the essential unit of the living organism.

'There have been many endeavours', he wrote, 'to establish analogies between animals and plants. All have failed since the idea of individual, as used for animals, is inapplicable to plants. Only in the lowest plants, in some *Algae* and *Fungi*, for instance, which consist of but a single cell, have we an individual in the animal sense. Plants, developed in any higher degree, are aggregates of fully individualized, independent, separate beings, namely the cells themselves. 'Each cell leads a double life; one independent, pertaining to its own development alone; the other incidental, as an integral part of a plant. The vital process of the individual cells, however, form the first indispensable and fundamental basis for both vegetable physiology and comparative physiology in general. Thus the primary question is, *what is the origin of this peculiar little organism, the cell?* [*Abridged.*]

Schleiden then develops Brown's views on the nucleus.

'Robert Brown', he says, 'with his comprehensive native genius, first realized the importance of a phenomenon which, though previously observed, had remained neglected. In many cells in the outer layers of Orchids he found an opaque spot, named by him the *nucleus of the cell.* He traced this phenomenon in the earlier stages of the pollen-cells, in the young ovulum, and in the tissues of the stigma. The constant presence of this nucleus in the cells of very young embryos struck me also. Consideration of the various modes of its occurrence led to the thought that it must hold some close relation to the development of the cell itself.'

From the cell—with its essential element the nucleus— Schleiden turned to a consideration of its origin. Here he blundered, thinking that cells arise by budding from the surface of the nucleus. Certain other activities of the cell, however, Schleiden observed with considerable accuracy. Notably he described the active movement of the cell substance, known now as 'protoplasmic streaming', in the tissues of plants (Fig. 147). This also had been observed by Brown (p. 244) and shown to Darwin.

§ 3. *Schwann* (1810–82) *and Animal Cells* (1839)

Theodor Schwann was a pupil of Johannes Müller. He was a simple-minded, pious man who showed a genius for research in several departments (pp. 393–5). His classic, *Mikroskopische Untersuchungen über die Ueberein-stimmung in der Struktur und dem Wachstum der Thiere und Pflanzen* ('Microscopical researches on the similarity in structure and growth of animals and plants,' 1839) is more searching than the work of Schleiden. It is mainly devoted to the investigation of the elementary structure of animal tissues which are more difficult of observation than those of plants.

Schwann opens with discussion of cartilage, in which cell outlines are observed more easily than in most tissues. Of cartilages he says that—

'the most important phenomena of their structure and development accord with corresponding processes in plants. These tissues originate from cells, which correspond in every respect to those of vegetables. During development, also, these cells manifest pheno-mena analogous to those of plants. The great barrier between the animal and vegetable kingdoms, viz. diversity of ultimate structure, thus vanishes. Cells, cell-membrane, cell-contents, nuclei, in the former are analogous to the parts with similar names in plants. Fundamentally, the form of the cell is that of a round vesicle, but flattening of cells against one another, the presence of inter-cellular substance between them in greater or less quantity, and lastly, thickening of cell-walls, have all been observed.' [*Abridged.*]

Schwann extends the discussion to the ovum or egg which is the beginning of the animal body. In some animals, as the hen, the egg is very large, being distended with food-substance—the yolk—and surrounded by a layer of protective albumen. In other eggs, as of the frog, the amount of yolk and albumen is much less. In yet others, yolk and albumen are reduced to a minimum, as in the microscopic egg of mammals, then recently discovered by von Baer (1828, pp. 471–2). Yet, as Schwann

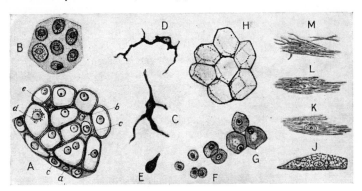

FIG. 144. Drawings by Theodor Schwann to illustrate the nature and origin of animal cells. All are highly magnified.

A, The first step in the origin of cartilage from cellular tissues. At the lower part the young cells are without cell-walls. In the upper part they have formed walls and are beginning to secrete cartilaginous substance. Nuclei and nucleoli are clearly visible. *B* shows a piece of maturer cartilage, in which the cells are embedded in a mass of cartilaginous material. *C*, *D*, and *E*, Pigment cells, such as are characteristic of the frog. In the lowest cell, which is contracted, the nucleus is concealed by the pigment. The upper two are more expanded and in them nuclei can be seen. *F*, *G*, Young cells from a developing feather. These cells may enlarge, secrete hard walls, and form the fine spongy tissue of the inner part of the shaft of the feather (*H*), or the cells may elongate, the protoplasm become granular, and finally break up into fibres (*J–M*). These form the tough fibrous matter of the outer part of the shaft of the feather. In either case the nucleus disappears and the cell dies. *F–M* show how structures of very diverse form can be differentiated from cells of the same type.

suggests, all these different types of egg are essentially cells. The egg may be enormously distended, but nevertheless the essential cell-elements of nucleus, cell-membrane, &c., are clearly traceable in it.

Further, the development of the egg into the young animal proceeds by division of the egg cell. This phenomenon is particularly evident in the earliest stages where the act of cell division is now usually referred to as 'segmentation'. This had been observed in invertebrates by Schwann's contemporaries, Siebold and Sars, in 1837. In 1838 segmentation was described in the mammalian ovum. Schwann himself saw the process in the hen's egg (1839) and treated it as a normal part of embryonic development. Soon after (1841) it was seen in the frog's egg, which has since become one of the classical objects for its investigation.[1]

From the egg, which is thus essentially a 'germ-cell', Schwann proceeds to investigate the adult tissues. His treatment of these passes at once far beyond that of Bichat. Of the tissues he distinguishes five classes on a cellular basis:

(a) Tissues in which the cells are independent, isolated and separate. Such is the blood.

(b) Tissues in which the cells are independent but pressed together. Such is the skin.

(c) Tissues in which the cells have well-developed walls that have coalesced to a greater or less degree. Such are cartilage, teeth, and bones.

(d) Tissues in which the cells are elongated into fibres. Such are tendons, ligaments, and fibrous tissue.

(e) Tissues which Schwann regarded as 'generated by the coalescence of the walls and cavities of cells'. Here he included muscles and nerves.

Schwann now passed to a general statement of his belief as to the cellular origin and structure of animals and plants. His conclusion may be expressed thus:

(a) The entire animal or plant is composed either of cells or of substances thrown off by cells.

(b) The cells have a life that is to some extent their own.

---

[1] Segmentation had been observed in the frog's egg by several early workers Swammerdam among them.

(c)   This individual life of all the cells is subject to that of the organism as a whole.

This general attitude is still valid.

'The question as to the fundamental power of organized bodies', wrote Schwann, 'resolves itself into that of the individual cells. We must consider the general phenomena to discover what powers exist in the cells to explain them. These phenomena may be arranged in two natural groups. First, those which relate to the combination of the molecules to form a cell. These may be called *plastic* phenomena. Second, those which result from chemical changes either in the component particles of the cell itself, or in the surrounding *cytoblastema*. These may be called *metabolic* phenomena.'

Two words here which Schwann himself invented require explanation.

Cytoblastema means the substance from which the 'cytoblast' or nucleus was supposed to be derived, that is to say the *protoplasm*. The word *protoplasm* (pp. 340–1) has now replaced it. The earlier term was, however, very useful in clarifying biological ideas and has had an interesting history.

Metabolic is an adjective which has a noun *metabolism*. It means, etymologically, 'liable to change'. The changes which it indicates are chemical and are those specially associated with life.

The changes which we call metabolic are ceaseless in living things. The processes of life involve the constant breaking down of complex substances—*katabolism*. The continuance of life involves their constant upbuilding—*anabolism*. These two events, which are related in the most intimate fashion, we still describe as *metabolism*. The idea which it represents is of very great biological importance. (The words *anabolism* and *katabolism* were not used by Schwann.)

§ 4. *Extension of the Cell Doctrine in the Plant Kingdom*

Schleiden fastened on the cell theory the false idea that new cells were derived by budding from the nucleus, which he therefore called the *cytoblast* (i.e. cell-bud). The error was left uncorrected by Schwann. Difficulties soon arose.

Karl Nägeli (1817–91), a versatile botanist who made contributions to several departments of biological thought, set himself to throw light on the origin of cells. He made microscopic examinations of cell-formation during the reproductive process and also at the growing points of plants. His researches were pursued upon a great variety of forms. The phenomena of cell-division, he found, were most easily seen in the lower Algae. In them the movement and behaviour of the living and transparent cell-substance could be watched. He soon saw reason to abandon the view that the nucleus buds off new cells (1842–6).

Further investigations on the subject were carried on by Hugo von Mohl (1805–72), professor of botany at Tübingen. He pointed out that the part of the vegetable cell just within the cell-membrane must be distinguished from the watery sap that occupies the centre. To this outer granular mucilaginous material he gave the name *protoplasm* (1846). The word has entered the general biological vocabulary. It indicates that part of the cell-substance which can alone, in proper sense, be said to be alive. In many cells the protoplasm is in constant movement.

Protoplasm means etymologically 'first formed'. Ancient theological writers designated Adam, the first created man, as 'protoplast.' The versatile Bohemian naturalist, Johannes Evangelista Purkinje (1787–1869), who had received a theological training, used the word in a communication of 1839 *On the analogies in the structural elements of animals and plants*. Felix Dujardin (1801–62) of Toulouse, a most acute and penetrating observer, entered in 1835 upon a critical examination of the relationship of certain protozoa. He used the terms 'sarcode' for protoplasm, and described many of its properties. It was von Mohl, however, who gave currency both to the word *protoplasm* and to the idea which it connotes. A series of very important and influential articles by him on the vegetable cell brought order into current views on the subject.

In the years that immediately followed, further important advances in the knowledge of the cell were made by Nägeli. By means of chemical examination he proved that protoplasm contains nitrogenous matter. He showed that in this it differs from certain other constituents of the cell, notably the cell-wall and the stored starch. He thus added a most significant factor to the conception of protoplasm.

§ *5. The Protozoa in Relation to the Cell Doctrine*

While Schleiden and Schwann, Nägeli and Mohl were elaborating the conception of the cell, Dujardin was investigating the old group 'Infusoria' (1835–41). He realized that it contained many unrelated forms, rotifers, worms, various algae, &c. The entire body substance of the remaining 'Infusoria' was, he noted, capable of contraction, movement, digestion, and other vital processes. The surfaces of the bodies of these Infusoria were, he observed, more or less clothed with minute movable hair-like processes or 'cilia'. These, by their movement, gave the organisms their power of progression. Dujardin saw, too, that the relatively 'structureless' character of the bodies of these creatures was in contrast to those of higher animals.

Several investigators were now reaching out to the conception of the Protozoa as organisms consisting each of but a single cell. It was Carl Theodor von Siebold (1804–84) who gave formal expression to this view in his text-book of comparative anatomy of 1845. In doing so he stressed cilia as the instruments by which many

Protozoa are able to move. He drew attention also to the presence and action of cilia in the organs of higher animals. In these, however, the ciliated cells being fixed, the movements of the cilia, instead of moving the cells, have a current-producing action. Twenty years previously Purkinje had seen cilia in vertebrates. At that date he was unable to express their movement as a form of cellular activity. This was now possible to von Siebold (1861). The cellular theme was rounded off in the same year by the comparative anatomist, Gegenbaur (p. 480), who showed, in supplement to Schwann (p. 343), that the eggs of vertebrates are to be regarded as cells.

The synthesis in the ideas of protoplasm, protozoa, and egg-cell was made by Max Schultze (1825–74) who succeeded Helmholtz as professor of anatomy at Bonn. He devoted himself to histology, which he studied in a wide range of animals. In 1861 he defined the cell as 'a lump of nucleated protoplasm'. His investigations extend to plants and to protozoa. In 1863 he introduced the conception of protoplasm as 'the physical basis of life'. He showed that it presents essential physiological and structural similarities in plants and animals, in lower and in higher forms, in all tissues wherever encountered.

A very important series of systematic advances, involving the conception of the cell, were made by Haeckel (pp. 487–9). Perceiving the difficulty of separating the animal from the plant kingdoms he introduced a third and coequeal group, the *Protista* (1866). The division has not proved to be tenable (p. 201), but the conception was useful in drawing attention to the great similarity of lowly animals and plants. Later Haeckel separated the sponges from the protozoa (1869). Extremely important was his clear expression of the great generalization that the animal kingdom is to be divided into unicellular and multicellular organisms, 'protozoa' and 'metazoa'. This formed an integral part of Haeckel's *Studien zur Gastraea Theorie* (1873–84, Fig. 168).

§ 6. *Extension of the Cell Doctrine in the Animal Kingdom*

The general interpretation of the tissues on a cellular basis in a large number of animal forms was especially the task of the Swiss, Albrecht Kölliker (1817–1905). He was a pupil of Oken, Johannes Müller, and Henle, and for many years a professor at Würzburg.

Kölliker happily initiated his work by applying Schwann's theory to the embryonic development of animals (see below). He treated the ovum as a single cell and the process of development as the result of cell-division (1844). The theme was further developed by Siebold (1849) and Remak (1852). Kölliker did an immense amount of good microscopical work and established histology as a separate discipline. He wrote the first text-book on the subject (1852). Nine years later he produced a model text-book of embryology.

The discoveries made by Kölliker in his chosen department are innumerable and his work has seldom needed correction. We note his account of the essential cellular nature of the non-voluntary muscles and his demonstration that nerve-fibres are no more than elongated processes of cells, the bodies of which are either in the central nervous system or in ganglia. The division of the nuclei during the process of segmentation of the ovum, on which Kölliker laid much stress, was first adequately observed by his colleagues, Leydig (1848) and Remak (1852). Kölliker regarded the nucleus as of especial importance as the transmitter of hereditary characters. His views on this matter, though seldom quoted, have been followed by almost all later workers on the subject. Kölliker took much interest in the problems of heredity and variation. He had not heard of Mendel's work, but he held the view that alteration in the characters of races takes place not gradually but by means of sudden and spontaneous changes, thus preceding De Vries.

Even more influential than Kölliker was Rudolf Virchow (1821–1902). He was for many years a professor at Berlin, and a prominent liberal alike in academic and in political matters. Virchow's liberal views found vent in the *Archiv für Pathologie* which he edited for fifty-five years. It was nominally devoted to the study of disease. But the test he applied was not so much whether an article referred to pathology as whether it was an important contribution to knowledge. Thus it contains a great variety of learning. Articles on oriental languages, on comparative anatomy and physiology, on anthropology, on medieval translations from Greek and Arabic, are interspersed between accounts of the histology of tumours, and on the infectious character of fever.

Virchow's main additions to biological thought are in his great *Cellular Pathologie* (1858). He analysed disease and diseased tissues from the point of view of cell-formation and cell-structure, much as Kölliker had analysed normal tissues. There are departments of pathology that Virchow explored so well that they have hardly been extended since his day. He set in motion the now familiar idea that the body may be regarded 'as a state in which every cell is a citizen'. Disease is a civil war, 'a conflict of citizens brought about by the action of external forces'.

In the *Cellular Pathologie* Virchow says:

'Where a cell arises, there a cell must have been before, even as an animal can come from nothing but an animal, a plant from nothing but a plant. Thus in the whole series of living things there rules an eternal law of continuous development. There is no discontinuity nor can any developed tissue be traced back to anything but a cell.'

Virchow crystallized the matter in his famous aphorism, *Omnis cellula e cellula* ('Every cell from a cell'), to be placed beside *Omne vivum ex ovo* ('Every living thing from an egg', Siebold's reading of Harvey, see p. 459), and *Omne vivum e vivo* ('Every living thing from a living thing', Pasteur, pp. 446–9). These are three of the widest generalizations to which biology has yet attained. They were all reached within the ten years around the middle of the nineteenth century.

Since Virchow and Kölliker, the study of the intimate structure of cells as distinguished from the tissues has become a separate and independent science, *Cytology*. It may be said to emerge into independence with the appearance of O. Hertwig's *Zelle und Gewebe* ('Cell and Tissue') in 1893. Cytology must be distinguished from Histology or the study of association of cells in tissues. The study of cell-division has become peculiarly important in relation to problems of heredity (chap. xv).

§ 7. *Nuclear Phenomena of Cell Division*

After the abandonment of Schleiden's theory that new cells were budded off by the nucleus, it became generally held that the nucleus disappears during cell-division to reappear in the daughter cells. This was at first the view of Kölliker. In the seventies new methods of staining microscopic preparations were introduced. These soon revealed that the seeming disappearance of the nucleus during division was illusory. It was due to certain profound transformations of the nuclear substance.

The elucidation of these changes has been the work of a great number of observers. Eduard Strasburger (1844–1912), professor of botany at Bonn, a pupil of Meckel, drew attention to the subject by his *Zellbildung und Zelltheilung* ('Cell-formation and Cell-division', editions from 1875 onwards). He described with great clearness the complex and important processes involved in division of plant cells.

The pioneer work for animals was performed by Walther Flemming (1843–1915) of Prague and Kiel. He, more than any other, initiated cytology as a special science. Much of his best work appeared in his *Zellsubstanz, Kern, und Zelltheilung* ('Cell-substance, nucleus, and cell-division', 1882).

Cell-division, as worked out by them, may be reduced to a schematic form. The division of a cell is led and controlled by the nucleus, and is essentially the same in plants and animals (Fig. 145).

In an ordinary resting nucleus there can be distinguished a fine network of material that stains deeply with certain dyes. As division approaches, this becomes arranged into a long thread that is more or less regular, continuous and spiral. The thread breaks up into a series of separate filaments, the *chromosomes*.

By the time that the chromosomes are distinguishable as separate bodies, other changes have occurred. The delicate membrane which normally surrounds the nucleus has disappeared. On either side of the nucleus a minute body or *centrosome* has become apparent,[1] while from each centrosome there radiate a series of lines forming a star-like *aster*.

The two asters extend until they meet around the equator of the nucleus. There the chromosomes collect. They then split along their length, and the twin halves pass, each in an opposite direction, along the lines of the asters toward opposite poles. Thus the split chromosomes become clustered in the neighbourhood of the two poles. Around each cluster a fine membrane forms. The daughter nuclei are now complete. In the protoplasm between the two daughter nuclei a partition appears.

---

[1] Most plants present no centrosomes.

Thus the cell is divided into two, each with its own nucleus.

It has been suggested that this complex process is needed to effect an equal partition of substance of mother

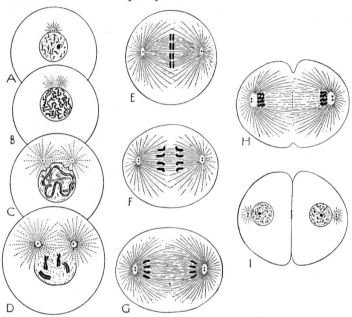

FIG. 145. Diagrams of process of mitosis in animals. *A–D*, Prophase. *E*, Metaphase. *F, G*, Anaphase. *H, I*, Telophase. Modified from E. B. Wilson.

nucleus between daughter nuclei. Its further 'meaning' has become more evident as the chromosome has been studied further in more modern times (chap. xv).

The nomenclature of mitosis tells much of the story of its discovery. The term *mitosis* (Greek 'thread') was introduced by Flemming in 1882. The useful terms *cytoplasm* and *nucleoplasm* to describe the protoplasm of the cell-body and the nucleus respectively are Strasburger's (1882). He also gave us the terms *prophase*

(Greek 'appearance before'), *metaphase* ('appearance next'), and *anaphase* ('appearance further') to describe the three consecutive stages of mitosis (1884). Flemming introduced in 1879 the terms *chromatin* (Greek 'colour') for the part of the nuclear substance that takes up certain stains. This part breaks up into the bodies now called 'chromosomes' (Waldeyer, 1888). Flemming again is responsible for *spireme* ('a skein', 1882); and *aster* ('star', 1892).

During the twentieth century the extent and exactness of our knowledge of cellular structure has increased apace. The centrosome—a word introduced by Theodor Boveri (1862–1915) in 1888—has been found a permanent constituent of the animal cell. It divides at an early stage of mitosis. Around it, there is found in animal cells an array of peculiar bodies rendered evident by their affinity for metallic salts. These attracted the cytologist Camillo Golgi (1884–1926) of Pavia in 1909, and have since been known as 'Golgi bodies'. Many other protoplasmic inclusions have been described. Some of them seem to be permanent, some transitory inhabitants of the cell-body or nucleus. The role of these bodies in the life of the cell is, however, still not clear.

§ 8. *Structure of Protoplasm*

Influenced doubtless by the molecular view of matter, there has always been a school that has sought to analyse the cell into lesser units. The names invented for these hypothetical bodies is legion. Herbert Spencer spoke of 'physiological units', Darwin of 'gemmules', Haeckel of 'plastidules', Weismann of 'biophores', Hertwig of 'idioblasts', De Vries of 'pangens'. In the twentieth century, with the rise of biochemistry, many have been disposed to refer protoplasmic behaviour to the activity of colloid molecules and especially to groups of amino-acids (pp. 389, 464).

All these terms assume a view of life that is untenable. They all assume that living things are such by virtue of being an aggregate of distinct minute bodies each of which is alike and each of which possesses the power of independent growth and division. Those who profess this simple faith draw comfort by throwing back the mystery of life on something that is so small as to be beyond vision. Such a solution of the problem can yield philosophical satisfaction only if certain other conditions be met. First among these is a demonstration that aggregates of non-living origin should exhibit certain properties which have been found only in living things and always in living things. These properties are metabolism, repair of injury, reproduction, heredity, adaptation to environment, and 'memory', that is to say, conduct determined by previous history. In fact, the parts of the cell can be given a meaning only in reference to the cell, just as the cell itself can be given a meaning only in reference to the organism. Kant knew nothing about cells, but he saw the cogency of this attitude long ago (pp. 215–18).

On the other hand, the question as to whether parts of the cell have any *particulate* as distinct from *independent* existence is not a question of theory but of observation. There is evidence that certain parts of the cell, among which are the chromosomes and the centrosomes of both animals and plants, and certain other bodies including those bodies known as 'plastids' (pp. 379–81) of plants, persist, divide, pass from cell to cell, and are derived only from bodies like themselves. The behaviour of some of these bodies, notably the chromosomes, is partly understood. A general discussion of the numerous other particulate bodies can be of little value in the present state of our ignorance.

The main physical fact about protoplasm itself is that it behaves as a liquid. Of this there is evidence of several sorts. The earliest and historically the most interesting is that associated with the name of Robert Brown. The so-called 'Brownian movements' had in fact been observed in the eighteenth century by Needham (pp. 441–2). Brown, however, was the first to describe them exactly in his *Microscopic Observations on the Pollen of Plants* (1828). This work, it may be remembered, was offered to the Rev. Mr. Farebrother by the surgeon Lydgate in the novel *Middlemarch* (begun 1869) by George Eliot (1819–80).

All microscopists are familiar with 'Brownian movement'. If very minute particles are suspended in a liquid, which is examined under high magnification, a constant dancing movement is seen. The particles seem alive. This was Brown's first thought, for he saw the movement first in living substance. Later he found it to be a purely physical phenomenon.

Brownian movement has become of greatly increased interest since the systematic investigation of the 'colloid state'. Its new study was initiated by Thomas Graham (1805–69, pp. 386–8) in 1861. More recently and in the twentieth century, Brownian movements have been shown to be related to molecular movement in liquids.

Brownian movements are visible under suitable conditions in all forms of naked living protoplasm. On the death of protoplasm the Brownian movements cease. This is because the protoplasm then ceases to be liquid.

Other evidence that living protoplasm is a liquid is that it tends to assume a spherical form in a state of shock. This was shown as long ago as 1864 by the German physiologist, Willy Kühne (1837–1900), a pupil of Claude Bernard and Virchow. He described the effect of an electric shock on those streams of protoplasm in plant cells which aroused the wonder of Darwin when demonstrated by Brown (p. 244).

It has also long been known that drops of watery liquid taken up or secreted by protoplasm take a spherical form. This is notably the case in the so-called 'food vacuoles' of protozoa. The spherical form is due to surface tension and is further evidence of the liquid nature of protoplasm.

High magnification reveals minute granules of various sizes in even the clearest protoplasm. The nature and classification of these is still in dispute.

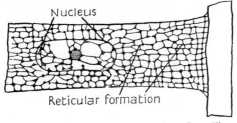

FIG. 146. Epidermal cell of earthworm to show foam-like structure. From Bütschli, 1892.

We may thus say that physically protoplasm is a complex colloidal system which commonly behaves as a viscous liquid and has particles suspended in it. Many investigators have, however, sought to demonstrate some further structure of protoplasm. Of these attempts the most important is the alveolar or foam theory first expounded in 1878 by the Heidelberg professor, Otto Bütschli (1848–1920). He found living protoplasm to have a honeycomb-like appearance. This he regarded as due to a mixture of two liquids of different degrees of viscosity. He tried to imitate this mixture mechanically, and he believed that his mixtures exhibited some of the reactions of protoplasm. In view of the excessively complex conditions that we know to prevail within the living cell, the idea seems very naïve. Bütschli, however, rendered a real service by firmly establishing the fact that protoplasm is a fluid.

There have been innumerable variations on Butschli's theme of the structure of protoplasm. All are open to a double criticism, one practical and the other theoretical.

On the practical side how far do the microscopical appearances represent actual objective conditions or mere optical effects? Or, again, are they artificial results of preliminary treatment? These are points that should be capable of decision.

On the theoretical side we may say that while there is no reason that protoplasmic activities should not be analysed separately, yet the results of such analysis cannot explain the living entity. Thus the fact that emulsions behave in some respects like living protoplasm does not get us nearer to understanding what living protoplasm is. It only helps us to understand how some of its behaviour is determined. If a man assaults another, his action may illustrate the laws of mechanics and may be analysable on the principles of Galilean physics. The analysis, however, will neither explain the feelings of the two nor determine whether or not the injured man will hit back or go to law. Similarly it helps little in understanding the conduct of the cell to be told that its protoplasm behaves according to the physical and chemical laws of an emulsion, for it behaves as a member of a tissue.

In fact, however, the knowledge that has emerged concerning the visible structure of protoplasm is that which has to do rather with particulate bodies, than with their relationship. The alveolar structure passes insensibly into the emulsional form, and both are included in the conception of the 'colloid state'.

§ 9. *Cellular Ageing, New Growths, and Tissue Culture*

The study of certain 'new growths'—of which the best known are cancers—has thrown some light on certain extremely important aspects of cell physiology.

Reliable statistics, now available from an enormous population, show that only a very small proportion of persons under forty develop cancer. Beyond that age the proportion rises as age advances. Cancer, in fact, characterizes the decline of life. This explains why it is so rare in wild creatures, for these normally die soon after passing their sexual prime. An interesting relationship to the cellular activities of the body is involved in this age distribution.

The cells of cancers descend from other cells or tissues of the body. They present an appearance resembling that of their normal ancestors. How, then, do they differ from them? The answer, in so far as any can be given, fits our knowledge of the age distribution of cancer.

Since Virchow (pp. 344–5), cancer-cells have been recognized to differ from normal cells not so much in structure—so far as that is known—but in conduct. They are isolated in the co-operative community of the tissues and organs of the body. The cells of the various tissues exercise an influence upon each other. According to the bodily needs they grow, they develop, they multiply. It is just this mutual balance and restraint of a group of cells that make us call it an 'individual'. But the cells that become a cancer multiply and give rise to tissues which subserve no function, which meet no demand of the cell community, which answer no physiological need, which obey none of the laws laid down by that power of co-ordination—entelechy, or whatsoever we may call it—that informs the body as a whole.

This defiance of the laws of the cellular community is accompanied by a feature which gives us a glimpse of the meaning of the power of co-ordination. The process of ageing, so that at last life ceases, is a function of the body as a whole, not of its tissues. It is a property only gradually imparted to the tissues. Thus embryonic tissue can be propagated for an indefinite time in a suitable medium, while normal adult differentiated tissue-cells can be propagated in media only with great difficulty or not at all. Cancer-cells partake in this respect of the nature of embryonic cells.

What is the power that gives cancer-cells this perverse quality? There is evidence that no parasitic organism is involved. The cumulative results of two generations of research point rather to cancer being related to those peculiar forms of stimulation of the tissues that are grouped together as 'irritation'. In some cases, at least, the carrier of the irritation is of a filterable nature (pp. 460–4).

Further, it is highly probable that at least two factors are involved in the disease. One factor is external, the irritant. Another factor is certainly internal, since experiments prove that it is possible, by very intensive hereditary accumulation, to increase liability or immunity to cancer.

Cancer presents us also with an interesting analogy with unicellular organisms. In multicellular organisms the fertilized egg continues to divide, and after the process has gone on for a time cells of two types are produced; one type constitutes the individual, the other type is for the maintenance of the species. This most important conception attracted the attention of August Weismann. It was developed in his famous essay *The continuity of the Germ-Plasm as the Foundation of a Theory of Heredity* (1885, pp. 543–7). Weismann distinguished between body-cells and germ-cells. He emphasized the continuing character of the germ-cells (Fig. 191), and pointed out that the mortal fate of the body-cells was the result of the division of labour within the multicellular body. The substance of these body-cells is to be distinguished from that of the germ-cells or *germ-plasm*. In certain unicellular organisms, on the other hand, the whole creature consists of germ-plasm. Such are, for example, the bacteria and certain protozoa. In such unicellular organisms there is no body as distinct from germ. Of these beings no part is pre-destined for death, no part need decay, ageing is unknown.

We can now see how cancer-cells present analogies to these lower forms of life. Something has happened to them whereby they have shed that mortality that is the lot of the other descendants of their cell ancestors. We may describe them as in a permanent embryonic state. They have come to resemble unicellular organisms. Like them they live a life that is for themselves alone, without regard to any community.

A new and promising mode of studying cells in general and those of new growths in particular was introduced in 1907 by Ross Harrison (1870–   ) of Johns Hopkins University. He found that if fragments of living tissues were placed in suitable media and kept under suitable conditions, the cells would multiply. The great technical difficulties—which vary for cells of different origin—were gradually overcome, largely by the efforts of the very skilled experimenter, Alexis Carrel (1873–1944) of the Rockefeller Institute, New York. Certain animal tissues, for example cells from the heart of a chick, first planted out in 1912, and still growing, breed true and produce cells similar to the original type. These results are extremely interesting in themselves. They seem, however, contradictory of certain others. Thus with some plants a small fragment or even a single cell, as with Begonia, may regenerate, not cells like itself, but a complete plant with all its tissues complete. A somewhat similar series of contradictions arises from certain mutilation experiments. At present these inconsistencies cannot be resolved.

§ 10. *Criticism of the Cell Doctrine*

The 'Cell Theory' is merely a general formula erected to cover a series of observations. It is thus open to criticism on the ground either (*a*) of the accuracy of the observations, or (*b*) of the adequacy of the description of the observations.

Naturalists think and speak of the cell as a separate entity cut off from other cells. It is a member of a community. Nevertheless in many cases it has been shown that cells of multicellular organisms are connected with their neighbours by protoplasmic bridges. Much attention was given to this matter by Strasburger (1901). It was soon found that not only in plants but also in animal tissues such protoplasmic bridges between cells are frequently demonstrable. They are encountered in the deeper layers of the skin, where they were seen as long ago as 1864 by Max Schultze (p. 352) who, however, interpreted them wrongly.

Certain unicellular organisms pass through a stage in which there is a naked mass of undivided protoplasm with a number of nuclei. An example is the parasite which gives rise to malaria. Certain multicellular organisms pass through a similar stage in the course of development. Notably this is the case with some insects and crustacea. Further, in some higher animals there are stages in development in which the cells act as though free to unite and their nuclei to multiply according to local needs. Moreover, in various diseased tissues, of which cancer is a type, the cells are often either multinuclear or united by bridges.

A series of facts of this order persuaded some that the so-called multicellular organisms cannot justly be compared to a community. These critics would treat the animal or plant individual as a continuous mass of protoplasm. This mass, they would hold, forms a single morphological unit, whether it appears as a single cell, as a cell with many nuclei, or as a system of cells. Such a view is to some extent supported by the results of experimental embryology (pp. 500–504).

Of the importance and value of these criticisms of the cell theory there can be no doubt. Nevertheless, the appearances on which they are based are exceptional and refer to special tissues, abnormal states, or particular phases of life. There is clearly a tendency in the higher multicellular forms to a definite and final separation of cells from each other. This is especially marked in the most specialized tissues. The most characteristic and highly specialized tissues are those of the nervous and muscular systems. The separateness of the units of these systems has been emphasized by the refinements of modern histology. That study seems to leave the cell theory as a useful, if inadequate, summary of a very important and significant group of facts.

During the twentieth century the main course of biological thought has been concentrated on the cell. The problems of heredity, of sex, of development, of the subordination of parts to the whole, of the essential nature of life, all have been reduced to cellular expression. These aspects we shall presently consider.

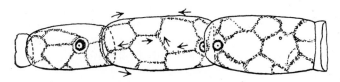

Fig. 147. Living vacuolated hair cells of potato. Magnified. Arrows indicate protoplasmic streaming. From Schleiden, 1838 (p. 335).

X

ESSENTIALS OF VITAL ACTIVITY

§ 1. *Physiology of Descartes and the Early Mechanist School*

THE widest generalization that we can make in the department of the biological sciences is a view of the nature of life itself. In this connexion the philosophy of Descartes has been a moulding force on the whole course of scientific thought (pp. 126–31).

The influence of Descartes on biological theory has been exerted through his treatise on physiology. He intended this to form part of his philosophical system and proposed its issue along with his *Discourse on Method*. It was the period of Galileo's conflict with the Inquisition. The great physicist had but recently been condemned (1632). Descartes, nominally loyal to the Church, decided to issue his *Discourse* (1637) without its physiological appendix. The physiological section did not appear until 1662, when its author had been in his grave for twelve years. Even then it was in a modified form and in Latin (*De homine*, 'On Man'). The true version appeared in French in 1664. It is the first book that seeks to cover the whole field now called 'animal physiology'.

Descartes had no extensive practical acquaintance with anatomy or physiology. He was fairly well read in these subjects, but he did not always understand what he had read. For instance, he did not fully appreciate either the nature or the implication of Harvey's discovery of the circulation (1628). Realizing, however, that it formed the best illustration at hand of the mechanical working of the body, he used it gladly though clumsily. Nevertheless his ingenuity was such that it would be difficult to exaggerate his influence on later biological thought.

Indeed that 'mechanistic' view of the nature of life, which is prevalent among biologists, had its origin with this great philosopher.

Although the general principles set forth by Descartes gained wide acceptance, yet the details of the mechanism which he presupposed were entirely imaginary and had no real existence. A strong point in the physiology of Descartes was the stress laid on the nervous system with its power of co-ordinating the various bodily activities. Thus stated, his view may seem very modern. Yet he was grotesquely wrong in his view of the way in which the nervous system performs its functions.

Descartes held, quite erroneously, that the nerves are hollow and provided with a system of valves at the points at which they branch. These valves he believed to be moved by a series of fibrils passing from the central nervous system along the nerves and terminating in the valves. He regarded as another factor in the activity of the nerves a certain subtle vapour, reaching them from the blood. This vapour, he believed, passing from the nerves into the muscles, distended the latter and thus brought their ends nearer together. Thus was produced what we now call a 'muscular contraction'. The movements of the vapour were, he thought, controlled by the valves in the nerves.

On the basis of these and of certain other imaginary contrivances, Descartes sought to elucidate all the workings of the animal body. The actions and reactions of animals could, he thought, be explained without the invocation of anything in the way of mind, consciousness, or feeling. He was thus a pure 'mechanist'.

An important part of the theory of Descartes was, however, the complete separation of man from all other animals. Man, he held, differs from them in his possession of a reasoning soul. This, he supposed to be situated in a structure in the brain, the *pineal body*. That body, we now know, is developed as an eye in certain fossil vertebrate animals and in one living lizard (*Sphenodon punctatum* of New Zealand). Its function in other vertebrates is not yet clear, but it is found in them all. Descartes held that animals have no soul and therefore no need for a pineal body.

This theory of Descartes was an ingenious attempt to explain the enormous difference between the activities of man and those of the animals. He held the difference to be one of kind. We now hold that, so far as biological science has anything to say on this theme, the difference is one of degree. It is as well to remind ourselves here that this attitude does not in itself exclude the revelation by other than purely biological methods, of a true difference of kind.

Soon after Descartes, the application of his mechanist theory was greatly extended by other workers. The most successful was the Italian professor, Giovanni Alphonso Borelli (1608–79), a pupil of Galileo. Borelli was an excellent mathematician and with his *De motu animalium* (Rome, 1680) he founded the science of muscular mechanics, according to the laws of statics and dynamics. These sciences, in their turn, had been founded by Stevin (1548–1620) and Galileo (1564–1642), on whose conclusions Borelli worked. Many of Borelli's results still stand. His success did much to persuade men of the value of the Cartesian physiology.

The suggestions of Descartes and of Borelli as to the mechanical workings of the animal body provided a fruitful stimulus to biological investigation. But the Cartesian mechanist views, in their original form, did not long remain acceptable to biologists. For generations, however, certain literary men with scientific tastes, but with little practical acquaintance with science, clung to the Cartesian doctrines. This was the case with the Paris physician, de la Mettrie (1709–51). In 1748 he published his famous essay, *L'homme machine* ('Man a machine'). It was burnt as atheistical.

De la Mettrie attempted to extend the mechanist view from animals to man, pursuing it into all departments of human activity. The work, though much read in its day, is of little biological importance. It was one member of a whole series of expressions of the same view. The great *Encyclopédie*, begun in 1751 and completed in 1772, provides further examples. Denis Diderot (1713–84), its main editor and contributor, exhibited the mechanist philosophy also in other works. The fact that both Mettrie and Diderot were 'spermatists' (pp. 506–7) has a bearing on their mode of thought.

§ 2. *Van Helmont* (1577–1644) *and the Beginnings of Chemical Physiology*

Even during the lifetime of Descartes, men of science perceived that purely physical devices were inadequate to explain all living activity. Chemistry, though still in its infancy, was also invoked. Away back in the sixteenth century, the Swiss, Philippus Aureolus Theophrastus Bombastus von Hohenheim (1493–1541), more compendiously known as 'Paracelsus', had made attempts to equate chemical action with bodily processes. The task was less unsuccessfully taken up by his follower, van Helmont, a contemporary of Descartes.

The Belgian, Jan Baptist van Helmont, was a medical man who spent his life investigating chemical processes. He was a very pious Catholic of mystical leanings, whose mysticism has deeply affected his literary style, rendering his works excessively obscure. He published little during his life. After his death, his son collected his writings— published and unpublished—and gave them to the world in a Latin volume which he called *Ortus medicinae* ('The fount of medicine', Amsterdam, 1648). The work employed such an outlandish terminology that it was little studied by biologists until the appearance of translations (into Dutch 1660, English 1662, and French 1671). It was thus, in its effect, contemporary with Descartes *On man* (1662–4).

Van Helmont, like Paracelsus, considered all physiological processes explicable on a chemical basis. He did not regard these processes as acting on their own initiative but as being each governed by one of a series of agencies, the *archaei*. This conception also he adopted from Paracelsus. Van Helmont postulated a regular hierachy of archaei in the various organs. The chief archaeus of all, he believed, supplies 'the reproductive power and is, as it were, the internal efficient cause. It has the likeness of the thing generated and equips the germ with the powers that determine its course of development.' It thus has analogies with the *entelechy* of Aristotle. Van Helmont was, in fact, trying to reconcile the vitalist and the chemical view of life. His theories are in opposition to those of Descartes, who regarded the seed as of the nature of a machine from which, if enough were known, the form and activity of the adult could be predicted.

Van Helmont introduced an entity subordinate to the archaei, for which he coined the term *blas*. The *blas humanum* is responsible for the specifically human functions of man's body, and other forms of *blas* preside over other physiological processes.

The physiology of van Helmont is otherwise based upon ferments and their action. With alcoholic fermentation as the type, he attempted to explain all physiological processes, and notably digestion, as due to the action of ferments. This is the most intelligible aspect of his work and is, to some extent, in accord with more modern teaching.

The followers of van Helmont are known as the 'iatrochemists' (Greek *iatros*, 'physician'). The nomenclature employed by them is often extremely bizarre, but they were working in a direction that has proved fertile for scientific advance.

The most prominent of the iatrochemists was Franciscus Sylvius (1614–72), a very able professor of Medicine at Leyden. His name is latinized from De la Boe. Sylvius devoted much attention to the study of salts. These he recognized as the result of the union of acids and bases, and he attained to the idea of chemical affinity—a very important advance. He sought to represent almost all forms of vital activity in terms of 'acid and alkali' and of 'fermentation'. Shedding the mysticism of van Helmont with its 'archaei' and its 'blas', he assumed that fermentation, which van Helmont had stressed, is a purely chemical process.

The school of Sylvius and its immediate successors added considerably to our knowledge of physiological processes—notably by the examination of digestive fluids. The action of these has some parallels to that of the microorganisms of alcoholic and other fermentation.

§ 3. *Plant Physiology in the Seventeenth Century*

The plant physiology of the seventeenth century was of very elementary character as compared to that of animals.

Van Helmont had demonstrated that the solid parts of a plant increase in weight apart from anything that they take from the soil. This action of plants was contrary to the current Aristotelian teaching that plants draw their food, ready elaborated, from the earth.

Malpighi held that the leaves form from the sap the material required for growth. He knew that elaborated food-substance is distributed from the leaves to the various parts of the plants. The sap, he wrongly thought, is brought to the leaves by the fibrous parts of the wood. He conceived an imaginary course for this nutrient sap. It goes downwards, he thought, into the roots and then again upwards to the organs above ground (1671–4). This view was fantastically developed by some of his contemporaries into a 'circulation of the sap', comparable to the circulation of the blood in animals. Nor was Malpighi much more fortunate in his conception of the process of plant breathing. He supposed that air is conveyed from the roots by the spiral vessels. The cause of this latter error was the close similarity which he thought he could discern between the 'spiral vessels' of plants and the 'tracheae' or air tubes of insects (cf. Figs. 78 and 89).

The earliest important experimental work on the physiology of plants was that of the French ecclesiastic, Edmé Mariotte (died 1684). This able physicist was a member of the scientific circle in France which developed as the Académie des Sciences (pp. 137–8). Mariotte observed the high pressure with which sap rises. He inferred from this that there must be something in plants which permits the entrance but prevents the exit of liquids. The pressure he compared to that of the blood-vessels, in a manner afterwards developed by Hales (p. 368). Mariotte held that it is sap pressure which expands the organs of plants and so contributes to their growth (*On the vegetation of plants*, 1676).

Mariotte was definitely opposed to the Aristotelian conception of a vegetative soul (p. 38). He considered that this conception fails to explain the extraordinary fact that every species of plant, and even the parts of a plant, exactly reproduce their own properties in their offspring. He was, so far as plants are concerned, a complete 'mechanist' (pp. 358–9), and therefore anti-Aristotelian. All the 'vital' processes of plants were for him the result of the interplay of physical forces. He believed, as a corollary to this view, that organisms can be spontaneously generated (chap. xii). This is a conclusion not often faced by advocates of mechanistic theory in modern times.

§ 4. *Stahl* (1660–1734) *and the Contest between Mechanism and Vitalism*

In the generation that was active in the last quarter of the seventeenth and the first half of the eighteenth century, there arose an important cleavage of physiological interests. The opposition between the 'mechanist' and the 'vitalist' came into clear view.

Georg Ernst Stahl (1660–1734) was professor at Halle and a fashionable and influential physician. He was much interested in Chemistry and saddled that science with the unfortunate theory of *phlogiston* which held its ground until the time of Lavoisier. In physiology Stahl set himself against the 'mechanism' of Descartes. To the French philosopher the animal body was a machine. To the German physician the word *machine* expressed exactly what the animal body is not. The phenomena characteristic of the living body are, Stahl considered, governed not by physical laws but by laws of a wholly different kind. These are the laws of the *sensitive soul*.

The sensitive soul of Stahl is, in its ultimate analysis, not dissimilar to the *psyche* of Aristotle (pp. 15, 38). Stahl held that the immediate instruments, the natural slaves, of this sensitive soul, are chemical processes. Thus his physiology develops along lines of which Aristotle could know nothing since he lived before the days of chemistry. Yet neither this nor the fact that Stahl did not regard himself as an Aristotelian can alter the fact that his theories are of essentially Aristotelian origin.

The views of Stahl are scattered through an almost incredible mass of writings. His physiological theories are, however, best set forth in his classical *Theoria medica vera* (Halle, 1707–8), which contains an important section 'Physiologia.'

Almost exactly contemporary with Stahl was a rival professor at Halle, Friedrich Hoffmann (1660–1742). He also was a skilled chemist and he was nearly as verbose as his opponent. In Hoffmann's view the body is like a machine. On the one hand, he separated himself from the pure mechanists of the school of Descartes and Boerhaave (see below) by claiming that bodily movements are executed under the influence of properties peculiar to organic matter. On the other hand, he separated himself from the Stahlian vitalists by denying the need to invoke the sensitive soul. He was deeply under the influence of the 'monadism' of the philosopher Leibniz, but attached great importance to the mechanism revealed by Harvey.

It is difficult to select from the huge mass of Hoffmann's writing. His doctrines are perhaps best expressed in his *Fundamenta physiologiae* (Halle, 1718). 'Life', he there says, 'consists in the movement of the blood. This circular movement maintains the integrity of that complex which makes up the body. The vital spirits which come from the blood are prepared in the brain and released therefrom to the nerves. Through them come the acts of organic life which can be reduced to the mechanical effects of contraction and expansion.'

An important protagonist in the controversy was Hermann Boerhaave (1668–1738) of Leyden, one of the ablest medical teachers of all time. Boerhaave was also skilled in chemistry and takes his place in the history of that science. He wrote a text-book of physiology, *Institutiones medicae* (Leyden, 1708), which ran through innumerable editions and was translated into many languages. It was the staple work on the subject until the appearance of the great treatise of Haller (pp. 370–2).

Boerhaave goes systematically through the functions and actions of the body. All are ascribed to chemical and physical laws. He does lip service to the influence of mind on the body. In practice, however, he ignores it and is as completely mechanist as Descartes. He still believed that something material passes down the nerves to cause movement, though he had himself published Swammerdam's experiment in disproof thereof (p. 164). The value of Boerhaave's physiological work is in its very clear and systematic exposition.

§ 5. *Hales* (1677–1761) *on the Physiology of Plants and Animals*

In the earlier eighteenth century, physiological writing was mainly confined to physicians. The attention of naturalists was concentrated on the discovery and arrangement of new forms of vegetable and animal life. Voyages of exploration kept them busy. Their work thus tends to be uniform and monotonous. From them on the one hand and from the speculative physicians on the other, Hales stands out for his independence of thought, his experimental ingenuity, and his power of lucid statement.

The Rev. Stephen Hales was educated at Cambridge in the mathematical and physical sciences. He became interested in botany, using Ray's *Flora of Cambridge* (p. 183) in his excursions. Most of his years were spent as vicar of Teddington in exemplary discharge of parish duties, and he led the uneventful life of a man of simple habits and serene mind.

Hales made many inventions of considerable importance. Best known were his 'artificial ventilators' which he succeeded in getting fitted to prisons, both English and French, then particularly insanitary. Great reduction in death-rate ensued. He made useful suggestions for the distillation of fresh from salt water, for the preservation of eatables, for cleansing harbours, for measuring the depth of the ocean by a mercurial pressure gauge, for winnowing corn, for preventing the spread of fires, for a thermometer for high temperature, and for the use of furze for fencing river banks.

The work of Hales on the functional activity of plants was the most important until the nineteenth century. His *Vegetable Staticks* (London, 1727) contains the record of a great number of ingeniously devised but simple experiments. Many of these are still repeated in the botanical laboratory. Their general idea is to explain the action of living plants on the basis of known physical forces. Thus he measured the amount of water taken in by the roots and the amount given off by the leaves, and so estimated what botanists now call *transpiration*. He compared this with the amount of moisture in the earth and showed the relationship of the one to the other. He made calculations of the rate at which water rises in the stems of plants, and he showed that this has a relation to the rate at which it enters by the roots and is transpired through the leaves. He measured the force of upward sap current in stems, *root-pressure* as it is now called. He sought to show that these actions of living plants might be explained as a result of their structure. With this in view he measured the absorption of water by substances with fine pores and the movements of water and other fluids in capillary tubes.

An interesting contribution by Hales was his demonstration that the air supplies something material to the substance of plants. This we know to be the carbon dioxide of the atmosphere. In his day, however, the gases of the atmosphere were hardly at all understood. The significance of this discovery was therefore overlooked. Following upon it, however, he showed with the aid of the air-pump, that air enters the plant not only through the leaves, but also through the rind.

The experiments and conclusions of Hales in animal physiology were as important as in vegetable physiology. He endeavoured to give an exact quantitative expression to the conception of the circulation of the blood.   He

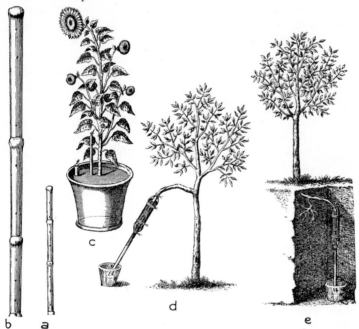

FIG. 148. From Hales, *Vegetable Staticks* (1727).

a and *b* exhibit the method adopted by Hales of showing the region of growth. A young stem is pricked with holes at known equal distances apart.   The distance between the holes is measured in the stem when older.   Growth takes place mainly in the middle part of the internodes.

c is a sunflower planted when younger in a pot on the top of which a metal plate is fixed.   Air and water are admitted through a tube fixed to this metal cover.   By comparing the dry weight of such a plant when it is grown, and the loss in the soil in which it grows, it can be proved that something material is absorbed into its substance from the air.

d is an apparatus for measuring the negative pressure resulting from transpiration.   The apparatus fixed to the branches is, in effect, a 'manometer'.

e is a similar apparatus for measuring root pressure.

showed that just as there is a *sap pressure* that can be measured, so there is a *blood-pressure* that can be measured, and he measured it. Moreover, he perceived that the pressure varies under different circumstances. It is different in the arteries and the veins. It is different during contraction of the heart from what it is during dilation. It is different in a failing and in an active heart. It is different in large and in small animals. All these differences Hales measured. He measured, too, the rate of flow in the capillaries of the frog. These experiments and conclusions of Hales were the beginnings of a quantitative development of the science of animal physiology.

It is specially characteristic of the work of Hales that he always sought to give an exact mathematical expression to his results. 'Science is Measurement' sums up his attitude.

§ 6. *Haller* (1708–77) *and the Doctrine of Irritability*

Albrecht von Haller (1708–77) was a Swiss of noble birth and ample means. After holding chairs in several German Universities, he retired to his native Berne. He exhibited unparalleled literary and scientific activity and stamped his views and methods of approach upon the whole range of biological research. He was certainly one of the most versatile men that has ever lived. He achieved distinction as a poet, botanist, anatomist, philosopher, physiologist, and novelist. He carried on a prodigious correspondence, was an exceedingly learned bibliographer, and was, perhaps, the most voluminous scientific author of his time. He took much interest too in the history of science. His great works on the bibliography of botany and of anatomy are still indispensable. They display judgment and penetration as well as system and almost incredible learning. Haller was a lover of his native mountains, and his writings on the natural beauties and the flora of the Alps are still of value.

Haller's greatest scientific achievement is his *Elementa Physiologiae* (Lausanne, 1759–66). This extensive work immediately replaced Boerhaave's *Institutiones medicae* (p. 366). It marks the modernization of the subject. The *Elementa* speaks the language of our own time, and as we peruse it we seem to have passed, at last, into a modern laboratory. Among the important subjects on which it throws new light are the mechanics of respiration, the formation of bone, and the action of the digestive juices. As regards embryology we note that Haller is an un-repentant preformationist and is opposed to the teaching of his contemporary Wolff (pp. 469–71).

Haller's most important contributions to biological thought are his conceptions of the nature of living sub-stance and of the action of the nervous system. These formed the main background of physiological thinking for a hundred years after his time. They are still integral parts of physiological teaching.

During the seventeenth and early eighteenth centuries the favourite doctrine of nervous action presupposed the existence of a subtle nervous fluid. Such was the view of Descartes (p. 359), Borelli (p. 406), and Boerhaave (pp. 366–7). An admirable experiment by Jan Swammerdam had made this view untenable. But Swammerdam's work was lost until published posthumously in 1736, and so the matter stood over till Haller's time.

Haller concentrated the problem on an investigation of the fibres. A muscle-fibre, he pointed out, has in itself a tendency to shorten with any stimulus, and afterward to expand again to its normal length. This capacity for con-traction, Haller, following a predecessor (Francis Glisson, 1597–1677), called *irritability*. He recognized irritability as an element in the movement of the viscera, and notably of the heart and of the intestines. The feature of irritability is that a very slight stimulus produces a movement altogether out of proportion to itself, and that it will continue to do this repeatedly, so long as the fibre remains alive. We now recognize irritability as a property of all living matter.

But besides its own inherent force of irritability, Haller showed that a muscle-fibre can develop another force. This second force comes to it from without, is carried from the central nervous system by the nerves, and is that by which muscles are normally called into action. Like irritability, it is independent of the will, and, like irritability, it can be called into action after the death of the organism as a whole. Haller thus distinguished the *inherent muscular force* from the *nerve force*. Both these forces he further distinguished from the natural tendencies to contraction and expansion, which, under changing conditions of humidity, pressure, and so on, produce changes in all tissues, living or dead.

Having dealt with the question of movement, Haller turned to consider feeling. He was able to show that the tissues are not themselves capable of sensation, but that the nerves are the channels or instruments of this process. He showed how all the nerves are gathered together into the brain. These views he supported by experiments and observations involving injuries or stimulation to the nerves and to different parts of the brain. He ascribed importance to the outer part or cortex, but the central parts of the brain he regarded as the essential seat of the living principle, the soul.

Throughout his discussion Haller never falters in his display of the rational spirit. He develops no mystical or obscure themes. Although his view of the nature of the soul may lack clarity, he separates such conceptions sharply from those which he is able to deduce from actual experience. He is essentially a modern physiological thinker.

§ 7. *Hunter as Vitalist*

During the earlier eighteenth century, owing to various causes, the interests of those who investigated plant biology and those who investigated animal activities drifted ever wider apart. Hales was almost alone in taking both departments into his wide purview. Toward the end of the century, with the all-embracing curiosity of Hunter and the philosophic probings of Goethe, the two studies drew together again. Men began again to consider life as a whole. Partly this expressed itself in the school represented by the 'Naturphilosophen' (pp. 215–23); partly, however, by a more naïve attitude which is well represented by John Hunter.

Hunter's older contemporary, Linnaeus (pp. 188–95), and his younger contemporary, Cuvier (pp. 227–31), used their knowledge of comparative anatomy as a guide to the arrangement of living things. These were not Hunter's ways. He, like Buffon (p. 293), was ever seeking the more general principles that underlie similarities or dissimilarities of structures. The most general principle of all, the principle from which *Biology* takes its name, is that mysterious thing called life. Hunter came no nearer to answering the question 'What is life?' than later biologists. In the course of his search, however, he reached some important conclusions.

Hunter considered that, whatever life may be, it is something held most tenaciously by the least organized beings, and is something that is independent of structure. These ideas lead to the conception of *protoplasm*, the substance, simple in appearance, inconceivably complex in fact, which seems the inseparable material factor without which life is never found. Hunter did not use the word protoplasm, which was invented fifty years after his death (pp. 340–1), but he was reaching out to this conception.

Life is normally exhibited, according to Hunter, in the various activities of living things, and notably in the power of healing and repair and renovation. This power is quite peculiar to living things, and cannot be paralleled in the non-living world. Nevertheless Hunter thought that vital activity can be suspended—as, for instance, in the egg. He was thus led to investigate what he considered the simplest forms of life. In doing so, he discovered what was from his point of view a 'latent heat of life' set free at death. Thus he found that sap removed from a tree congeals at freezing-point but that the living tree itself may be reduced far below freezing-point before the sap loses its viscous or liquid quality. The suggestion was interesting though the phenomena, as we now know, are susceptible of other explanation. He made similar experiments on other forms of life to illustrate the same point. He also came to regard the heat given off by germinating seeds as evidence of something of the nature of a latent heat of life.

§ 8. *The Balance of Life*

Light on the vital activities of plants was thrown by the chemist, Joseph Priestley (1733–1804). In his *Experiments and observations on different kinds of air* (London, 1774) he demonstrated that plants immersed in water give off the gas which we term 'Oxygen'. He observed, too, that this gas is necessary for the support of animal life.

Priestley's contemporary, the French chemist, Antoine Lavoisier (1743–94), made quantitative examinations of the changes during breathing. These displayed the true nature of animal respiration and proved that carbon dioxide and water are the normal products of the act of breathing.

In the meantime Jan Ingenhousz (1730–99) was introducing the highly important concept of the balance of animal and vegetable life. Ingenhousz was an engineer, of Dutch origin, educated at Louvain, Leyden, Paris, and Edinburgh. He worked in London with Hunter, and was in contact with all the scientific movements of his day. He travelled widely and wrote on a variety of topics. Nearly all his communications exhibit an original outlook.

In 1779 Ingenhousz published in London his *Experiments upon vegetables, discovering their great power of purifying the common air in the sunshine and of injuring it in the shade and at night*. It contains a demonstration that the green parts of plants, when exposed to light, fix the free carbon dioxide of the atmosphere. He showed that plants have no such power in darkness, but that they then give off, on the contrary, a little carbon dioxide.

This discovery is the foundation of our whole conception of the economy of the world of living things. Animal life is ultimately dependent on plant life. Plants build up their substance from the carbon dioxide of the atmosphere together with the products of decomposition of dead animals and plants. Thus a balance is kept between the animal and the plant world. The balance can be observed in the isolated world of a small aquarium.

All this is obvious enough now. But the importance of the work of Ingenhousz was at first insufficiently appreciated, despite the labours of Lavoisier on the nature of air. Not until vegetable physiology came to be taken up in the light of the cell theory was it realized that this power of plants to fix carbon dioxide is one of the most significant of all the activities of living things (pp. 413–15).

A discerning Swiss protestant pastor, Jean Senebier (1742–1809), of Geneva, had however perceived some of the implications of the discovery of Ingenhousz. He stated the matter clearly in his *Mémoires physico-chimiques sur l'influence de la lumière solaire* (Geneva, 1782). Senebier was much under the influence of his townsman Charles Bonnet (p. 211), who had been feeling his way toward the same idea. He was also the medium through which the doctrines of Spallanzani (p. 442) became better known. Finally, Senebier stimulated his fellow Genevan, Nicholas Theodore de Saussure (1767–1845)—the third man of science of that name—to work on the chemistry of plant respiration. The subject is clearly expounded and assumes its modern form in de Saussure's *Recherches chimiques sur la végétation* (1804).

§ 9. *T. A. Knight* (1759–1838) *and Tropisms*

A new field of research in plant physiology was opened up by Thomas Andrew Knight. This English country gentleman was a correspondent of Sir Joseph Banks (pp. 240–1). With the purely practical end of improving agriculture he made a surprising number of important scientific contributions. He is best remembered by the device known as *Knight's wheel*. Germinating plants are attached to a rapidly rotating disc which may be in a horizontal or vertical position. If the plane of rotation be vertical, the line of action of gravitational force is constantly changing while the line of action of centrifugal force remains constant with reference to the axis of the plant. Under these circumstances the stem of the plant grows along the radius toward the centre of the wheel, while the root grows away from the centre.

The reaction of the plant to gravity, which can thus be eliminated and replaced by centrifugal force, is known as *geotropism* (Greek = 'earthward turning') a word introduced by Sachs (1868). It is spoken of as 'positive' for the root and 'negative' for the stem. Geotropism is a phenomenon of great importance in living things in general and in plants in particular. It is the prototype of a whole series of similar reactions that have since been discovered in living things, both plant and animal.

These movements are now known by the more general

term of *tropisms*. They may be defined as the simple and non-voluntary reaction of an organism or part of an organism by movement, growth, or bending in response to an external stimulus. We thus have *phototropism* (reaction to light), *heliotropism* (reaction to the sun), and *thermotropism* (reaction to heat). There are also chemical tropisms, one well-known form being the movement of the spermatozoids of ferns and other vascular cryptogams towards malic acid. Some naturalists, notably Jacques Loeb (1859–1924), have ascribed many of the phenomena of growth and development to tropisms. The theme has been particularly stressed in connexion with the conduct of the plants and lower animal forms, but also in the modern department of Experimental Embryology (pp. 500–4). In the higher animals tropisms pass insensibly into reflexes (pp. 422–4).

§ 10. *Liebig* (1802–73) *and the Chemistry of Vital Activity*

The work of Priestley, Lavoisier, Ingenhousz, Senebier, and Knight was appreciated by the English chemist, Sir Humphry Davy (1778–1829). He combined their views with little change in his *Elements of Agricultural Chemistry* (London, 1813). The real creator of the chemistry of vital activity was, however, the commanding German teacher, Justus von Liebig (1802–73).

Liebig was professor of chemistry first at Heidelberg and afterwards at Giessen. He was convinced that all vital activity could be explained as the result of chemical and physical factors. Over the door of the University Laboratory which he founded he had inscribed the dictum *God has ordered all His Creation by Weight and Measure.* His great achievement was his application of chemical knowledge to the phenomena exhibited by living things. He did much to introduce laboratory teaching, and certain apparatus which he invented is still in constant use.

Liebig improved methods of organic analysis and notably he introduced a way of determining the amount of urea in a solution. This substance is found in blood and urine of mammals, and was long regarded as the first organic compound to be 'synthetized', that is, built up from elemental materials.[1] It is of very great physiological importance, for it is regularly formed in the animal body in the process of breaking down the characteristic nitrogenous substances known as 'proteins' (pp. 385–90).

Along with Friedrich Wöhler (1800–82) who was working towards the synthesis of organic substances, Liebig wrote a paper (1832) in which he showed, for the first time, that a complex organic group of atoms—a *radicle*, as it is now called—can remain as an unchanging constituent through a long series of compounds, behaving throughout like an element. The discovery is of primary importance for our conception of the chemical changes in the living body. From 1838 onwards, Liebig devoted himself to attempting a chemical elucidation of living processes. In the course of his investigations he did pioneer work along many lines that have since become well recognized. Thus he taught the true doctrine, then little recognized, that all animal heat is the result of combustion, and is not 'innate'. He also classified articles of food with reference to the functions that they fulfilled in the animal economy into fats, carbohydrates, and proteins.

Very important was Liebig's teaching that plants derive the constituents of their food, their carbon and nitrogen, from the carbon dioxide and ammonia in the atmosphere, and that these compounds are returned by the plants to the atmosphere in the process of putrefaction. This development of the work of Ingenhousz and of others, made possible a philosophical conception of a sort of

---

[1] It was not really synthetized as the process involved HCN, which, at that time, had not itself been synthetized.

'circulation' in Nature (pp. 374–6). That which is broken down is constantly built up, to be later broken down again. Thus the wheel of Life goes on, the motor power being energy from without, derived ultimately from the heat of the sun. The wheel turns on and on. Whether it will ever stop depends upon its motor at the centre of the solar system.

It was very unfortunate that Liebig conceived putrefaction as a purely chemical as distinct from a vital process. It took Pasteur long years to displace this view (pp. 444–6).

§ 11. *The Chlorophyll System*

From the time of Aristotle the idea had prevailed that plants absorb their nourishment exclusively by their roots. Van Helmont had shown, however, that something was also drawn from the atmosphere (p. 363), but it was long before his lead was followed. The ultimate destruction of the ancient fallacy proceeded along two lines. Firstly there was the investigation of the formation of the carbohydrates, and secondly that of the nitrogenous substances. The former we consider here; the latter in the next section.

By far the major part of existing living matter is contained in green plants. These provide the ultimate source of aliment for the entire animal kingdom. The economic significance of the sources from which the substance of plants is replenished cannot, therefore, be exaggerated. A most important source is carbohydrate, especially in the form of starch, the formation of which is associated with the green matter itself.

In the seventeenth century Leeuwenhoek (pp. 166–9) saw starch granules in the tissues of plants and he portrayed their remarkable concentric form. We now know that starch is built up in the plant from the carbon dioxide absorbed from the atmosphere; that starch formation is a function of the living cell intimately connected with the green substance; and that the process is active only in the presence of light. The name *Chlorophyll* (Greek = 'leaf green') was attached to this green substance by the two French chemists, Pierre Pelletier (1788–1842) and Josephe Caventou (1795–1878) in 1817.

A step toward the modern position was made by the French physician and botanical experimenter Henri Joachim Dutrochet (1776–1847). It was already known from Ingenhousz (p. 375) and others that the plant as a whole gave off oxygen and absorbed carbon dioxide. Dutrochet recognized (1837) that only those cells that contain green matter are capable of absorbing the carbon dioxide.

Perhaps the most influential botanical teacher of the nineteenth century was Julius Sachs (1832–97), a pupil of Purkinje (pp. 340, 418). Sachs long professed botany at Würzburg. After applying himself to morphology he turned to physiological investigation. From 1857 onward he was immersed in problems of nutrition. He soon became convinced that the chlorophyll is not diffused in tissues but is contained in certain special bodies—*chloroplasts* as they were later (1883) named by Schimper (p. 275). Sachs showed also that sunlight plays the decisive part in determining the activity of chloroplasts in absorption of carbon dioxide. Further, chlorophyll is formed only in the light. Moreover, in different kinds of light the process of carbon dioxide assimilation goes on with different degrees of activity. The views and discoveries of Sachs were set forth in his great treatise on botanical physiology (1865).

Important for the discussion of starch formation is the process by which the gases of the atmosphere come into contact with the tissues of plants. In animals this process is more evident, especially in such as mammals in which the aeration of the lungs is an active process. Plants, however, were long in giving up their secret.

The *stomata* of plants, or little openings on the surface of the leaves, had long ago been figured by Malpighi (pp. 152–7). Dutrochet in 1832 called attention to the fact that the stomata communicate with intercellular spaces within the substance of the leaf. Nevertheless the manner of entry of gases into the leaf system remained in dispute for more than half a century. It was not until the 'nineties that the stomata were generally recognized as the normal channel of gaseous entry.

The critical experiment that determined opinion toward Dutrochet's view was the outcome of the demonstration by Sachs that the appearance of starch in the chloroplasts follows immediately the absorption of carbon dioxide. Thus the presence of starch can be used as an index of carbon dioxide absorption. Leaves and parts of leaves were coated with wax so as to occlude the stomata. By this beautifully simple experiment it was found that starch is developed only in the uncoated parts (1894).[1]

In the course of these discussions, the structure of the stomata and the nature of the chloroplasts became much better known. This was largely the work of Sachs himself seconded by Nathaniel Pringsheim (pp. 532–3). The chloroplasts have often, if not always, the power of growth and division independent of the cell in which they dwell. To such an extent is this the case that it has even been suggested that they are, in effect, separate organisms. Chloroplasts are merely one of a class of self-multiplying cell inclusions (p. 349), named 'plastids' in adoption of a term of Haeckel (1866). Plastids are found mainly, but not exclusively, in plants.

---

[1] Some aquatic plants and certain cryptogams as well as certain fossil forms are devoid of stomata. Plants without stomata absorb gases through their general surface.

The formation of starch in the chloroplast is essentially a vital phenomenon. It takes place only in the living chloroplast and in the living cell. Chlorophyll is not a simple chemical substance. Since 1864 it has been known that it consists of a mixture of at least two green substances together with at least two yellow pigments. The work of the following sixty years did not suffice, however, to differentiate the function of these constituents.

It was shown in 1882 by the physiologist T. W. Engelmann (1843–1909) that all parts of the spectrum do not activate the chloroplast equally. The red is the most efficient, the violet less, and other parts of the spectrum hardly at all.

The actual chemical process of starch formation is still in dispute, but it is certain that there are intermediate products of which formaldehyde is believed to be one. The process is reversible, since the carbohydrate can be oxidized into carbon dioxide and water. The process is specially active in growing seedlings. The formula is usually thus represented:

$$CO_2 + H_2O + \text{Light Energy} \rightleftharpoons HCHO + O_2$$

§ 12. *The Nitrogen Cycle*

We turn now to a consideration of the origin and fate of nitrogenous substances in the plant. The earlier nineteenth century chemists Gay-Lussac, Thénard, Davy, Liebig, were well aware of the importance of nitrogen in the substance of plants. Liebig persuaded men of science that nitrogen was taken up by the roots in the form of ammonia compounds and nitrates. He steadied the whole physiological position by rejecting the old idea of absorption of humus and declaring that 'carbon dioxide, ammonia, and water contain in themselves all the necessary elements for the production of all living animal and vegetable matter. Carbon dioxide, ammonia, and water are

also the ultimate products of their processes of putrefaction and decay' (1840).

The French chemist and mining engineer, Jean Baptiste Boussingault (1802–87) applied himself persistently
and, in the end, successfully to the nitrogen problem.
During the 'fifties he proved that plants absorb their
nitrogen not from the atmosphere but from the nitrates of
the soil. He showed further that no organic or carbon-
containing matter is necessary in soil for the growth of

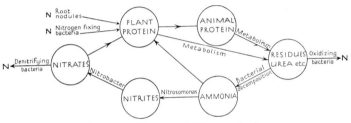

FIG. 149. Diagram of Nitrogen Cycle.

plants, provided that nitrate be present. Thus the carbon
in plants must all be derived from the carbon dioxide of
the atmosphere. He applied quantitative methods in these
experiments (1860) and was correct in his results so far
as concerned the plants on which he worked.

Two decades had not gone by before the question of
nitrogen nutrition was raised in another form by the
famous historian of chemistry, Marcellin Berthelot (1827–
1907). This distinguished man did much important work
on the synthesis of organic compounds. His investigation
of plant nutrition fits into this series of his researches. In
1876 Berthelot showed that free atmospheric nitrogen can
be fixed by electric discharges. Next he proved that the
presence of certain sugars aids this process. In 1886 and
the following years he demonstrated that bacteria acting
in clay soils are able to fix nitrogen.

It was now evident that the matter of nitrogen fixation

was of high economic importance. It was soon taken up by many other workers whose discoveries have been of the utmost bionomic importance and have modified our view of the 'nitrogen cycle' (Fig. 149).

The nature of this cycle is best brought out by a diagram. Its existence has been known since Liebig. Many details have since been elicited. But over and above the familiar circulation of nitrogen it came to be realized that certain plants can obtain this element in a way unknown to Liebig. The discovery is perhaps the most significant addition made since his time to our knowledge of the economics of life on land. Many other comparable physiological cycles are now recognized.

In ancient times it was a rule of agriculture to alternate leguminous with corn crops. Virgil in his *Georgics* advises the farmer to do this. The rule was not wholly forgotten during subsequent ages. In 1686 Malpighi investigated the germination of leguminous plants and figured minute nodules on the roots of the seedlings (Fig. 19). His observations were repeated by de Candolle (1825), Treviranus (1853), and others. The nodules were further found to contain bacteria (1866). The growth of these bacteria was investigated by several observers (1878–86) while Berthelot's work was going forward.

Despite all these concurrent investigations, the real significance of the phenomena presented by the Leguminosae, was not fully grasped until the appearance in 1888 of a classical paper by the two German investigators Hermann Helriegal (1831–95) and H. Wilfarth. They proved that these plants can absorb atmospheric nitrogen; that they do so by means of their root nodules; that their rate of nitrogen absorption is proportional to their nodulation; and that they can live and grow independently of either ammonia or of nitrates.

Not only is this matter of great economic importance,

but it has, in more recent time, been brought into line with a comparable series of phenomena in other plants. The agents of nitrogen fixation in the Leguminosae are bacteria. Other organisms in other plants are associated with the first growths of the roots of seedlings and often with the growing points of roots throughout life. Thus the organism that we recognize as a higher plant, is in fact, a colony of organisms of very different types.

To complete the conception, we should develop the chemistry and bacteriology of soil. These subjects, however, had not reached the stage at which a brief summary was possible until well after the nineteenth century.

§ 13. *The Chemistry of Protoplasm*

There is an immense literature on the chemical character of protoplasm as the 'physical basis of life'. Strictly the subject is insoluble since protoplasm can only be investigated when it has ceased to be the basis of life. We may learn what protoplasm takes in and what it throws out. But living protoplasm is beyond the reach of the chemist's activities. It is protoplasmic products and dead protoplasm that have been the subject of his researches.

Dead protoplasm consists of a very complex mixture of numerous substances. Of these the bulkiest is water. The others are largely made up of the complex nitrogenous group known as *proteins* and their derivatives, of the *lipoids* or fatty bodies, and of the *carbohydrates* or starchy substances. The distinctness of these three types was made definite by Justus von Liebig in his great textbook *Organic chemistry applied to agriculture and physiology*, (1840). The conception of protoplasm itself as a chemical substance was made familiar by T. H. Huxley in his well-known essay *The Physical Basis of Life* (1869). Neither used the nomenclature with which we are nowadays familiar.

The Dutch chemist, Gerard Johann Mulder (1802–80) obtained a certain complex substance to which he attached the formula

$$C_4 O H_{62} N_{10} O_{12}.$$

He believed that this was the essential constituent of all organized bodies and named it *protéine* (1838). Later he worked with Liebig who soon found that there was no such definite compound. The word, however, was retained for the nitrogenous products of which it was a mixture. These came ultimately to be known as *proteins*.

The words *carbohydrate* and *lipoid* had a similar but slower development to that of *protein*. The three terms were employed collectively for the first time with reference to protoplasm by the Leipzig pathologist Ernst Wagner (1829–89). The conception of living protoplasm as an enormously complex mixture of those three types of substance dates from the appearance of Wagner's *Handbuch der allgemeine Pathologie* (1862).

Living protoplasm is liquid (pp. 349–52). Nevertheless an elementary acquaintance with its behaviour shows that it exhibits a considerable degree of 'viscosity', that is, it has some of the properties of a sticky or of a jelly-like substance. Modern views of the intimate structure or composition of living protoplasm have become closely linked with a comparison of its behaviour to that of other substances in the *colloid* ('glue-like') state. The study of the colloid state is one of the many areas in which the old sciences of chemistry and physics have become indistinguishable from each other.

The basis of our knowledge of *colloid* substances is due to the chemist Thomas Graham (1805–69). He performed the researches with which his name is associated from 1850 onward while Master of the Mint in London. The term 'colloid' was already in use, but he applied it to a particular state of matter. He distinguished soluble substances in general into the two great classes, *colloids* and *crystalloids*.

The apparatus used by Graham was of the simplest character possible. There can be no doubt, however, either of the originality of his methods or of the great importance of his results. He observed that certain substances (*a*) pass very slowly into solution, (*b*) do not crystallize, and (*c*) cannot diffuse or diffuse very slowly through organic membranes. Of these substances glue is the type, hence the name *colloid*. In this class are starch (compare starch paste), white of egg, gelatine (the basis of most table jellies). Opposed to these in all three respects are the *crystalloids*.

Graham distinguished sharply between colloid and crystalloid substances, but was aware that certain substances— silica for instance—could exist as either colloid or crystalloid. He recognized too that instability was a characteristic of colloids. Moreover, he perceived that most colloids are of organic origin. He almost foresaw certain modern views of the nature of vital activity in his conception that the surface energy of colloids 'may be looked upon as the probable primary source of the force appearing in the phenomena of vitality'.

The knowledge of the essential nature of colloids was but little extended until the twentieth century. Investigators of our own generation have given a physical interpretation to the differences between the colloid and crystalloid states. A solution of substances in either state may be looked on as a suspension of solid particles in another medium. The difference depends on the size of the particles, those of colloids being much larger than those of crystalloids. Thus the difference is of degree rather than of kind, though there are practical difficulties in accepting this theoretical conclusion.

In colloid solutions the dissolved particles are from

about 2/1,000,000ths of a millimetre in diameter up to about fifty times that diameter. Much of the recent progress in the knowledge of the colloid state has been due to the use of the *ultra-microscope*, invented in 1903. By its means it is possible, in colloid solutions, to see light reflected from the floating particles. These particles can, moreover, be observed in a state of Brownian movement.

A valuable conception has been introduced by W. B. Hardy (1864–1934) of Cambridge into the view of colloids (1899 onwards). It affects our conception of the events which follow on the death of protoplasm.

Colloidal solutions, such as those of gelatine or white of egg, can be 'fixed' by certain substances into a more or less solid state. This may happen in one of two ways. Either the solid elements may form a fine sponge-like framework enclosing minute droplets of liquid, or the solid elements may aggregate into larger separate particles forming an excessively fine emulsion (Fig. 150). The method of fixation determines which of these states shall result. The microscopic appearance will depend on which state is produced. This is the case with protoplasm. Thus the structure of dead protoplasm cannot be regarded as any real index of the state of living protoplasm.

This criticism does not, of course, apply to larger protoplasmic bodies such as chromosomes (p. 346), plastids (p. 349), asters (pp. 346–8), and the like which can be seen during life.

Among the colloids, biologically the most important is the vast and varied class known as proteins. They are absolutely necessary to the building up of protoplasm. Dead protoplasm largely consists of them. They are not only essential for growth and repair of living substance, but they can be and are used as a source of energy and of heat, though the carbohydrates and fats share this function with them. Chemically the proteins are all built up of very large molecules. The modern chemistry of the proteins is based on the work of the great German chemist Emil Fischer (1852–1919) from 1882 onwards.

Emil Fischer demonstrated that proteins are built up of linkages or condensations of numbers of molecules of the substances known as *amino-acids*. The members of this very peculiar class are characterized by the presence in

A   B

FIG. 150. Diagrams to illustrate 'phases' of colloidal systems. The black is the solid, the white the liquid element. *A* represents an ordinary *hydrosol* in which very minute solid particles are freely movable. Such a preparation was made by Faraday early in the nineteenth century with extremely finely divided gold. His preparations still preserve their character. *B* represents a *gel*. It has an alveolar or honeycomb structure. The liquid drops are imprisoned by more or less solid walls. Such a preparation is given by a strong solution of gelatine when cooled to the ordinary jelly form.

each of one or more NH_2 ('amino') groups and one or more $COOH$ ('carboxyl') groups. The former gives them basic qualities, the latter acid. According as one or the other predominate, the amino-acid acts as a base or acid.

A favourite theory of the nature of protoplasm regards it as a mixture of amino-acids. These can become immeasurably complex by associating with each other in varyingly intimate ways. A modern mechanist view of life pictures all vital activity as a continuous change and interchange of the conditions and relations of amino-acids. These, it is held, act through local changes in the degree of viscosity. It is a fact, in accord with this view, that protoplasm does at times change its viscosity in a regular and systematic manner as, for instance, in the process of cell division. Many other phenomena of the living cell have been interpreted as due to changes in degree of viscosity.

Another aspect of protoplasmic activity is that of enzyme action. The word enzyme (Greek 'in yeast') was introduced by Willy Kühne (1878) to distinguish a class of organic substance which activates chemical change. Such enzymes can act on an indefinite amount of material without losing their activating power. The living body produces a large number of enzymes. These are remarkably specific in their action.

Within the protoplasm, though not of it, are numerous materials, the so-called 'food substances', which are often of relatively simple composition. Under this heading are to be included sugars and their derivatives, fats, and the 'reserve' proteins (p. 386).

The problem of the nature of protoplasm thus resolves itself into that of the nature of the matrix in which a vast variety of controlled reactions are taking place, and the ways in which the matrix can influence these reactions. The chemical processes at any moment within a single cell are of many and varied types. In spite of the smallness of cellular dimensions, these must somehow be spatially separated from one another.

A very striking fact concerning the reactions within the protoplasmic complex is the ease with which they occur. Synthesis of sugars is still not possible without cyanogen —a substance hitherto detected in nature only in comets —but in the green cell it is both customary and rapid. Many enzymes, capable of comparable processes, have been extracted from plants. When extracted, they will accelerate reactions which are otherwise excessively slow. Enzymes must be almost as numerous as the reactions which correspond to cellular activity. The cells of yeast have already yielded scores of enzymes. Of their origin very little is known.

FIG. 151. Drawings made for Goethe to illustrate the morphology of leaves. (See p. 219.)

A. A monstrous rose. The stamens are replaced by petal-like expansions. The pistil is represented by a continuation of the stem on which are petal-like and leaf-like structures together with others of a character intermediate between the two.

B. A seedling bean to illustrate the homology of cotyledons with other leaves. A plant, *Welwitschia*, has since been discovered which does not develop beyond this four-leaf stage.

C. Young chestnut branch just emerged from bud. It shows transition from scales, through leaf-like scales, to complete leaves.

FIG. 152. Jahannes Müller (1801–58)

XI

RELATIVITY OF FUNCTIONS

§ 1. *Johannes Müller* (1801–58) *and the Law of Specific Nerve Energies*

A NEW spirit was introduced into Biology in Germany by the gifted but short-lived Johannes Müller. He ranks among the greatest biologists of all time.

Müller was the son of a shoemaker. He showed early promise, studied Medicine, filled the chair of Anatomy and Physiology at Bonn, and was called thence to Berlin. His last twenty-five years were crowded with teaching and writing. The general direction of modern physiology and morphology has been largely determined by him. His character and bearing were of a dignified and lofty asceticism, and he approached his labours as a prophetic call.

In Müller's *Handbook of Physiology* (1834–40), the results of comparative anatomy, chemistry, and physics were for the first time systematically brought to bear on physiological problems. His researches on the chemistry of the animal body touch on those of Liebig at many points. His most important physiological work, however, dealt with the action and the mechanism of the senses. His explanation of colour sensations, his account of the internal ear, his description of the structure and action of the organs of voice, were all important starting points for modern physiological research.

The doctrine specially associated with Müller's name is the 'principle of specific nerve energies'. This teaches that the kind of sensation, following the stimulation of a sensory nerve, depends not on the mode of stimulation but on the nature of the sense organ with which the nerve is linked. Thus mechanical stimulation of the nerve of the eye produces luminous impressions and no other; stimulation of the nerve of hearing gives rise only to an auditory impulse, and so on.

Müller's doctrine of specific nerve energies is of such importance that it is well to consider some of its implications.

What do we know of the world in which we live? Only what our senses tell us. But how do our senses convey anything to us? That no man can answer. All we know is that certain external events somehow initiate specific disturbances in certain nerves, that these nerves convey the disturbances to the brain or central nervous system and that a sensation then arises. We have a glimmering of understanding of the mechanism by which the external event elicits a specific nerve impulse. But as to how that impulse becomes a sensation, which alone is what we experience, and how that experience can give rise to something which so alters a nerve or series of nerves that it induces action—of these things we are as ignorant as Aristotle. There are reasons to believe that ignorant we shall remain and that here is a veil which never can be rent by mortal man.

But consider further. Such external events as we experience, we know only by their action on our senses. Nevertheless from one and the same event we may receive utterly different sensations. Thus, an electric stimulation of the optic nerve will give rise to a visual sensation; the same stimulation of the olfactory nerve yields a sensation of smell; of the auditory nerve, a sensation of sound. Further, different events may give rise to the same order of sensation. Thus it matters not whether the optic nerve is stimulated electrically, thermally, or mechanically; in each case the sensation will be visual. If our optic nerve were grafted to our auditory organ and our auditory nerve to our optic organ, we should find ourselves transported to a world so strange that we cannot form the remotest conception of it. To beings with senses different from ours, the world would be utterly different.

The law of specific nerve energies is thus fundamental for our view as to the range of validity of scientific method, and indeed of experience as a whole. That law is a standing criticism of the 'common-sense' view that the world is just as we see it and that its contents, and particularly the living things in it, can be exactly and completely understood by us.

The later part of Müller's short life was given to morphological research. He was one of the founders of the modern science of embryology (p. 477). There is hardly a group in the animal kingdom on the knowledge of which he has not left his mark. Moreover, he worked out the microscopic anatomy of glandular and cartilaginous tissues, and thus prepared the ground for his pupil Schwann (pp. 336–8). Again, his introduction of the microscope into the scientific study of disease was carried further by another gifted pupil, Virchow (pp. 344–5). There are many other special departments which Müller initiated.

Müller was a convinced vitalist. He laid emphasis on the existence of something in the vital process that was and must remain insusceptible of mechanical explanation or physical measurement. This doctrine, however, occasionally misled him. Thus he held it impossible to measure the velocity of a nervous impulse. Yet that velocity was measured by his own pupil, Helmholtz, some ten years later.

§ 2. *Karl Ludwig* (1816–95) *and Mechanism*

Less universal as an intellect than Johannes Müller, but certainly as important as a teacher, was Karl Ludwig (1816–95). He was, perhaps, the most successful of all teachers of physiology. Much of the important physiological research between about 1870 and 1910 was the work of his pupils. He published little in his own name.

Among the many lines of investigation initiated by Ludwig, the most remarkable depended on his introduction of new technical methods. He had an exceptionally wide knowledge of the physical sciences, and he excelled in mechanical skill and in ingenuity of device. The most important invention or rather adaptation of Ludwig is the mechanically rotating drum or *kymograph* (Greek 'wave writer'). This instrument is now in general use for the permanent record of all sorts of continuous movement. The self-recording barometer is a familiar example. The kymograph led to a much wider application of the method of automatic record. Ludwig applied it specially to indicate the movements of breathing as well as the variations of the pressure of blood in the arteries. It has since been in constant use for recording vital movements and changes of many kinds, including the transient electrical disturbances associated with nerve impulses.

An instrument invented by Ludwig is the mercurial blood-gas pump. Its purpose is to separate from the blood the mixture of gases which it contains. This apparatus was indispensable for the investigation of the physiology of breathing by Ludwig's pupil Pflüger (pp. 409–10), and modifications of it are still in use.

Ludwig devoted much attention to secretion. Here especially his work gave support to mechanist views. He succeeded in showing that the process of secretion can be so transformed as to do external mechanical work. Thus secretion can be made to fit a more modern conception of energy.

Ludwig explained many other physiological events on a physical or chemical basis. His methods were followed and developed by his pupils, and he is largely responsible for the mechanistic view of the nature of life that was and is prevalent among the leading exponents of the science of animal physiology.

§ 3. *The Early French Experimental Physiologists. Claude Bernard* (1813–78)

French science in general has been more isolated than English, German, or Italian. At the beginning of the nineteenth century the supreme influence in French biology was Cuvier, of whom it has been said that 'he mastered not only his subject but also his opponents'. His destined successor was M. F. X. Bichat (1771–1802). He died at thirty-one, too early to develop his conclusions.

Bichat's analysis of the body into 'tissues' (p. 331) has survived only in name; in essence it has been rejected. Bichat's basic idea was that the life of the body is the resultant of the combined and adjusted lives of the various tissues of which it is upbuilt. He held to the conception of a definite vital force. The activities of living things he regarded as a result of the conflict of this force with physico-chemical forces. The latter have full play at death, but not till then. In giving a separate life to each tissue Bichat, though turning away from the vitalism of Stahl (pp. 365–6), and still more from the mechanism of Boerhaave (pp. 366–7), was really reviving the old conceptions of Paracelsus and Helmont.

Bichat's spiritual successor was François Magendie (1783–1855). Recognizing 'vital force', he regarded it as beyond the reach of experiment. But the physico-chemical elements of the body he considered to be fully within his field, and he threw himself with enthusiasm into their investigation. He indulged in a perfect orgy of experiments that were technically most skilful though seldom philosophically thought out. His most important result was the exact demonstration of the doctrine, almost contemporaneously with Charles Bell (p. 417) that the anterior nerve-roots of the spinal cord convey the impulses of movement, while the posterior convey impulses to the brain that are translated into sensation (Fig. 155).

The pupil and successor of Magendie was Claude Bernard. He was a severe and powerful thinker with an Olympian intellectual aloofness. Inheriting the technique and principles of Magendie, he utterly surpassed him in the deep and searching manner in which he devised his experiments.

One of Bernard's greatest discoveries, the elucidation of which occupied over ten years, was that the liver builds up from the nutriment brought to it by the blood, certain highly complex substances which it stores against future need. These substances, and notably that known as *glycogen*, it subsequently modifies for distribution to the body according to its requirements.

Now Wöhler in 1828 had synthetized urea from ammonia and cyanic acid (p. 378). Urea was recognized as a final degradation product formed by the body in breaking down substances derived from food. It had also become recognized that the source of bodily energy is this breaking-down process. Bernard, by his work on glycogen, demonstrated that the body not only can *break down* but also can *build up* complex chemical substances. This it does according to the various requirements of its various parts.

Bernard thus destroyed the conception, then still dominant, that the body could be regarded as a bundle of organs, each with its appropriate and separate functions. He introduced what we may call a 'physiological synthesis', a conception that the various forms of functional activity are interrelated and subordinate to the physiological needs of the body.

No less important, as bearing on this conception, was Bernard's work on the physiology of digestion. Up to his time, an elementary knowledge of the facts of digestion in the stomach constituted the whole of digestive physiology. While Bernard was working on the glycogenic function of the liver, one investigator had suggested that the secretion sent forth by the organ known as the 'pancreas', or sweetbread is capable of emulsifying fats. This, it was held, is its function in digestion. Another worker demonstrated that pancreatic juice acts also on starchy matter in food.

Bernard now showed that digestion in the stomach is 'only a preparatory act'. The pancreatic juice, passing into the intestine, emulsifies the fatty food substances as they leave the stomach, and splits them into fatty acids and glycerin. He showed further that the pancreatic juice has the power to convert insoluble starch into soluble sugar, and that it has a solvent action on such 'proteins' as have not been dissolved in the stomach.

A third great synthetic achievement of Bernard was his exposition of the manner of regulation of the blood-supply to the different parts of the body. This we now call the 'vaso-motor mechanism'. In 1840 the existence of muscle-fibres in the coats of the smaller arteries was discovered. Bernard showed that these small vessels contract and expand, thereby regulating the amount of blood supplied to the part to which they are distributed. This variation in calibre of the blood vessels is, he showed, associated with a complex nervous apparatus. The reactions of the apparatus depend upon a variety of circumstances in a variety of other organs. Thus he again provided an illustration of the close and complex interdependence of the various functions of the body upon each other.

Bernard's clear conception of the interdependence and reciprocal relations of the organic functions led him to a very valuable generalization. He perceived that the characteristic of living things, indeed the test of life, is the preservation of internal conditions despite external change.

'All the vital mechanisms,' he held, 'varied as they are, have only one *object*, that of preserving constant the conditions of life in the internal environment'.

The phrase stamps Bernard's belief that the living organism is something *sui generis*, something quite different from everything in nature that is not living. The organism has an *object*, and it uses a mechanism for attaining that object. Is this conception infinitely removed from that of Aristotle?

What is the internal environment of an organism? Bernard was thinking chiefly of the blood. But if we think of a part in terms of cells we see the environment of the cell made up of four main factors:

(*a*) The neighbouring cells and cell products.

(*b*) The substances that are brought to it by the blood.

(*c*) The substances that it throws off and that are removed from it by the blood.

(*d*) The nervous impulses that come to it, the physical nature of which is still but very little known.

To all of these we shall be returning.

§ 4. *Energetics*

A characteristic phenomenon of life is certainly movement. This is something over and above the molecular and atomic movement, or the mechanical change to which all matter is subject. Organic movement is of relatively large and visible masses, is relatively rapid, and is conditioned by an 'end' or 'purpose'. Whatever our view of the nature of life, when we seek to describe its exhibitions we are forced to use terms which imply such end. Aristotle recognized this. He invented the term *entelechy* or 'indwelling purposiveness' to describe the governing principle of the organism.

At first sight it might be thought that the vegetable kingdom is exempt from the rule that movement is characteristic of life. This, however, is not the case. Many fixed plants pass through a phase in which they move freely. Movement is characteristic of the sexual process in all plants. It is also characteristic of the phenomena of cell division (pp. 346–8). Moreover, in cells of fixed plants there is an active streaming of living protoplasm (p. 335, and Fig. 147). Movement, in fact, is as characteristic of living protoplasm as is metabolism, of which movement is at once an exhibition and a product. Movement, like metabolism, may, under some circumstances, be reduced to a minimum.

Physicists have worked out the general principles of the transformations of the various forms of energy. These are expressed most conveniently in terms of their heat equivalents. Landmarks in the history of the subject are the enunciation in 1824 of his famous 'axiom' of the dissipation of energy by Sadi Carnot (1796–1832); the determination in 1840–3 of the mechanical equivalent of heat by J. P. Joule (1819–89); and the formulation in 1845 of the law of the 'conservation of energy' by J. R. Mayer (1814–78) and Hermann Helmholtz (1821–94). Until these doctrines were available, there could be no true science of energetics, either for the biological or non-biological world.

A fundamental principle of the science of energetics is that, in any system, the capacity for doing mechanical work tends always to decrease, except in so far as it is increased from without. The principle is of great importance whether we consider mechanical systems or living bodies. It also has its implications in cosmology and philosophy.

The energy of living things is derived from chemical changes. These may be expressed as the oxidation of food and of the products of food. The energy of living things is utilized in various ways and need not and often does not pass through the form of heat.

A form of energy, associated with liquids, is that known as 'surface energy'. When a liquid, such as the protoplasm of an *Amoeba* is in contact with another liquid, such as water, the surface between them has the properties of a stretched film. It can therefore do work if the tension of this film be decreased, just as the stretched india rubber of a catapult can do work when released. Attempts to explain the action of protoplasm have been made on this basis.

For these attempts *Amoeba* has often been taken as the exemplar. The behaviour of this organism can be reduced to terms of protoplasmic flow. The flow in this and in other protozoa, as well as movements of protoplasm in higher beings, could, it was believed, be explained as due to changes of surface energy. These changes are in their turn due to chemical changes within the protoplasm, which alter or suddenly destroy the surface tension at different points.

The great mathematical physicist J. Clerk Maxwell (1831–79) compared the transactions of the material universe to 'a system of credit'. Each transaction consists of the transfer of so much credit or energy from one body to another. This act of transfer or payment is called *work*. There is no part of the universe accessible to us where the ledger has been closed. The transactions are ever going on. In the world of life these transactions are very active, are largely chemical in nature, and are constantly in opposite directions. The transactions in the living organism may be summed up in the word *metabolism*.

So far as has yet been discovered, the actions of living things accord with the general laws of physics and chemistry. They have not been shown to break the laws of energy. It is true that living things are peculiar systems which cannot be imitated, produced, or paralleled, and that some of their processes—and notably the progress from youth to age—are irreversible.[1] Nevertheless, so far as can be seen, living things obey the laws of thermodynamics and other rules of physics. Many biologists therefore refuse to recognize or discuss the existence of anything of the nature of a 'vital action'. In the present stage of research, the conception of 'vital action' has no place in the physiological laboratory, since physiologists are occupied in examining functions separately and not in examining organisms as wholes. None of these reasons need prevent the biological philosopher from forming his own views according to the evidence.

It must be remembered that the amount of energy consumed by the cell need have no relation to the external work done. Some cells, such as those of the muscles and of the secretory cells of the glands, do considerable external work; others, as the nerve-cells or the fertilized egg-cells, hardly any. But both types demand a good supply of oxygen. The work done by the nerve-cells and the egg-cell must, therefore, be internal. These cells need energy for the maintenance and development of their own structure. The demand and search for and the utilization of energy for the maintenance of internal structure is altogether characteristic of living things.

In this connexion it is appropriate to invoke a famous illustration by Clerk Maxwell. An established law of thermodynamics is that, in a system in which temperature and pressure are uniform, and so enclosed that neither change of volume nor passage of heat is permitted, no inequality of temperature or pressure can be produced without expenditure of work.

But if we look within the system and if we consider it not as a uniform mass but as a group of molecules, we find that these conditions of uniformity are not fulfilled.

---

[1] For some organisms and under some conditions this statement requires qualification.

'If we conceive', says Clerk Maxwell, 'a being whose faculties are so sharpened that he can follow every molecule in its course, such a being, whose attributes are still as essentially finite as our own, would be able to do what is impossible to us. The molecules in a vessel full of air at uniform temperature are moving with velocities by no means uniform. Now suppose that such a vessel is divided into two portions, A and B, by a division in which is a small hole, and that a being, who can see the individual molecules, opens and closes this hole, so as to allow only the swifter molecules to pass from A to B, and only the slower to pass from B to A. He will thus, without expenditure of work, raise the temperature of B and lower that of A in contradiction to the law that we have considered.'

'Clerk Maxwell's demon' (1871) may be regarded as the vitalistic crux. Does our conception of the behaviour of inorganic matter provide that uniformity in diversity which is demanded for any adequate conception of the stream of living things throughout the ages? Will it explain the appearance of purposiviness of the developing ovum? Can it account for consciousness and will? Does it cover all the phenomena of life and especially of mind? It seems to the author that an answer to these questions in the negative does not necessarily contradict or infringe the laws of thermodynamics. He therefore considers the something that corresponds nearly to the old 'vitalism' may still be regarded as a scientific approach to biological phenomena.

§ 5. *Muscular Action*

Among the higher animals, movement is very evident. It is always brought about by the action of muscles. These structures were long ago subjected to microscopical analysis. The process began with Leeuwenhoek and Stensen in the seventeenth century and continued through a long line to Bowman (1840), Kölliker (1851), and beyond. These observers have shown that muscles are essentially collections of innumerable minute elongated fibres which are definitely cellular and are living. Muscle-fibres are so constructed that during 'contraction' their ends, and consequently the ends of the muscle itself, are approximated (Fig. 93).

The word 'contraction' must not mislead the reader into supposing that either the muscle fibre or the muscle becomes smaller in total bulk. Such is not the case. No change of bulk takes place during muscular contraction. What happens within the muscle-fibre during contraction? This has been the subject of microscopic investigation since Leeuwenhoek. The optical difficulties are so great and their interpretation so various and so unsatisfactory that almost nothing was learned during the nineteenth century as to these structural changes. They are a matter of current discussion.

As regards chemical aspects of the question we are in better case. One important chemical conclusion in connexion with the 'contraction' of muscle is that the sugar— with which it is kept continuously supplied from the stored glycogen of the liver—is changed into another form of carbohydrate, lactic acid. The process is not essentially an oxidation, for it takes place freely in the absence of oxygen.

A second important conclusion is that *recovery* from contraction is an *active* chemical process. About one-fifth of the lactic acid produced during contraction is normally oxidized—'burnt up'—during recovery or relaxation. The remaining four-fifths of the lactic acid is reconverted into glycogen and stored for further use.

The history of our knowledge of lactic acid and its relation to muscle tissue is intricate. Lactic acid was originally isolated from sour milk in 1780 by the Swedish chemist Scheele (1742–86). Its formation in milk was ascribed to micro-organisms by Pasteur in 1857 in his epoch-making

Mémoire sur la fermentation appelée lactique. The same acid had been found in meat extract by Liebig in 1832 and synthetized in 1850. In the latter year Helmholtz showed that muscular contraction resulted in the formation of an acid substance. In 1867 Adolf Fick (1829–1901) of Cassel, a pupil of Ludwig, showed that muscular contraction was associated with carbohydrate metabolism. In 1871 it was proved that the muscles were the site of the chemical regulation of heat in the animal body. In the same year Hugo Kronecker (1839–1914), another pupil of Ludwig, demonstrated that the acid formed during muscular contraction is similar to that in sour milk. The role of the muscles in heat regulation was settled during the next decade in Ludwig's laboratory at Leipzig.

The action of the muscles accords with the ordinary laws of mechanics. In antiquity this was perceived with some clearness by Galen. He knew of such mechanical laws as those of the lever and the pulley set forth by Archimedes. Galen applied them to muscular action (Fig. 156).

In the seventeenth century, the impetus given to the science of mechanics by Galileo encouraged attempts in the same direction. The principle of the parallelogram of forces was applied to muscles in general by Niels Stensen (1667), and to the muscles of respiration in particular by John Mayow (1668). Giovanni Alfonso Borelli (1608–79) of Bologna, a pupil of Galileo, placed muscular mechanics on a firm scientific footing. His great *De Motu animalium* (Rome, 1680) seeks to treat all the movements of the body, both voluntary and involuntary, on a mechanical basis.

The mechanics of muscular movement has attracted the attention of human anatomists since Borelli. Mechanical principles have been invoked in the examination of the more rigid parts on which the muscles act. The skeleton, and its accessory structures, the cartilages, are formed in many respects along the lines best suited to meet the muscular stresses and strains that they have to bear. Thus the structure of the bones is often close to that which would have been given them had they been designed by a modern engineer (Fig. 153).

Ever since Galen, much has been made of this evidence

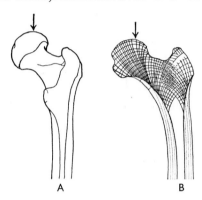

A   B

FIG. 153. *A*, Upper part of human thigh bone viewed from behind. The thigh bone has to support the entire weight of the body, the direction of action of which is shown by the arrow. *B*, Vertical section of the thigh bone. The lines show the direction of stresses and strains. These correspond with the development of bony tissue. This part of the thigh bone is thus constructed much as it would have been by an engineer with due regard to economy of material and the nature of the material with which he had to work.

for design in the body. The argument, however, cuts both ways, for there are many parts of the body the design of which is, in fact, very inadequate. Take for example, the eye. That organ is focused by alteration in the shape of the lens—the result of its elasticity. But the elasticity of the lens, like that of a piece of caoutchouc, gradually wears out. With advancing age we all have to take to spectacles. Had the lens been focused, like that of a telescope or microscope, by movement backward and forward, no such artificial device would have been needed. No good working optician would send out an instrument so clumsily designed as the human eye.

We may note that the investigation of muscular mechanics has been almost exclusively upon the human body. It has been in the main a tribute which the science of biology has paid to the art of surgery. To complete the conception of the body as a muscular mechanism, it would be necessary to treat the subject on an evolutionary basis. The 'entelechy' (=indwelling purpose) or vital force or whatever we call it that shapes our bodies has to work with imperfect instruments. Those instruments, both in their values and in their shortcomings, can be understood only through their evolutionary history. It is this historical element which marks them as vital products and separates them from those of the inorganic world.

§ 6. *Respiration as Combustion*

The importance of breathing to the animal organism cannot be missed. It provides the crude test of life, and we do not die until we 'breathe our last'. The word used for 'spirit' (Latin *spiro*, 'I breathe'), the theoretical basis of life, contains this idea in most languages (pp. 15–16). The importance of breathing was recognized in the physiological system of Galen (pp. 62–3). Real advance in the knowledge of the subject was, however, deferred until Chemistry became a science in the second half of the eighteenth century.

In the nineteenth century, the bare chemical facts of respiration being ascertained and their resemblance to those of combustion, the organism came to be compared to a steam engine. In such a machine, the draught of air leads directly to oxidation of the fuel. The accompaniment of oxidation is heat which is transformed into work. The parallel is now quite familiar. It was drawn in 1842 by the German physician J. R. Mayer, founder of the doctrine of Conservation of Energy (p. 401).

Arising out of the discussion of the relation of work to oxidation in the animal body were two fundamental questions:

(*a*) What is the essential nature of the combustible material?

(*b*) Where in the body is it consumed?

An answer to the former question was reached by Liebig and his colleagues. Food substances were classed as proteins, fats, and carbohydrates (p. 386). Of these only the first contains nitrogen. It reappears mainly in the urine and mainly as urea. The fats and carbohydrates are completely oxidized in the body into water and carbon dioxide. It thus became possible to calculate the amounts of the food substance absorbed from the ratios and amounts of the nitrogen compounds and carbon dioxide thrown off and the oxygen consumed.

The answer to the latter question was longer in coming. The perfecting of the mercurial blood-pump by Ludwig and his pupil E. F. W. Pflüger (1829–1910) was the main factor (p. 396). This instrument extracts from blood the contained gases, which can then be analysed. It was thus established that oxygen absorbed during breathing is taken up in the form of a chemical combination with the red colouring matter in the blood corpuscles. This colouring matter, known as *haemoglobin*, gives up part of its oxygen as the blood passes to the tissues, and then returns to the lung for a fresh charge. The discharge of oxygen in the tissues is accompanied, as is well known, with a change in the colour of the blood from scarlet to purple. Similarly carbon dioxide is taken up from the tissues and discharged in the lung.

For long there was doubt as to whether and how far the oxidation took place in the tissues or the blood. By

1872 Pflüger had shown that it took place only in the tissues, whither the oxygen is conveyed by the blood, as is also the nutriment. Thus the study of respiration was brought into relation with that of nutrition.

Before Pflüger, the relationship of respiration to animal heat was regarded as a blind mechanical process. Thus Liebig held that the rate of oxidation and consequent heat production is proportionate to the amount of food absorbed from the alimentary canal and oxygen introduced into the lung, just as the heat of a furnace is a function of the fuel consumed and the draught created. In support of this he showed that increase in the amount of nitrogenous food absorbed is followed by proportional increase in the amount of nitrogen excreted. This view, of animal respiration merely as combustion, is untenable, but here, for the moment, we leave it.

The conception of the nature of respiration in plants was only gradually brought into line with that of animals. By the end of the first third of the nineteenth century it was well known that green plants in light absorb carbon dioxide and give off oxygen, while in the dark they absorb oxygen and give off carbon dioxide (pp. 380–2). The two processes were held to be of comparable character and became known as 'day respiration' and 'night respiration'. Liebig went so far as to deny that plants respired at all in the proper sense.

Sachs in the 'sixties was the main agent in the dispersal of the confusion that thus arose. The major facts he established in connexion with plant respiration were four in number:

(*a*) Respiration in plants, as in animals, involves absorption of oxygen and oxidation of assimilated substances. This leads to the formation and exhalation of carbon dioxide. The process is masked but not suppressed by the process of starch formation and the accompanying assimilation of carbon dioxide and discharge of oxygen by the chlorophyll apparatus.

(*b*) The respiratory process of plants becomes more active with rapid growth or metabolic change.

(*c*) The streaming movements of protoplasm and other active exhibitions of movement by plants are decreased and finally cease if oxygen be withheld.

(*d*) If nutrition be suspended, weight is lost. This is ascribable to respiration.

Thus the respiratory process of plants was brought fairly into line with that of animals. The process, as with animals, was freely compared to the combustion in an engine. Moreover, the process of combustion was regarded by Sachs as essentially a consumption of fats and carbohydrates which were separately grouped in the hypothetical protoplasmic molecule.

§ 7. *Respiration as Relative to Physiological Needs*

During the 'seventies, largely owing to the work of Pflüger, it transpired that the view of respiration as combustion is inadequate to explain the phenomena. If no food be taken, the amount of nitrogen excreted falls greatly, but there is little fall in the amount of oxygen consumed. This means that fat and carbohydrate are being oxidized, while protein is being conserved. Such is, in general, the policy of the animal body.

The ratios in which protein, fat, and carbohydrate replace one another were next determined. These are proportional—within wide limits—to the energy which they liberate by oxidation within the body. Thus the rate of oxidation is not determined merely by the rate of food supply but is regulated according to the energy requirements of the body—and is moreover so regulated with great accuracy.

It should be remembered, however, that even with complete starvation of nitrogenous material, the consumption of nitrogen does not cease. It may fall, however, to the extraordinarily low level of about 8/1,000,000ths of the body weight per day (1911). This minute amount of nitrogen is obtained by the organism through the break-up of its own tissues.

Just as the rate of oxidation is not determined by the food supply, so also it is not determined by the supply of oxygen. Again, within wide limits, Pflüger found that the amount of oxygen consumed is altered neither by rise nor fall of oxygen concentration. It is determined by the requirements of the organism—and again the determination is very exact.

Similarly—within wide limits—the body determines its own temperature, whatever the temperature of the surrounding medium. Pflüger showed that if the surrounding temperature of a warm-blooded animal be lowered, the cold evokes, *via* the nervous system, a rise in heat production. The body temperature is maintained at the same level. This maintenance is thus a factor determining the energy requirements.

The progress of the physiology of plants was on essentially the same lines. Everywhere consumption has come to be seen as no blind process, but as related to the physiological needs of the organism as a whole. The process is less evident in the plant than in the animal only because the plant is less integrated under a directing mechanism, because, it is in short, less an 'individual'.

We might go through the whole gamut of functions and find them all subservient to the preservation of a norm. The science of physiology, like every other science, has many 'loose ends' which have not yet been traced to their connexion with the central theme. But wherever research has been long maintained upon a definite line, it leads back to this conception of bodily activity regulated in subservience to physiological needs. It is a theme that has been expounded by many writers, but by none with more learning and eloquence than by the English philosopher and physiologist, J. S. Haldane (1860–1936). The conception gives us the justification of biology as a separate science that can never become a department of Chemistry and Physics.

'Up to a certain point we can understand living organisms mechanically. We can, for instance, weigh and measure them and their parts, and investigate their mechanical and chemical properties. This enables us to predict certain points in their behaviour. But when we look more closely it becomes quite evident that the knowledge we gain hardly touches any fundamental physiological problem. We cannot escape from the relativity of the phenomena we are dealing with.

'The only way of real advance in biology lies in taking as our starting point, not the separated parts of an organism and its environment, but the whole organism in its actual relation to environment, and defining the parts and activities in this whole in terms implying their existing relationships to the other parts and activities. We can do this in virtue of the fundamental fact, which is the foundation of biological science, that the structural details, activities, and environment of organisms tend to be maintained. This maintenance is perfectly evident amid all the vicissitudes of a living organism and the constant apparent exchange of material between organism and environment. It is as if an organism always remembered its proper structure and activities; and in reproduction organic 'memory', as Hering (1870) figuratively called it, is transmitted from generation to generation in a manner for which facts hitherto observed in the inorganic world seem to present no analogy. We can discover and define more and more clearly by investigation these abiding details of structure and activity, distinguishing accidental appearances from what is really maintained; and this process of progressive definition is the work of the biological sciences.' (J. S. Haldane, 1931.)

§ 8. *Vital Activity of Plants and Animals Approximated*

During the eighteenth and early nineteenth centuries,

the study of plants had tended to become widely separated from that of animals. Thus a true science of 'Biology' in the sense of Lamarck and Treviranus (p. 298) became more and more distant. From about 1860 plant and animal biology drew nearer to each other. It is worth tabulating some of the reasons for this. The approximation has had and is having an important influence in moulding biological thought.

(*a*) The methods of respiration and of nutrition in plants and animals have been shown to be basically similar despite great apparent differences (pp. 410–11).

(*b*) The chlorophyll apparatus is specially characteristic of plants, but there are plants that have no chloroplasts and there are a few unquestionably animal species that have these structures. Moreover the chloroplasts can multiply independently of the organism in which they are contained, and the containing organism, in some cases, independently of the chloroplasts. It has even been shown that certain lower forms, which usually contain chloroplasts, can, under suitable conditions, thrive through an indefinite number of generations after their loss (*Euglena*).

Plants are divided from animals by a very indefinite line, and the chlorophyll apparatus is found on both sides of that line. Its extensive occurrence in the plant world is partly correlated with the stationary habit. The larger plants cannot go forth to gather their food, they fasten on that which comes to them.

(*c*) The conception of 'protoplasm' has had much influence. That conception was directed by the discovery that the living substance of plants is indistinguishable in general appearance and in many of its activities from that of animals. In all essentials the vital unit, the cell, was seen to be the same for plants and animals (pp. 334–41).

(*d*) The processes of cellular division (mitosis) and

sexual cell union (conjugation) are, on the whole, more easily traced in plants than in animals. Gradually they have been revealed as essentially identical in the two kingdoms (pp. 334–9).

(*e*) The discovery of the sex processes, especially of the flowerless plants (Cryptogams), by Hofmeister and his followers (pp. 525–32) led to the conception of an alternation of generations in plants (p. 539). This was freely compared to the well-known phenomena of alternation of generations in animals. Modern views of alternating 'haploid' and 'diploid' cell generations have made the analogy still closer.

(*f*) The general conception of a 'physiologia' of life as a whole, involving the interdependence of plants and animals, has been developing ever since Liebig. It expresses itself in such formulae as the 'nitrogen cycle' (p. 383).

(*g*) The evolutionary view gave a new conception to what may perhaps be called the *economics* of life. There arose the tendency to examine the relations of living things to each other. This led to the discussion of how communities, especially plant communities, adjust their forms and behaviour to factors, such as moisture, heat, light, nutriment, and so forth. Looking back on the history of modern botany, one can see much of the work on absorption, nutrition, metabolism, and growth, as well as on geographical distribution, converging on to this topic (pp. 382–5).

Ecology, as this aspect of biology is named, has been especially extended to a consideration of the mode of association of plants into communities and the relation of such communities to one another. The same approach may be used to animal communities in relation both to each other and to plant communities.

(*h*) The essential phenomena on which the modern doctrine of heredity is based have been drawn in equal

‑ .(41).‑ ‑

measure from the animal and the plant kingdom. Their similarity in the two is generally admitted (ch. xv).

All these movements began to make possible a true science of Biology, a science that should treat life as a whole. Of that science Darwin was one of the earliest and perhaps the ablest exponent. It is certain that from about 1860, the time of his greatest activity, most departments of biology begin to take on a new aspect.

The aspect in which plants differ most fundamentally from animals is in the matter of integration. As we ascend the animal series we encounter an ever increasing tendency for the general co-ordination of the organism by a nervous system. No such tendency is visible in plants. It is to this action of the nervous system that we now turn.

§ 9. *The Sensori-motor System*

In the case of plants, the investigation of vital functions has chiefly centred either round the cell, or round the organism as a member of a community. With animals the subject has become differently orientated. The developments of experimental physiology have here awarded a position of peculiar dominance to the nervous system.

By the time of Haller (pp. 370–2), the naked eye anatomy of the nervous system had become quite familiar. It was made more exact by several anatomists in the later eighteenth century, prominent among them being Vicq d'Azyr (p. 210).

A new physiological phase was opened by Luigi Galvani (1737–98) of Bologna, whose name is preserved in the 'galvanic battery'. Galvani showed (1791) that if a nerve be subjected to a certain type of stimulation, the muscle to which it leads will contract. The electrical nature of Galvani's method was revealed by Alessandro Volta (1745–1827) of Pavia, who is commemorated in our system of 'voltage'. In the fifth decade of the nineteenth century the Berlin Professor, Emil du Bois-Reymond (1818–96), pupil and successor of Johannes Müller, showed that a nervous impulse is always accompanied by the passage along the nerve of a change of electrical state. He and other investigators demonstrated moreover that chemical changes in the muscle accompany contraction. These chemical changes are initiated—'lit up' we might say,—by the nervous impulse.

Meantime Magendie (1783–1855) and Sir Charles Bell (1774–1842) had worked on the double spinal roots from which the segmental nerves of the body arise. They showed that of these roots, one conveys only sensory elements, while the other conveys only motor elements (Fig. 154). Magendie (p. 397) was the teacher of Bernard and founder of the first journal devoted to physiology (1821). Through this publication the investigation of the action of individual nerves became familiar to biologists of the nineteenth century.

In the first half of the nineteenth century there appeared many comparative studies on the nervous system. Cuvier based his classificatory system in part upon the nervous reactions (p. 232). He had himself explored the nervous system of Molluscs, Starfish, and Crustacea. His influence may be traced in a detailed work on the anatomy of the vertebrate nervous system prepared under the superintendence of Magendie (1825). The subject was extended by many less distinguished workers.

The solid researches on a variety of invertebrates by the 'Naturphilosoph' Carl Gustav Carus (1789–1869), Professor at Leipzig, and the very refined dissections of insects by the English amateur, George Newport (1803–54), should have drawn attention to the fact that the bodies of many invertebrate groups are no less dominated by their nervous organization than are those of the vertebrates. Yet Franz von Leydig's (1821–1905) important text-book of *Comparative Histology* (1857), while stressing invertebrate forms, did little to emphasize the dominance of the nervous system. Not until the appearance of T. H. Huxley's *Manual of the Anatomy of the Invertebrated Animals* (1877) was full stress laid on the complete nervous control in all except the lowest members of the animal series.

Despite the lead of Huxley, the nervous anatomy and especially the nervous physiology of the invertebrates remained neglected. Nevertheless the internal structure of the nervous system in the higher animals was investigated with very great detail. It was found to be almost inconceivably complex. The investigations were greatly helped by the introduction of new technique, at which we may now glance.

The early anatomists, from Vesalius onward, recognized that the central nervous system consists of two main parts—the grey and the white matter. It was perceived that in the brain the grey matter is mostly on the surface, while in the spinal cord it is mainly central in position.

Soon after the foundation of histology as a special science, it was observed that white matter consists of masses of enormous numbers of fibres, while grey matter contains also numerous cells. This was known to Purkinje (1835, pp. 334, 340) was and formally set forth by Henle (p. 450) in his *Allgemeine Anatomie* (1841). It was however, more than forty years before Kolliker proved that *all* nerve-fibres are nothing more than enormously elongated processes given off from nerve-cells with which they retain continuity (1889). These nerve-cells are to be found either in the central nervous system itself or in the various ganglia (Fig. 154).

This discovery of Kölliker gave meaning to a series of facts that had long been under observation. As far back as 1851 the half French, half English investigator,

Augustus Volney Waller (1816–70) had demonstrated that if a nerve be cut, only the fibres peripheral to the injury will degenerate. The reason is that they are cut off from the nerve-cells of which they are branches and whose metabolism they share. As the degenerative character of nerve-fibres is microscopically easily determinable, 'Wallerian degeneration' placed in the hands of investigators an experimental means of tracing the course of nerve-fibres through the nervous system.

Another system of conducting comparable investigations has also been developed. In 1873 the Pavia Professor, Camillo Golgi (1844–1926), introduced a method of depositing metallic salts within various cell structures. These deposits are very evident under the microscope. Ten years later Golgi succeeded in applying this method to the central nervous system. He showed that the cells in that system tend to resemble irregular polygons from the angles of which project processes, *axons*, the essential parts of the nerve-fibres which ultimately end in a complicated system of branches, *dendrites*. The dendrites form twig-like 'arborizations' round other dendrites linked to other cells. Ultimately the system ends in terminal cells associated with sense organs, glands, or muscles.

The method of Golgi has been developed for the sense organs as well as for muscles and glands by many investigators. Of these the most prominent has been the Swede, Magnus Gustav Retzius (1842–1919). In application to the structure of the central nervous system itself, remarkable work was done by Ramón y Cajal (1852–1934) of Madrid, almost the only important scientific investigator that Spain produced for at least three centuries.

§ 10. *Localization of Nervous Functions*

Such researches stamped upon biology the conception of an immensely complex series of systems for the transportof nervous impulses. These systems, if intact and working well, determine the activities, the reactions, the whole life of the organism. Most significant work has been done during the last half century in the light of this conception.

Very important for our view of the activity of the higher organisms was the localization of the functions of the central nervous system. During the nineteenth century work on this was entirely upon the higher vertebrates, having as ultimate aim the elucidation of the nervous and mental phenomena of man. Some of the earliest experimental evidence was adduced by the Parisian surgeon, Julien Legallois (1770–1814). He proved (1811) that an injury to a certain spot in the *medulla oblongata*, a structure that lies at the junction of brain and spinal cord, causes cessation of breathing. Here then lies a centre controlling this function.

The Viennese, Franz Joseph Gall (1758–1828), who had long been working on the functions of the brain, was in Paris when Legallois made his researches. Gall now produced in collaboration his monumental treatise on the brain and nervous system (1811). In it the structural distinction between the white and the grey matter is clearly set forth, as well as the excessively intricate character of the various tracts (in one of which Gall is still commemorated) which go to make up the brain and spinal cord. Gall's attempt to map out the surface of the brain, the 'cortex', according to function, gave rise to the pseudoscience of Phrenology.

Not till long after Gall's death was cortical localization again taken up by investigators of repute. In 1861, however, the French surgeon, Paul Broca (1824–80) demonstrated in a post-mortem room at Paris a relationship between loss of speech and injury to a definite area of the cortex. The work was soon carried into the experimental field.

In 1870 a very versatile naturalist, Gustav Fritsch (1838–91), and a student of insanity, Eduard Hitzig (1838–1907), working together at Berlin, found that stimulation of certain parts of the cortex regularly produced contraction of certain muscles. The Englishman, David Ferrier (1843–1928) followed this up by demon-

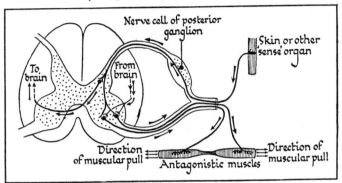

FIG. 154. Diagram to illustrate simplest form of Reflex. An afferent impression from a sense organ to the spinal cord may give rise to an afferent impulse by a purely intra-spinal process. This impulse may be of the nature of a complex and balanced muscular act involving a whole system of muscles, some of which may be antagonistic to each other. All this may take place not only unconsciously, without any intervention from the higher nerve centres in the brain, but even in an animal from which the brain has been removed. On the other hand, channels exist (and are indicated in the diagram) for passage of impressions to and impulses from higher centres. These higher centres in many cases control and modify the resulting muscular or other action to a greater or less degree.

strating that other areas of the cortex, which do not evoke muscular activity, are nevertheless functionally differentiated (1876).

Following on researches of this type, the surface of the brain was mapped in great detail. Special areas were associated with movements of different parts and different organs. Others were related to various forms of sensory discrimination such as sight, sense of position, weight, taste, and the like. Others involved the use of language.

Thus arose a naïvely simple psychological system that proved quite inadequate in the following century.

Influential in determining modern views of the action of the nervous system have been researches on the nature of 'reflex action', that is, non-voluntary movement in response to a sensory stimulus. The conception may be traced in physiological writings of the seventeenth and eighteenth centuries from Descartes onwards. The term 'reflex action' was invented (1833) by the English physiologist Marshall Hall (1790–1857).

By experiments, chiefly on cold-blooded vertebrates, Marshall Hall demonstrated in the spinal cord the nervous centres for a variety of reflexes. His papers, culminating in his *Memoirs on the Nervous System* (1837), mark a period in the history of nerve physiology. The study of reflexes has resulted in the localization of functions in the grey matter of the spinal cord much as with the grey matter of the cortex.

Since Hall's time there has been vast extension of the conception of reflexes. Besides the simple nervous arc (Fig. 154) there are more complex arcs which depend for their action on an elaborate mechanism. Beside 'spasmodic' events, as sneezing, coughing, scratching, &c., many of the ordinary acts of life, standing, walking, breathing, &c., are expressible as reflexes. The attempt has also been made by Pavlov and others to press even the 'instincts' into the same category, and the cortex has been shown to have the power of establishing new reflexes. The school that has been thus occupied sought to explain all the reactions and indeed the whole life of the higher organisms on a purely objective basis without reference to volitional elements.

§ 11. *Nervous Integration*

If the simple reflexes of animal bodies are tested, it will

be found that they clearly serve certain ends. Lightly touch the foot of a sleeping child and it will withdraw it. Tickle the ear of a cat and it will shake it. Exhibit savoury food to a hungry man and his digestive process will at once get to work, his mouth will 'water'. These instances might be multiplied an hundredfold. Such reflexes are admirably adapted to their ends. Many will continue in an animal in which the brain has been removed, provided that the spinal cord be still intact. Nevertheless, in the higher animals, and especially in man, the reflexes are controllable to a greater or less extent by the will.

But to leave the question at that would give a false idea of the extremely complex functions performed by the central nervous system. Thus, the spinal cord which, to the naked eye, is a longitudinal and little differentiated nervous mass, is, in fact, a collection of nerve-centres which have historically, both in the individual and in the race, been formed by the union of a series of separate segments. Each segment in this system governs certain functions or movements of the body, and the activity of each segment is related in various ways to the activity of the other segments. There is thus a very complex process of 'integration' which runs right through the nervous system.

Even during the nineteenth century, growing knowledge of the bodily functions of chemical and physical nature revealed that these activities are far more largely under nervous control and discipline than was formerly conceived to be possible. Thus, the main factor in the activity of any part is its blood-supply, but the blood-supply is determined, as Bernard showed (p. 397–99), by state of contraction of the vessels of supply, which are in the their turn under nervous control. Similar relations prevail for the state of nutrition of muscles, for the action of the sweat glands of the skin, for the mechanism of childbirth, and for a thousand bodily states. The regulation and control of all these events, processes, and states came to be called *integration*.

The investigation of nervous integration is especially associated with the name of Sir Charles Sherrington (1861–1955) of Oxford. The picture formed of the nervous apparatus is that of a machine in which some parts work spontaneously, automatically, and with complete uniformity; others, though mainly automatic, are susceptible of various degrees of alteration and adjustment; others need intermittent or constant attention and demand for their functioning fresh supplies of energy at longer or shorter intervals: while, finally, others have hardly yet taken a fixed form and are improvised as occasion demands. Thus the nervous system is a system of systems of every degree of independence.

These systems, each with a certain individuality of its own, date from every stage of evolution, the more ancient being, as a rule, the more automatic and the less dependent on other systems. The most ancient, the chemical messenger or 'so-called' 'hormonic' system, we share with the simplest living things which consist of but one cell. Very recent are the factors in the nervous system that are specially developed in man as contrasted with the higher apes. Such are those associated with the delicate co-ordination of sensory impressions and motor impulses involved in such acts as speaking, reading, writing, and the like. Each of these systems, high or low, ancient or recent, has its own place in the body. For some the exact controlling centre is demonstrable, and some lower systems can function without the aid of any other save those which control their nutrition. All this accords with the Aristotelian maxim, 'The first to live is the last to die'.[1]

----

[1] The phrase is not in Aristotle's work, but the sense is there and association with him is habitual.

§ 12. *Beginnings of Comparative Psychology*

When we watch the conduct of any being we ascribe its actions to sensations, emotions, motives, thoughts, comparable to those which we ourselves experience. In the case of human beings this forms an admirable working hypothesis and is indeed the theory which carries us through life. It is the basis of our laws, our customs, our very society. It has held human communities together through the ages. The only alternative doctrine is what philosophers call *solipsism* (Latin *solus* = 'alone' and *ipse* = 'self'), the view that self is the only object of real knowledge and, in the extreme, the only thing really existent. If any individual should push this view beyond a mere philosophic tenet and act as though he believed it, we should regard him as eccentric and antisocial. If he acted consistently with his belief, we should treat him as insane.

When we contemplate higher animals, dogs for instance, we ascribe to their minds some at least of the qualities of our own. The method works within limits. In practice a dog shows signs of knowing his master, manifests affection for him, exhibits hunger, passion, content, and so forth. The application of this 'commonsense' method is more or less possible according as the being with which we deal is more or less like ourselves. When, however, we pass to insects, in which the nervous system is organized on a wholly different though no less complex plan than our own, or to protozoa which exhibit no separate nervous system, the task becomes impossibly hard. The 'subjective' method that reads our minds into other beings is here merely misleading.

With the advent of evolutionary views naturalists turned to comparative studies. Comparative anatomy, as being the easiest, was the first cultivated. Comparative physiology came much later into the field. A consideration of comparative psychology on a scientific, that is on an objective, basis was perforce deferred until physiology could provide an adequate technique. Since the mind of beings other than oneself can only be known through their conduct, some exact, some physiological method of investigating conduct had to be devised.

By a truly extraordinary mental ellipsis, a school of thinkers now developed the conception that since mind can only be known through behaviour, therefore mind has no existence. As though we should say that there was no Phidias since we can know him only through his sculptures! From the philosophical point of view this extreme 'behaviourist' position seems hardly worthy of answer, for the one thing we *do* know about ourselves is that we think. All else, even our own behaviour, is inference. We can, however, discuss behaviouristic interpretation from the purely scientific point of view that does not go behind phenomena except to other phenomena. That is to say, we may adopt the scientific method of limiting our field of study and may decide to consider only the phenomena of behaviour without forming any hypothesis concerning its motive. This is a perfectly sound attitude, provided that its necessary limitation is recognized. From this phenomenological point of view, behaviourism is a necessary corollary to biological 'mechanism'. Nevertheless, such behaviourism is not, as some behaviourists seem to think, inconsistent with vitalism. There are certainly philosophical reasons for rejecting 'behaviourism', but rejection of 'mechanism' is not one of them.

The extreme mechanist doctrines of some of the early followers of Darwin created a great stir in their day. They are, however, of little importance for the history of science, since they were not based on experimental evidence. But a new period was opened in the early 'nineties by two investigators who were philosophically poles asunder. These were C. Lloyd Morgan (1852–1936), of Bristol, and Jacques Loeb (1859–1924), first of Strasbourg and later of the University of California.

The basis of Lloyd Morgan's comparative psychology is, in effect, a return to the essential principles of science. These, it is said, were laid down in the fourteenth century by the great Franciscan opponent of the Papacy, William of Ockham (1270–1349), to whom is ascribed the dictum 'entities are not to be multiplied beyond necessity'. This 'law of parsimony', sometimes called 'Ockham's razor', is fundamental for the logic of science. To this principle Lloyd Morgan was appealing when he wrote:

'In no case may we interpret an action as the outcome of the exercise of a higher psychical faculty, if it can be interpreted as the outcome of the exercise of one which stands lower in the psychological scale' (1893).

But while this was being written, experiments were in progress, the interpretation of which reduced the lower psychological scale to the level of physico-chemical rules. Indeed *chemotropism* in the case of spermatozoids of ferns was already well known. Moreover the American school, led by H. S. Jennings (1868–   ), was making similar observations on protozoa, which showed that certain behaviour of these creatures also is determinable by physical and chemical means. The conception was developed in great detail by Loeb and his followers and was extended to a large variety of animals, chiefly invertebrates. These investigators showed that many actions and attitudes that might be thought 'voluntary' in the human sense are closely controlled by physical conditions. The contemporary work of the isolated French amateur, Henri Fabre (1823–1915), also demonstrated that a multitude of seemingly purposeful acts of insects are performed without reference to even the most elementary form of reason, and even without regard to the interests of the performer or of the species.

In the meantime Lloyd Morgan had developed his conceptions of the nature of the psychical faculties. He distinguished three levels of mental activity. Of these the lowest *sentience*, a vaguely conscious state, is possessed by all animals, and is described in man by the term 'affective'. Above this stands *effective consciousness* which we must suppose possessed by such creatures as can profit by experience. There is evidence to suggest that this too is present in all animals, even in protozoa, though it is more obvious in the higher orders. Lastly there is the third level, *self-consciousness*, which can be present only in a small number of the higher creatures.

This scheme, with various modifications and qualifications, still holds the field with most thinkers, and has provided the outline for the doctrine of 'emergent evolution'. The basic criticism to which it is subject from the point of view of science is that its ultimate criterion, self-consciousness, is not susceptible of measurement or expression in terms other than itself. This is merely to say that self-consciousness is not susceptible of scientific analysis, which is a very different thing to proving that it has no existence. The latter view was, however, taken by some extremists attracted by doctrines arising from work on 'conditioned reflexes'. Such were some of the disciples of Pavlov.

§ 13. *Conditioned Reflexes*

Many of the simple reflexes, or automatic movements in response to stimulus from the environment, appear very early in life, and a number are present before birth. The animal exhibits these *innate* reflexes, without any regard to its individual experience as an individual in the world. The absence or exaggeration of reflexes, normally innate or early acquired, implies some injury or disease of the nervous system. A number of these reflexes are, in fact, of value to medical men in testing for nervous disorders.

Beside these innate reflexes, common to the species, there are also a number of acquired or *conditioned* reflexes, the development of which depends upon the history of each particular individual. They are called 'conditioned', since their nature depends on the conditions of their establishment. It will, however, be found impossible to draw an absolute distinction between some conditioned reflexes and some reflexes that have been acquired very recently in evolutionary history.

The study of conditioned reflexes is the work of the last twenty years and is associated closely with the name of the Russian physiologist Ivan Pavlov (1849–1930). This distinguished investigator had developed a method of inquiry into the mechanism of the highest nerve centres without appealing to the consciousness of the organism. The technique of investigation of conditioned reflexes is largely determined by Pavlov's method of measuring accurately the flow of saliva in dogs—a somewhat slender basis, it must be confessed, on which to found a new philosophy and psychology. If a hungry dog is shown food, his mouth waters—the saliva flow is increased. If a bell be rung each time food be given, he comes at last to secrete saliva at the sound of the bell, irrespective of food. The showing of the food is the *unconditioned stimulus*, the ringing of the bell is the *conditioned stimulus*.

Our experience with the dog accords in fact with our own and is, in itself, nothing new. Our own dinner bell arouses us to the fact that we have an appetite. The new feature is the manner, at once exact and objective, in which these reflexes can be studied. The chief difficulty in their study consists, as in many scientific experiments, in limiting the number of 'variables'. If we are investigating, let us say, the effect of pressure on a gas, we must be sure that the temperature remains constant. If we are investigating the effects of temperature, we must take steps to secure uniformity of pressure. Such precautions are especially difficult in studying the acquired reflexes. Pavlov and his many followers took immense pains to eliminate the sources of error due to 'multiplicity of variables', which in this case means multiplicity of stimuli. Their work, like much successful scientific work, has been mainly concerned with the perfection of methods, It is, however, only with the results that we are here concerned. Pavlov had shown that each conditioned reflex is associated with a definite part of the cortex of the brain. If that part is removed, the reflex disappears.

An important aspect of the conditioned reflexes is the manner of their inhibition. Any sort of disturbance, curiosity, anxiety, noise, fear, change of light or temperature may interfere. How well we know this with children at their meals! The removal of these interferences had been Pavlov's chief task. Special laboratories were so constructed that the experimental animals were under absolutely uniform conditions and did not even see the experimenter. Thus conditioned reflexes have been experimentally established in relation to many organs, the pupil of the eye, the movements of so-called voluntary muscles, even the beat of the heart and the process of breathing. The ease with which the various kinds of conditioned reflexes can be established is very various. Moreover, the strength or weakness of the reflex depends largely on the intensity and duration of the requisite stimulus.

Pavlov's point of view was that, through the medium of conditioned reflexes, any part of the nervous system may be coupled up with any other part. When we consider that the nervous system contains many millions of nerve-cells we can realize the truth that no creature uses more than a small fraction of its cerebral powers. Pavlov himself held that however complex mental activity may be, it is in essence compounded of successions of acquired and modifiable connexions of one neurone with another. Pavlov recognized in the brain no other, no higher function than this. He accepted no *science* of psychology beyond this development of neurology. For him, 'freedom', 'curiosity', 'purpose', were but conditioned reflexes, and religion but one of the higher conditioned reflexes. Such a view is an extreme development of mechanist doctrine. It is inconsistent with vitalistic theory and is equally inconsistent with any of the philosophical doctrines, such as emergent evolution or holism, which had been elaborated, during the nineteenth century, as attempts to cover the phenomena of life.

§ 14. *Mind as Conditioning Life*

During the nineteenth century there was an enormous extension of scientific interest in the analytical study of animal function by means of physical experiment.

The exponents of this science of physiology have applied themselves mainly to the higher animals. They have devoted themselves almost exclusively to an examination of the parts or functions in the adult or developed state. The results have been portentous in bulk, complexity, and interest. Fundamental for the details of scientific medicine, they are less useful in helping us to a conception of the organism as a whole.

The animal body is, as it were, a vast and complex maze. The physiologist enters it and he wanders there as long as he will. But his close and detailed report on its paths and walls helps but little toward the exposition of the design as a whole. A bird's-eye view would be more productive. Such a glimpse, though blurred and distant, has been better obtained by the 'general physiologist', who has devoted himself to the examination of protoplasm and especially of the life, movement, and habits of the lower organisms; by the embryologist who has sought out the beginnings of the organism; and by the biochemist who has analysed the physical and chemical character of the products of vital activity.

The physiologist, in his special duties, is well nigh bound to consider isolated functions. He selects respiration, nutrition, muscular movement, the action of the nervous system, or the like. But the performance of each of the functions of each of the systems is inextricably linked with the performance of the functions of all the other systems. We are always looking for metaphors in which to express our idea of life, for our language is inadequate for all its complexities. Life is a labyrinth. But a labyrinth is a static thing and life is not static. Life is a dance, a very elaborate and complex dance! The physiologist cannot consider the dance as a whole. That is beyond his experimental power. Rather he isolates a particular corner or a particular figure. His conception of the dance, as thus derived, is imperfect in itself and, moreover, in obtaining it he has disturbed the very pattern of the dance. The shortcoming of his method becomes fairly evident when he seeks to relate his corner to another in a far distant part of the dance.

Moreover, even should he seek to treat the organism as a whole, he is still almost bound to consider it as an 'individual', complete and separate in itself, shut off from its environment and its history, born, as was Minerva, armed and fully equipped from the head of Jove. But in fact living beings are not so. There is every degree of dependence on their fellows among organisms. 'Individuality' comes into prominence only in the more differentiated groups. The term is almost inapplicable to plants in which physiology is, in effect, the physiology of a community and is a study not far, in its conceptions, from that of bionomics. The very idea of the 'individual' involves an historical record which the science of physiology has hitherto almost ignored.

The special development and isolation of the science of animal physiology have been largely conditioned by its relation to medical studies. The seminal biological ideas of the nineteenth century were unquestionably in connexion with Evolution, Biogenesis, Heredity, and the Cell Theory. The work of those who professed physiology was but little directed by any of these themes except the last. The illustration of Evolution has been provided almost entirely by the field naturalist and the comparative anatomist. The phenomena accompanying the beginnings of organisms have been illuminated by the cytologist and the experimental embryologist. The activities of the cell and its protoplasm have been elucidated mainly by botanists. The study of Heredity and Sex has demanded specially trained naturalists. 'Physiology' retained a peculiarly lonely position among the biological sciences during the nineteenth century.

Since the dawn of the twentieth century there has certainly a been breakdown of this isolation. Physiology alone is, however, of its nature incapable of presenting any picture of the mode of action of the organism as a whole, though modern doctrines of the workings of the nervous system have given some explanation of many forms of animal behaviour.

Yet the functions of the nervous system, like those of other systems, are relative to the other functions of the body. Not only is respiration, for example, regulated by the nervous system, but the nervous system itself is regulated by the character of the respiration. Raise the amount of carbon dioxide in the blood, and the respiratory movements are first stimulated and finally diminished *via* action on the respiratory centres. It would be possible to show that the same is true of any system or part of a system in relation to any other. What picture, then, can physiological processes give us of the interrelated complex of activities that we can call an organism?

The physiologist has found that his science can be best prosecuted on the higher animals because the functions of these creatures are best differentiated. If he wishes to study movement, respiration, nutrition, nervous action, he finds in the higher animals separate organs devoted to these processes. Such organs he cannot so easily or cannot at all find in the lower organisms. In the lowest of all, the Protozoa, every process is carried on in a single cell.

But the most distinctly and clearly developed characteristics of the highest animals are their mental powers. To discuss these in the mechanistic nomenclature adopted by physiology is mere contradiction in terms. The one thing that we really know is our own thoughts, and external things—including the science of physiology—we know only in relation to these. How then can external things be said in any sense 'to explain' our thoughts? It is more intelligible to invert the process and to say that phenomena —including those of physiology—are parts of our thinking, than to say that our thinking can be built up of phenomena.

But if we emphasize the conception of science as dealing with phenomena—'things which appear'—we reach a *modus vivendi* both for a conception of mind and for the findings of science. Having agreed that science shall deal only with phenomena, we expressly exclude our own mind, which is not an appearance at all but that to which appearances happen. Science must keep to the phenomenal level. On that level she must prosecute physiological study. But no amount of that study will truly represent an entity in which is any element of mind. Is that element of mind found in other organisms than myself? Unless the solipsist view (p. 425) be taken, this question must be answered in the affirmative.

FIG. 155. The first attempt to interpret graphically in mechanical terms the action of muscles and ligaments (Vesalius, 1543).

Fig. 156. Louis Pasteur (1822–95)

XII

BIOGENESIS AND ITS IMPLICATIONS

§ 1. *Early Ideas of Infection and Spontaneous Generation*

MANY savages think that the properties of any kind of matter can be transferred to neighbouring matter. They see that decaying animal or vegetable substance corrupts neighbouring material, or again, that hot or cold things make neighbouring things hot or cold. Thus for them corruption and incorruption, heat and cold, are 'contagious', that is, pass from object to object by reason of contact. Generalizing, they think many other things contagious—a man's cleanliness and uncleanliness, his holiness and unholiness, his power and his weakness. This belief is described by anthropologists as 'sympathetic magic'. To the savage, the passage of disease from person to person is a normal part of a world full of such magic.

Man rises in the scale of civilization. Religion crystallizes out from the mass of vague beliefs. Gods now govern the fate of man. An epidemic outbreak in the tribe is a most impressive event. The gods have dealt the blow. It is the unknown that is most dreaded, and 'the pestilence that walketh in darkness' has ever been more feared than 'the arrow that flieth by day'. The Bible strikes a human note when the appearance of the very Angel of Death at the threshing-floor of Araunah comes as a relief to the stricken King of Israel.

Greek, medieval, and renaissance physicians studied epidemics. The Greek man of science remained very sceptical concerning the infectious nature of these diseases, and indeed concerning infection in general, for that suspect conception came from a magic-ridden world. It was in the later Middle Ages that physicians, acting on promptings from the book of Leviticus, came to realize that epidemics are simply outbreaks of disease of a peculiarly infectious type. Such disease, it was observed, is almost always associated with fever. Theories were formed to explain the evident association of infection and fever. A common view was that these conditions were allied to the phenomena of fermentation or leavening. We still speak of 'zymotic' diseases (Greek *zymē*, 'leaven').

Ideas on the subject of infection were given some exactness by the Veronese physician, Hieronimo Fracastoro (1484–1553). He did much to separate the different types of fever, and he sought to classify the infectious diseases. He was an ardent follower of Lucretius (*c.* 98–55 B.C.), the Roman atomic philosopher, whose work had been recently rediscovered. Following his master, Fracastoro expounded the idea that infection of all kinds, including fermentation, is the work of minute 'seeds' (*seminaria*) or germs. This view remained more or less latent in men's minds until Pasteur's time.

The really salient fact in the seventeenth century conception of zymotic disease was, however, not its 'seminal' origin but its relation to fermentation. Fermentation and zymotic disease have indeed much in common. Both are associated with rise of temperature, both are infectious, both can be propagated indefinitely without diminution of the original focus. Moreover, it seemed to early observers, who interpreted the facts wrongly, that both were capable of being 'spontaneously generated', that is, of arising without the existence of any predecessor like to themselves, of being, in fact, without parents.

Now this power of spontaneous generation suggested to the men of the age that both fermentation and infectious disease partake of the nature of life. Certain living things, it was held, could, under certain conditions, be spontaneously generated. Moreover, all living things have this power of indefinite propagation. Thus the early scientific study of infectious disease came to turn on the current conceptions of spontaneous generation and of fermentation.

The misleading doctrine of spontaneous generation is often laid at the door of Aristotle. It so fell out that when that doctrine was first questioned, the authority of Aristotle—or rather the contemporary misunderstanding of him—was a very real obstacle to scientific advance. It is also true that Aristotle believed in spontaneous generation and gave it a place in his biological scheme (p. 43). But his error was shared by every naturalist until the seventeenth century. Indeed it is hard to see how these men, with the knowledge at their disposal, could take any other view. It is but just, therefore, that the father of biology should be acquitted of any special fault in misleading his successors in this matter.

With the advent of the general use of the microscope in the second half of the seventeenth century, new tendencies set in. On the one hand, exploration of minute life showed many cases of alleged spontaneous generation to have been falsely interpreted. Thus plant galls had been regarded as spontaneously generated. Nevertheless, Malpighi showed that these curious growths are related to insect larvae (p. 157). On the other hand, the microscope revealed minute organisms which seemed to appear out of nothing. Thus Leeuwenhoek saw excessively small creatures in infusions of hay and other substance (p. 169). Such infusions, perfectly clear when first prepared, become in a few days or even hours cloudy with actively moving microscopic forms. These seemed 'spontaneously generated'.

§ 2   *Redi* (1621–97), *Needham* (1713–81), *and Spallan-zani* (1729–99)

The first scientific treatment of the question was made toward the end of the seventeenth century. Francesco Redi of Florence, at once poet, antiquary, physician, and naturalist, proved by experiment that, if living causes be excluded, no living things arise. Using no microscope, his work failed to convince those who based their belief in spontaneous generation on microscopic appearances. Nevertheless, Redi's experiments are faultless so far as they go, and his arguments unanswerable so far as they apply to flesh-eating flies. He checked his operations by what are now called 'controls'.

In his *Esperienze intorno alla generazione degli inettis* ('Observations on the generation of insects', Florence, 1668), Redi tells that he

'began to believe that all worms found in meat were derived from flies, and not from putrefaction. I was confirmed by observing that, before the meat became wormy, there hovered over it flies of that very kind that later bred in it. Belief unconfirmed by experiment is vain. Therefore I put a [dead] snake, some fish, and a slice of veal in four large, wide-mouthed flasks. These I closed and sealed. Then I filled the same number of flasks in the same way leaving them open. Flies were seen constantly entering and leaving the open flasks. The meat and the fish in them became wormy. In the closed flasks were no worms, though the contents were now putrid and stinking. Outside, on the covers of the closed flasks a few maggots eagerly sought some crevice of entry.

'Thus the flesh of dead animals cannot engender worms unless the eggs of the living be deposited therein.

'Since air had been excluded from the closed flasks, I made a new experiment to exclude all doubt. I put meat and fish in a vase covered with gauze. For further protection against flies, I placed it in a gauze-covered frame. I never saw any worms in the meat, though there were many on the frame, and flies, ever and anon, lit on the outer gauze and deposited their worms there.' [*Slightly abbreviated.*]

It is odd that, despite these admirable experiments, Redi continued to believe that gall insects were spontaneously generated. This subject was taken up by Vallisnieri (1661–1730) who again demonstrated that the larvae in galls originate in eggs deposited in the plants (1700). Vallisnieri compared the process of gall formation, as well as infection of plants by aphides, to the transmission of disease. Other investigators showed that fleas and lice— to this day popularly thought to be 'bred by dirt'—are, in fact, bred only by parents like themselves.

The later history of the doctrine of spontaneous generation was long and stormy. The battle raged back and forth. On the one hand, it was shown that by screening, boiling, or chemically treating a medium, the appearance of minute organisms was either delayed or altogether avoided. On the other hand, cases were repeatedly adduced in which organisms did so appear, despite all precautions.

About the middle of the eighteenth century the controversy reached an important stage in a discussion between John Turberville Needham and Lazzaro Spallanzani. This debate was virtually repeated a century later between Pasteur and his opponents.

Needham was an English Catholic priest, who made some interesting contributions to science. In 1748 he published what was in effect a repetition, made in conjunction with Buffon, of the experiments of Redi of the previous century. Needham's treatment of the problem was, however, more refined than Redi's since he aimed at excluding even the most minute microscopic organisms.

Needham came to the opposite conclusion for microscopic organisms to that reached by Redi for maggots. He boiled mutton broth and placed it in a corked phial so well closed with mastic that it was 'as good as hermetically sealed'. He sought thus to exclude the exterior air, that it might not be said that animalcules arose from air-borne germs. The flask was opened after a few days. It was swarming with animalcules. 'The very first drop yielded multitudes perfectly formed, animated and actively moving'. He followed this up with observations on infusions of other animal and vegetable substances. All gave the same phenomena with little variation.

Needham was answered by Lazzaro Spallanzani of Scandiano, an investigator of very great experimental skill. Spallanzani, who was in orders, bore the title 'Abbot' (Abate), and was professor successively at Reggio, Modena, and Pavia. He made important contributions to several departments of biology. Some of his experiments on spontaneous generation bore a close resemblance to those of Pasteur.

Spallanzani was dealing with very minute organisms. He perceived that the early stages must be minuter still. Thus the problem resolved itself into excluding forms of life so small as to be beyond the reach of his microscopic vision. In his *Saggio di osservazioni microscopiche relative al sistema della generazione dei signori Needham e Buffon* (Modena, 1767), he says:

'I sought to discover whether long boiling would injure or prevent the production of animalcules in infusions. I prepared infusions with eleven varieties of seeds, boiled for half an hour. The vessels were loosely stopped with corks. After eight days I examined the infusions microscopically. In all there were animalcules, but of differing species. Therefore long boiling does not of itself prevent their production.' [*Abbreviated.*]

Spallanzani then tried excluding air. He placed infusions in five series of flasks. One series was left open. The other four series were first sealed with the blow-pipe and then raised to the boiling-point. The duration of boiling for these four series was respectively a half-minute, a minute, a minute and a half, and two minutes. The flasks were left for two days. The open series swarmed with microscopic organisms. Of the sealed series, that boiled for a half-minute contained smaller organisms, while the remainder contained only excessively minute forms.

With this clue Spallanzani now raised the time of exposure to heat. At last he found that if a sealed flask be subjected to the temperature of boiling water for between one-half and three-quarters of an hour, no development of organisms ensued so long as the flask remained sealed.

The experiments of Needham and Spallanzani were much discussed. Some thought the flaw in Spallanzani's method lay in the necessity of air for the development of organisms. Heating the air in the sealed flasks, they considered, had spoiled it for purposes of generation. Spontaneous generation in the presence of unspoiled air seemed still to be a possibility.

Several workers in the early nineteenth century repeated Spallanzani's experiments with improved technique. Thus Schwann (pp. 336–9) also proved that no putrefaction ensues in fully sterilized broth to which none but air previously heated can have access. He was convinced, however, that such heated air rests unchanged for vital purposes since he was able to show that it will serve for animal respiration (1836–7).

The weak point of such experiments was that they sought to prove a universal negative. One positive result, and the rest became worthless. Thus some still remained convinced of the reality of spontaneous generation. Of these the most prominent was the naturalist. Félix Archimède Pouchet (1800–72). In 1859 he brought out a large work, *Hétérogénie*—his name for spontaneous generation. Its elaborate details of cases and conditions of the supposed process are now mere curiosities. The fallacies inherent in his methods were fully demonstrated by Pasteur.

§ 3. *Pasteur* (1822–95) *on Fermentation*

Pasteur was by training and profession a chemist and especially a physical chemist. When he began his first great series of researches, the question of the optical action of crystals was much before the scientific world. His first scientific successes were obtained on the optical action and varieties of the tartrates. These are produced in the course of fermentation, and his researches carried him from crystalline substances to the ferments which make them, from experimental physics to great industrial problems. Much of his subsequent work had practical issues, but he never faltered in the purity of his scientific enthusiasm.

Pasteur was, however, by no means the first to demonstrate that fermentation is associated with organisms. Thus in 1837 Schwann revealed that yeast consists essentially of a mass of plant-like beings. He showed that the presence of these is a condition of the acoholic fermentation of sugar.

In 1854 Pasteur became professor of chemistry at Lille, a great industrial centre. There he set himself to that study of fermentation which bore him on to the study of disease. He made the first important step in 1857, when he gave to a local society a paper on the souring of milk. This *Mémoire sur la fermentation appelée lactique* is a landmark in the history of Biology. He had discovered in sour milk a substance that proved to be a ferment that produces lactic acid. He isolated the ferment, added it experimentally to milk and watched it act. It was, in fact, a mass of bacteria of a species nowadays quite familiar.

At this stage Pasteur was called to Paris, where the rest of his scientific life was passed. He had hitherto been primarily a chemist. His work on sour milk led him to acquire skill with the microscope. His development of microscopic technique made him one of the founders of bacteriology, and indeed one of the great biologists of all time.

At this period the current views of fermentation were controlled by Liebig (pp. 377–9). That eminent chemist regarded fermentation as a peculiarity of organic matter, and the fermentational change as 'of the nature of *death*'. Pasteur's experiments were to reveal the essential part played in it by *living* organisms. From the first, the teaching of the two men was opposed. For Pasteur, fermentation was a vital phenomenon and demanded the presence of living organisms.

From the souring of milk Pasteur turned again to the investigation of the most familiar fermentation, that of sugar into alcohol in the making of wine. He found that for the formation of wine, essential elements were yeasts or moulds which he detected on the skins of the ripening grapes. If these were absent, deteriorated or in wrong proportions, the fermentation was either arrested or on abnormal lines. Pasteur sought to trace the life-history of these organisms. He perceived that their mode of growth was controlled by their conditions of life.

Thus Pasteur passed to the diseases and defects of wine. Most of these, he found, were due to abnormal ferments. Study of the ferments soon suggested that there is a great variety of organisms associated with fermentation. Not all are of the type of yeasts. Some are quite different in form. These are yet smaller and grow often in chains, and are bacteria. Pasteur's investigations were summarized in his historic *Études sur le vin* (Paris, 1866).

Pasteur was now quite convinced that fermentation,

decomposition, and putrefaction are all vital processes. It is life that brings about these changes. Innumerable minute organisms are perceptible in decomposing material. They are the cause of the process of decay. The germs of them are brought by the air.

§ 4. *Biogenesis versus Abiogenesis*

A great difficulty was here encountered in the demand that Pasteur's views made on the germ-bearing capacity of the air. His critics were not slow to avail themselves of this. According to Pasteur, they said, the air must be one solid mass of germs! Some held that the organisms found in fermenting and decomposing matter were the result, not the cause, of these processes. The old idea of spontaneous generation still had its advocates.

By 1859—the year of publication of the *Origin of Species*—Pasteur was engaged in acute controversy on the question of the 'Origin of Life'. Discussion turned round what was then regarded as the lowest form of life, the bacterium. Were bacteria ever spontaneously generated, or were they not? If a supposedly sterilized flask of broth 'went bad', and bacteria made their appearance in it, was it certain that they had come from without? May they not have generated spontaneously within the broth? Must all life, even lives so low, so minute as these, be the work of life? Life must begin somewhere; then why not here, at this lowest stage?

Pasteur, like his predecessor Spallanzani, had now the task of proving a universal negative. The task is impossible in formal logic, but science is not formal logic. In a few years, with a magnificent series of studies, Pasteur so cut away the foundations of the doctrine of spontaneous generation that it crumbled into ruin. Biologists now hold that all living things are themselves derived from living things and always have been, in so far as properly designed experiment can test the matter. The crux is the germ-carrying power of the air.

The most effective criticism of the experiments of Spallanzani had been that in boiling his sealed flasks, he had also altered the contained air. Early investigators of cells were naturally interested in the origin of minute forms of life. Despite Schwann (pp. 336–9), the question was still essentially in the state in which Spallanzani had left it.

Now if, as Spallanzani and Pasteur believed, the beings associated with fermentation come from the air, it should be possible to find them there. Pasteur therefore made air-filters of gun-cotton and drew air through them. The cotton was then dissolved. The deposit that settled at the bottom of the vessel revealed numbers of rounded and of elongated microscopic bodies. These resembled organisms already observed in fermenting substances. On the other hand, gun-cotton from filters through which air already filtered had been drawn, revealed no such structures. The unfiltered air then contained these bodies.

But it could still be said 'Yes, these structures that you demonstrate are in the air, but they are not alive. It is not they, but something far more minute and subtle, that gives rise to the swarms of living things that we see in fermenting and putrid substances.'

To the question thus posed, Pasteur found a complete answer. A wonderfully simple but triumphant experiment clinched the matter. An infusion of a fermentable substance is introduced into a flask. The neck, very narrow and long but left open, is drawn out into an 'S' shape. Flask and contents are raised to boiling-point for a long time. The source of heat is now withdrawn, but the flask is left undisturbed in still air. It so remains for days, weeks, months. Its contents do not ferment. Now the neck is severed. The fluid within is thus exposed to the fall of atmospheric dust. In a few hours it is fermenting. Organisms are demonstrated in it under the microscope (*Mémoire sur les corpuscles organisés qui existent dans l'atmosphére*, 1861, Fig. 157).

The success of this experiment marks the downfall of the doctrine of spontaneous generation. In the two-thirds of a century that have since elapsed, it has been shown in various ways that if due precautions be taken to exclude

FIG. 157. Pasteur's crucial experiment to prove that fermentation or putrefaction is the result of the action of air-borne organisms. The S-shaped flask contains a putrescible fluid such as meat broth. The flask containing the broth is subjected to prolonged heating to destroy all organisms. It is then left in position with the mouth open. Days, weeks, months, even years, may pass without sign of putrefaction. No organisms reach the broth, since any that enter the open mouth fall on the floor of the neck and remain there. Sever the neck of the flask so that organisms can fall from the neck directly on to the surface of the fluid and thus multiply. In a few hours putrefaction sets in. This is shown by the formation of a film of scum on the surface just below the severed neck. Microscopically the broth is seen to be teeming with organisms.

living organisms and their eggs, spores, or seeds, no fermentation, putrescence, or other production of minute life ever takes place. It is all a question of the adequacy of the precautions. This adequacy is a question of technique.

The germ-carrying power of the air was reduced to more exact expression by the English experimenter, John Tyndall (1820–93). He devised beautiful methods for determining the physical conditions of aerial purity. He also demonstrated the errors in technique that had led some observers to oppose Pasteur. Tyndall's work (1876–81) was well presented and widely read. Its appearance marks the final abandonment by men of science of the doctrine of spontaneous generation.

Biologists are now united in rejecting this belief, at least for organisms as they now exist upon earth. Whether spontaneous generation has ever taken place with other and yet lower forms of life or elsewhere than on this earth is a question on which science cannot, at present, express any opinion.

The word *biogenesis* (Greek, 'life-origin') was coined by T. H. Huxley in 1870 to express 'the hypothesis that living matter always arises by the agency of pre-existing living matter'. The opposite hypothesis is *abiogenesis*.

§ 5. *Early Work on Microbic Origin of Infectious Disease*

Long before Pasteur, it had been suggested that the phenomenon of infectious disease might be ascribed to minute organisms and to their passage from one host to another. In one infection this had been actually demonstrated. The condition known as 'the itch' is both caused and carried by a minute creature related to the spider. This 'itch mite' is just visible to the naked eye. The microscopists of the seventeenth and eighteenth centuries had described it, and had cured the disease with ointments injurious to the mite. They had even conveyed it experimentally. Moreover, certain diseases of plants were well known to be due to and carried by gall insects, aphides, and other insects (pp. 157, 441–2).

An important pioneer of the germ theory of disease was the amateur microscopist, Agostino Bassi (1773–1856), of Lodi near Milan. In 1835 he gave the first demonstration that a vegetable micro-organism could be a cause of infection. After prolonged investigation, Bassi proved that a certain disease of silkworms was transmissible from moth to worm, and that the transmitting material was a minute fungus.

In 1840 the Berlin anatomist, Jacob Henle (p. 452), set forth in detail his theory that infectious disease in general is caused and conveyed by invisible forms of life. During the next two decades the idea was taken up by several other workers. A few minor superficial maladies were shown, with a high degree of probability, to be related to micro-organisms. The microbic doctrine of disease was, however, still vague and generally regarded as a scientific heresy.

In the 'fifties and 'sixties the French silk industry was suffering acutely from an epidemic among the silkworms. It looked as though the trade might perish altogether. The industry depends on the success of cocoon formation, as the caterpillar becomes a chrysalid. The caterpillars were forming hardly any silk and were dying, or languishing to die as chrysalids. Such few moths as were hatched showed signs of disease.

There were many theories concerning this disastrous state of affairs. Certain microscopists had found numerous oval corpuscles in the worms and moths. Their association with the disease was, however, very vague.

Pasteur took up the question at the request of the French Government. After a long investigation he did at last succeed in showing that the corpuscles were the cause of the disease. The matter was exceedingly complicated, and he made many false starts. He was, in fact, dealing not with one disease but with two. He showed that each is always associated with its own special organism (*Études sur la maladie des vers à soie*, Paris, 1862).

The general nature of infection was still a mystery, but here were two infectious diseases in silkworms shown to be associated with microscopic organisms. The hint was very valuable. Primarily, it enabled Pasteur to recommend measures that saved the silk industry of France. Secondarily, it led to discoveries which have opened out entirely new departments of biology, initiated a new era in medicine, and given a new view of the world of life.

The next infection on which light was thrown was anthrax. This is a very fatal disease of cattle, and is sometimes transmitted to man. As with silkworm disease, rodlike bodies had been found in the tissues of sick animals.

The matter had gone farther. A French observer, Casimir Davaine (1812–82), after long research, had shown that bodies which he called *bacteridia* are always found in the blood of sick animals in numbers proportional to the gravity of the attack. He showed that such blood conveys the disease, when inoculated in amounts so minute as 1/1,000,000th of a drop (1863–8). A German observer, C. J. Eberth (1835–1927), had proved that the substance conveying the disease could be separated from the blood. He had filtered the blood of animals suffering from anthrax, and injected the filtrate into healthy animals which had nevertheless failed to develop the disease (1872).

The bacteria had also begun to attract the attention of botanists. Among them, Ferdinand Cohn (1828–98) of Breslau applied himself most successfully to the subject. He recognized the great number and variety of bacteria, and sought to classify them according to their effect on the medium in which they dwelt. He reserved a place for those which cause disease. Cohn's work on bacteria in the period 1872–6 clarified the current attitude toward these organisms. He is rightly regarded as the father of bacteriology.

Bacteria are a group of lowly plants, perhaps allied to the fungi, but entirely without any trace of sexual processes. Like fungi, they are without chlorophyll and therefore cannot build up organic substance from the carbon-dioxide of the air (pp. 375–8). They depend for their sustenance upon the substances in which they live. This existence is described as 'saprophytic' (p. 323). The derivation of the word implies that they 'grow in putrid substance'. Bacteria are the great agents of decay of all kinds, owing to the fact that they set up rapid and profound chemical changes in the organic substances which support them. They are also necessary for various vital processes. Bacteria fall into two great classes, according to their mode of life. The *aerobic* bacteria flourish only in the presence of oxygen; the *anaerobic* cannot so flourish. It happens that the organism of anthrax, which for a bacterium is large and conspicuous, is aerobic.

§ 6. *Koch* (1843–1910) *on Anthrax*

The achievement of unravelling the life-history of the anthrax bacillus belongs to Robert Koch. He had a poor education and little culture. He studied medicine at Göttingen. Though an inconspicuous pupil at the time, it is yet probable that he was inspired by the theories of his teacher, Henle (1809–85), who was himself a pupil of Johannes Müller. In 1872 Koch settled as a country doctor not far from Breslau. He was thrown entirely on his own resources. Anthrax broke out among the cattle of his district, and he began to study it microscopically.

Koch found that in anthrax the spleen was specially affected. He examined fragments of it microscopically and saw the anthrax bacilli as Davaine and Eberth had seen them (p. 451). He found that the disease could be conveyed to the mouse. Propagating it for twenty generations, he recovered the characteristic organisms from the last.

But the key to Koch's progress in the study of the disease was his discovery that the organism can be grown outside the body. He used as soil the body fluids of the animal, and especially blood serum. He ascertained with

exactness the conditions under which growth of the bacilli
will take place.

For investigating growth at different temperatures,
Koch employed a primitive sort of incubator. It was so
arranged that the temperature could be adjusted. A fragment of spleen of a mouse infected with anthrax was sown
in a drop of blood serum. This was kept at body temperature for eighteen hours and examined hourly. The
drop finally presented microscopically a very characteristic
appearance.

In the centre of the preparation, where air could not
penetrate, the bacilli, being aerobic, suffered from want
of oxygen. They were there growing very slowly and
were mostly separated from each other. Farther out,
where the oxygen supply was better, the bacilli had grown
longer and many were placed end to end. These organisms, like many of their kindred, when actively multiply-
ing tend to grow thus in threads. Yet farther out, the
threads were so long that they became twisted and bent.
The nearer to the edge, the longer the threads. In those
threads in contact with the outer air, little round refring-
ent bodies had formed. These were the 'spores'. They
were very highly resistant to such destructive influences
as heat, chemical agents, drying, &c. This fact is of the
utmost significance for the history of the disease, as
Koch fully realized (Fig. 158, p. 463).

In 1876 the unknown Koch wrote to Cohn at Breslau
that he was prepared to demonstrate the life-history of the
anthrax bacillus. Cohn immediately invited him to do so
in his laboratory. A few days later Koch fulfilled his
promise. Several men of science were specially invited as
witnesses. The results were published under Cohn's
auspices. Koch's great paper, *Die Ätiologie der Milz-
brandkrankheit, begründet auf die Entwickelungsgeschichte
des Bacillus Anthracis* ('The causative factors of anthrax

based on the development of the Bacillus Anthracis'), marks the beginning of our exact knowledge of bacterial infectious diseases.

Koch was now able to devote himself to science. He proceeded at once to develop technique. In a short time he had made the two contributions that have enabled bacteriology to develop as a separate science. First he raised the microscopic visibility of bacteria, and second he improved the method of cultivating them.

As regards visibility. The eye does not readily distinguish the shape and arrangement of bacteria under the microscope. Stains had been occasionally used to raise their visibility. The aniline dyes were now available. Koch introduced them into bacteriology and elaborated methods of staining suitable for a variety of organisms. The staining methods commonly employed by modern bacteriologists are more or less directly derived from those of Koch. Perhaps the greatest triumph of these methods is the discovery by Koch himself of the elusive tubercle bacillus in 1882.

As regards methods of cultivation. Early workers on bacteria were constantly confused by preparations that contained more than one species. Many elaborate methods were introduced to obtain growths consisting of but one species—'pure cultures' as they are now termed. Koch adopted, after many trials, the beautifully simple device of a transparent medium of jelly-like consistence. The fluid from which the organism is to be grown can then be diluted indefinitely before it is used to infect the medium, and the very smallest quantities can be used. Thus colonies of bacteria can be raised, each from very few or even from a single organism. Moreover, anaerobic organisms can be grown within the medium. The system of cultivation in transparent media introduced by Koch in 1881 has remained in use to this day without substantial alteration.

A direct result of the use of Koch's technical methods has been the discovery of the causative organisms of a large number of diseases. Among these are cholera, diphtheria, glanders, Mediterranean fever, plague, pneumonia, and typhoid. The bacteria of all these are now well known.

§ 7. *Immunity*

Koch and his German colleagues thus applied themselves with success to the development of bacteriological technique and to the discovery of the causative organisms of disease. It is a department which we need follow no further. In the meantime Pasteur and the school that he had founded were studying the powers that the body possesses for protection against disease. These studies were of the highest practical importance. Here, however, we have only to consider certain biological aspects.

It was well known that many infectious diseases, such as measles, scarlet fever, and the like can commonly be contracted but once. The sufferer having recovered becomes 'immune'. Further, Pasteur had before him the artificial establishment of immunity to small-pox by vaccination which is, in effect, a mild form of the disease. Pasteur sought a parallel to this. He found it in the disease called 'fowl cholera'.

He prepared cultures of the germs of fowl cholera and noticed a fact that has proved of great significance. If he kept the cultures too long and then inoculated fowls with them, the fowls developed only a mild attack. Into such a fowl, on recovery, he inoculated an ordinary virulent culture of the disease. The animal withstood it. It was 'immune'. The discovery (1880) opened up a new department of biology.

Pasteur now turned to anthrax, where he found a somewhat similar state of affairs. If a virulent culture of anthrax bacilli be kept at a temperature of 42° Centrigade (108° Fahrenheit) for eight days, it loses much of its virulence. It may then safely be inoculated into susceptible animals. Further, he found that animals thus treated become immune to the disease. He demonstrated this on cattle, and ultimately protected millions of them. He carried out successful researches of a similar character in other diseases of which the best known is hydrophobia.

The question of the nature of immunity early arose. Two rival theories appeared. One of them, associated with the name of Elias Metschnikoff (1845–1916), may be called the 'solidist theory', the other the 'humoral theory'.

According to Metschnikoff, infective organisms that find their way into the body are engulfed and digested by certain cells which he named 'phagocytes' (1884, Greek, 'devouring cells'). He based his views, in the first instance, on invertebrates, and set them forth in detail in his *Leçons sur la pathologie comparée de l'inflammation* (1892). The process that he describes certainly takes place, but his work threw light chiefly on local inflammation. The general nature of immunity cannot be explained by it.

The humoral theory was stimulated by a very important discovery in connexion with the phenomena of immunity, made in 1890 by the Prussian army surgeon, Emil von Behring (1854–1917). He showed that it is possible to produce immunity against tetanus and diphtheria by injecting serum from an animal that has been infected and has recovered. The serum, then, in these cases at least, is the carrier of the immunity. It bears with it the *antitoxin*—a word introduced by Behring.

The subject of immunity has turned out to be enormously complex. Its greatest exponent was Paul Ehrlich (1854–1915), discoverer of a chemical cure for syphilis

(1909). The body, under various forms of excitation, constantly throws into its blood-stream substances directly or indirectly inimical to the organisms of diseases. They are definitely specific for the disease for which the body elaborates them. Nor is the body prepared only against attacks of disease. There are poisons of unknown chemical composition against which similar immunity may be acquired. Further, it must be remembered that whenever we recover from an infectious disease, we must have developed some degree of specific immunity against it. If we did not, the infecting organism would have got the upper hand and we should not have recovered.

No theory yet advanced is adequate to explain all the known facts of immunity, or even a large proportion of them. The details of the science of immunity are thus not of great importance for us here. What is of importance is that the study reveals to us the complexity of the conditions involved in the 'internal environment'. Just as the functions are active according to physiological needs, so are they active according to pathological needs. In disease, as in health, the organism must be considered as a whole. The test of life is ever this—Does the organism seek to preserve its own norm? We can no more consider a disease apart from the sufferer than we can an organ apart from the organism. We are back on that wonderful balance that life alone maintains.

§ 8. *Biology and Disease*

Pasteur and Koch brought the phenomena of disease well within the range of biological science. Many contemporaries and successors extended their work in many departments. The special application of the knowledge of disease to the life of man has given a practical direction to these researches. With their extension in the domain of medicine we are not directly concerned. Yet it is well

to recall that if the problems of disease are to be regarded
in a philosophical manner, they must be treated on a
biological basis.

The sciences are usually developed along the lines of
the material needs of the day. Special departments are
established not so much because of a pressing intellectual
necessity as because of the practical demands of the times.
Thus develop such sciences as forestry, economic zoology,
or the study of immunity, which become intensively
investigated without regard to their general relation to
the main body of knowledge. Nevertheless, we may sometimes turn from the more special aspects which these
subjects represent, and consider their position as part of
a great scientific unity. Let us, in this spirit, consider
Medicine as a department of Biology, recalling at the
same time that if Biology is now an independent discipline,
it first became so, historically, as a department of Medicine.

There are various departments of medical research
which, in modern times and with the rapid advance of
biological knowledge, have been placed in the hands of
those who follow the study of biology in the academic
sense. To a few of these lines of work we may briefly refer.

The study of Protozoa has its own special technique,
and entered long ago upon the experimental stage. Many
important and common disorders are of protozoal origin.
Among them we may recall malaria, certain forms of
dysentery, syphilis, and sleeping sickness. The investigation of these and of other diseases is mainly the concern
of specially trained biologists who have had experience of
allied forms of Protozoa, and who understand the complex
life histories of the groups as a whole. Moreover, much
light has been thrown on human protozoal diseases by
the study of the protozoal parasites of animals. Thus has
arisen the study of the specialist science of *Protozoology*.

Many diseases are, as we now know, conveyed by

Insect parasites. Thus malaria and yellow fever are conveyed by certain species of mosquito; typhus by certain species of louse; plague by certain species of flea; relapsing fever by certain species of tick; sleeping sickness by certain species of fly. Many other instances might be adduced. Close attention to the distribution and life-history of such insects has illuminated many obscure and little-understood features in the nature of all these diseases. There are some insects, moreover, which without being carriers of a specific disease, yet distribute infection impartially by reason of their habits. Such is the domestic fly, which has typhoid, many forms of sepsis, and the very fatal infantile diarrhoea to his discredit. Because of facts such as these, a special knowledge of the life-history and habits of insects is demanded for the advancement of Medicine. Thus has arisen the study of *Medical Entomology*.

Again, a number of diseases have their origin in the action of metazoan parasites, chiefly of the type of flatworms (Trematoda and Cestoda) and round worms (Nemathelminthes). These enter usually through the digestive or urinary tract, but there is almost no part of the body that is not, from time to time, infested by them. Many of these forms are parasitic during parts of their lives in other animals besides man, and many of them have life-histories of excessive complexity involving some of the classic cases of alternation of generations (pp. 523–30). These forms require very knowledgeable interpretation. Thus has arisen *Helminthology* (Greek *helmins*, 'a worm') as a special science.

More obviously related to medical studies, as they are usually understood, are such sciences as Bacteriology, Physiology, Anatomy, and Comparative Anatomy. Moreover, of late years most of the main biological themes have appeared in a medical aspect. The problems of heredity, of the nature of sex, of the natural term of life, and of life as a cycle, of generation, of the nature and origin of species as illustrated by man, of the possibility of modifying specific characters—all these have their medical aspects. Further, it has long been known that certain diseases affect certain animals and not others, and this has become of importance both for the study of the spread of disease and for its experimental investigation. The domestic animals have been specially studied in this connexion, and so we are getting a real science of *Comparative Pathology*.

§ 9. *Some Failures of the Theory of the Microbic Origin of Disease*

The practical results of the study of the microbic origin of disease are so impressive that we are apt to forget the shortcomings in our knowledge of the nature and life of micro-organisms. Thus from the biological point of view the first question that we should like answered is, 'How does an infectious disease originate in the first place?' Instead of answer there is almost complete silence.

All through the animal and vegetable kingdoms we encounter the phenomenon of *parasitism*. The parasite lives in or on its host. In that limited habitat, its organs are lost and it becomes degenerate. The flea, living on the surface of the body, loses its wings. The mistletoe can no longer root in earth. Many parasitic plants have lost their chlorophyll. Intestinal parasites have lost their sense-organs, and among them are forms without digestive canal and even without power of movement. It is assumed that the bacteria are such degenerate organisms, though we are without knowledge of their ancestry.

How does such a form become parasitic? We are entirely in the dark. We are struck with the fact that diseases often attack only one particular species, but why the fowl, for example, alone suffers from fowl cholera, and man alone from typhoid fever, it is hard to say. Why should not allied forms be attacked by such diseases? We can but guess.

A disease must have started at some time. Without raising the question of the origin of life itself, it is evident that there must be points in history at which disease germs of each type enter a body for the first time, and cause a disease for the first time. We hear of historical or contemporary records of previously unknown diseases. But no one has ever watched the process of the first appearance of a disease from the bacteriological point of view. We have, therefore, no real knowledge as to whether any particular disease is new or not.

Again, if a disease is traced through history, it is evident that its type changes. Scarlet fever, measles, influenza, are very different in one year from what they are in another. Why? Is it that the infective organism has changed or is it that the host has changed? We know nothing on these matters, and the answers that are given do but darken counsel with words.

On these fundamental points our age is as ignorant as other ages were. So long as this is so, it is but blind optimism that can speak of 'the secret of disease' having been 'wrested from nature'. It is no small part of the function of science to define the limits of knowledge. Unjustified optimism is as much the enemy of science as is unreasoning credulity.

§ 10. *Viruses*

Towards the end of the last century there gradually came into view an aspect of infectious disease that promised to throw light on fundamental biological problems. It was demonstrated that some diseases are caused by organisms smaller than bacteria. They can pass through filters whose mesh is so fine that bacteria are withheld by them. Thus in 1892 it was demonstrated that tobacco plants infected with 'mosaic disease' yield a juice which, when thus filtered, can still convey the infection. In 1897 the German bacteriologist, Friedrich Loeffler (1852–1915), showed that the agent causing foot-and-mouth disease of cattle is similarly a filter-passing organism.

These minute organisms in their collective state are known as *viruses*. They cause many diseases in animals and plants. In man virus diseases include smallpox, measles, mumps, chicken-pox, yellow-fever, influenza, rabies and poliomyelitis (infantile paralysis). Many virus diseases have been shown to be transmitted by insects; thus the mosquito *Aëdes aegyptii* infects man with the deadly virus of yellow fever, and it has been demonstrated that a number of plant diseases may be conveyed by insects, notably by aphids. Even bacteria themselves are susceptible to virus diseases which are known as *bacterio-phages*. The existence of these last suggests a hopeful way of treating bacterial disease. This hope has been partly fulfilled.

The organisms of viruses are so minute that the ordinary light microscope is of little value in elucidating their structure. The smallest particles that can be separately discerned by the use of the light microscope are of the order of 1/4,000th of a millimetre in diameter; but even the largest of the virus organisms have a diameter of no more than 1/3,000th of a millimetre. The other virus organisms range down to 1/100,000th mm. Nevertheless the structure of these minute beings has been made in-directly visible by special methods such as the use of the short wave-length rays of ultra-violet light, and by elec-tron microscopy. Under certain conditions, however, the presence of a virus in a cell may be made evident by the presence of much larger non-living structures, the *in-clusion bodies* visible with the ordinary microscope. These were first recognized in hydrophobia. The nature and action of *inclusion bodies* is a subject of current discussion.

Viruses are not known to live or multiply in any medium other than the living cells of the host. Nevertheless they show the prime criterion of life, that of self-

FIG. 158. Bacilli of Anthrax, from a culture, highly magnified.
The rod-like organisms are growing typically in chains. Some of the rods have white clear spots in them. These are the highly resistant 'spores'.

reproduction, using and absorbing foreign material and converting it into their own substance.

The tiny bulk of the virus organisms leaves room for only a very few hundred protein molecules. These must, by their interchanges, provide for the physiology of living things with its unique contribution of chemical flexibility and physical stability. These minute beings not only vary but they must adapt themselves to their environment by some degree of adaptation to heat, drought, poisons, etc. Without some element of such adaptation the term 'life' is meaningless. Some have suggested that the viruses exhibit a passage from the world of not-life to that of life, claiming that a few hundred molecules, each made up of numbers of variously linked amino-acids afford immense possibilities of variation (p. 389)

Is life then the mere concomitant of a happy juxta-position of events in non-living matter. The onus of demonstration lies on the mechanist, for the only known origin of living things is from other living things. There are, however, less formal and more philosophical reasons for thinking that life cannot be given a purely chemico-physical expression. Some of these we have already considered, others we shall later consider.

XIII

DEVELOPMENT OF THE INDIVIDUAL

§ 1. *Seventeenth-century Embryologists*

WITH the advance of comparative anatomy in the sixteenth and seventeenth centuries (pp. 204–10) came the beginning of the study of development, *embryology* as it is now called. Coiter broached the subject (p. 109) and Fabricius of Aquapendente (pp. 109–10, 202) published the first illustrated embryological works. His *De formato foetu* ('On the form of the foetus') saw the light in 1600. His *De formatione ovi et pulli* ('On the formation of the egg and the chick') appeared posthumously in 1621. The former deals well with later stages of the embryo, when it has assumed an appearance similar to that of the new-born young. The latter, in treating of the earliest stages, breaks new ground but contains many grave errors. Among these is the representation of the parts as distinctly formed at a far earlier date than is the case. Fabricius, it must be remembered, had no microscope. Some of his errors passed to an important school of later writers, who held that the young animal is from the first 'preformed' in the egg (Fig. 159).

During the seventeenth century, there appeared a number of treatises on embryology. The majority of them were befogged by metaphysical speculation. The most famous is the *De generatione animalium* (1651) forced from the reluctant William Harvey in his old age. Despite the strange conservatism of parts of this work, it contains some important conclusions. The most significant is set forth on its allegorical title-page (Fig. 160). There a rather comical looking Jove opens an egg—a very Pandora's box. From it emerge many creatures, embryos of man, mammals, reptiles, fish, insects, and even plants. On the egg is inscribed *Ex ovo omnia* ('All creatures come from an egg'). The phrase is prophetic, for Harvey had not seen the minute ova of viviparous creatures such as

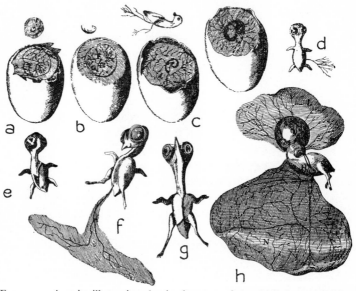

FIG. 159. A series illustrating the development of the chick from Fabricius ab Aquapendente (1621). In *b* and *c*, of the third and fourth days of incubation, the chick is wrongly represented as having almost its adult form.

mammals. Harvey does not, in fact, attach the exact meaning to the word *ovum* that it now has and, in the language of the day, the phrase must have been almost meaningless.[1] His discourse, however, has this of genius, that its content is greater than its utterer knew. Save for those beings that arise by budding or by fission, all living

[1] It may be that, in calling up the image of an egg, Harvey had in mind the *seminales rationales*, the primary rudiments of all things emanating from the mind of God, as set out by St Augustine and later medical thinkers? This would be by no means in discord with Harvey's mode of thought.

things are derived from eggs or from something analogous thereto.

Harvey's treatise is also important as being averse to the false doctrine of 'preformation'. For the most part it is very Aristotelian, but Harvey's philosophical prejudices, though they sometimes led him astray, placed him in this matter on the side of truth. Observationally, too, Harvey's work stands out in the accuracy of his description of the egg, in his account of the ovary and oviduct of the hen, and in many points in the early history of the incubated egg.

Embryological study was placed on a firm basis by Malpighi (p. 152). His *De formatione pulli in ovo* ('On the formation of the chick in the egg', 1673) and *De ovo incubato observationes* ('Observations on the incubated egg', 1689) foreshadowed some of the more important general lines of subsequent embryological research and gave an exact foundation of knowledge on which to build.

Despite his observational acuity, Malpighi believed that he could discern the form of an embryo in an unincubated egg. Nevertheless, the egg which he was examining was not really unincubated. No hen had brooded on it, but it had been laid two days previously in an unusually hot Italian August. This embryo provided a basis for the extraordinary doctrine of *preformation* and its yet more extraordinary development, the theory of *emboîtement* or encasement (p. 506). On such a shaky foundation have been built whole systems of biological philosophy. Nay, the very hope of salvation of men such as Bonnet was erected upon it! (p. 211).

Apart from the embryo as a whole, Malpighi describes stages in the development of certain of its organs. He is most successful with the heart. Some of his figures of that organ are so good that they would fit a modern textbook without much alteration (Fig. 76, p. 152).

The generative organs of mammals were accurately described (1668–73) by de Graaf and Stensen, though the knowledge of their minute ova was long deferred (pp. 473–4). The *spermatozoa* or essential male elements, were figured for the first time by Leeuwenhoek in 1679 and

FIG. 160. Part of allegorical title-page of William Harvey, *De genera-tione animalium*, London, 1651.

soon after by several others (Fig. 172). The discovery gave rise to strange views as to the nature of generation (pp. 506–7).

Not until the mid-eighteenth century was the knowledge of the developing embryo pushed farther than Malpighi had left it. The reason for the arrest was the general acceptance of the doctrine of preformation. This removed the embryological motive. If all the parts of an organism be preformed from the beginning, it were easier to investigate them in the grown animal than in the embryo.

The next important landmark in the history of the subject is the work of Wolff.

§ 2. *Wolff* (1738–94) *and his Successors*

The German, Caspar Friedrich Wolff, produced at 21 his remarkable booklet, *Theoria generationis* (Halle, 1759). Five years later Catherine the Great offered him a post in Russia. There he published a series of observations on the formation of the intestine in the chick (St. Petersburg, 1786). Neither work attracted much attention till the nineteenth century, when J. F. Meckel (1761–1833) issued a German translation of the *Observations* (Halle, 1812). Meckel, like Wolff, worked at Halle, where he pursued comparative studies with enthusiasm. He has left his name on several anatomical structures. He published a monograph of the egg-laying mammal, the duck-billed Platypus, then recently discovered. Meckel was a befogged thinker but a very popular teacher, and his advocacy of Wolff's writings gained them wide attention.

Wolff showed that the organs of plants—leaves, stipules, roots, &c.—are developed by differentiation from uniform tissues at the tip of the growing shoot or root. He was the first to indicate that this growing point consists of an area of undifferentiated tissue. The demonstration had as corollary that the growth of the bud could not be a mere process of 'unfolding', since during the earlier stages the elements of the bud—such as leaves or parts of the flower—are not present. These elements appear only gradually. They first emerge as tiny prominences in undifferentiated tissue at the growing point (Fig. 161).

The process of emergence is, in appearance at least, a true *epigenesis* (Greek, 'origin upon'), an event in which something appears which was not there before, even in rudiment. *Epigenesis* was thus opposed to the then prevailing view of *preformation* (p. 506) which in the language of that day was often denominated *evolution* (literally 'unfolding'), a source later of some confusion.

Wolff set himself also to examine the formation of the parts of the flower. He showed that these appear in the first instance exactly as do the leaves, and that the rudiments of leaves are indistinguishable from the rudiments

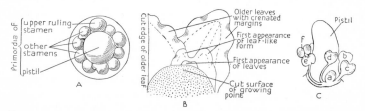

FIG. 161. Diagrams of development of leaf and flower-bud after Wolff, 1759. Wolff was such a poor draftsman that his figures are difficult to understand unless redrawn. This has been done here, using his outlines.

The diagrams, inferior though they are, illustrate the process of 'epigenesis'

A and *C* represent early magnified stages of the flower of the common bean, *A* being from above, and *C* from the side. In *C*, *a* is an anther not yet raised on its filament, *b* to *f* are more developed anthers, *f* being definitely cleft into two chambers. *B*, Vertical section through growing point of young cabbage.

of floral parts. This suggested that the parts of the flower are of the nature of 'modified leaves'.

Wolff made a somewhat similar series of demonstrations in the developing chick. He showed that the abdominal organs and notably the intestines, like the parts of the plant, develop from apparently homogeneous tissue that at first shows no trace of its ultimate fate.

Wolff's name is still attached to a structure peculiar to the embryo, the development of which he was able to trace from such undifferentiated tissue. It is known as the 'Wolffian body' or 'primitive kidney'. It disappears in later life, and does not function in any living vertebrate except certain fish. Since Wolff's time, however, a most primitive excretory system, from which the Wolffian body has doubtless arisen in the course of evolution, has been demonstrated in that lowly and primitive 'chordate', *Amphioxus*.

There is an idea in the work of Wolff which yielded important results in the nineteenth century. Arising, doubtless, out of his researches on the growing points of plants, he compares the first rudiments of organs in animals to 'leaflets'. Later, in describing various folds and layers in the embryo, he compares them to 'leaves', or 'layers'. This is the beginning of the so-called *germ-layer theory*, afterwards developed by von Baer and his successors.

§ 3. *Von Baer* (1792–1876) *and the Mammalian Ovum*

The subject of development was taken up early in the nineteenth century by the Russian, Heinrich Christian Pander of Riga (1794–1865). He published a beautifully illustrated monograph on the development of the chick (Würzburg, 1817). A close personal friend of Pander was von Baer, who introduced the definitive modern stage in the study of development.

The Esthonian, Karl Ernst von Baer, after wandering through several universities, settled with his friend Pander at Würzburg. Von Baer was called as professor to Königsberg in 1817. He there prosecuted embryological studies inspired by Pander. In 1834 von Baer accepted a chair at St. Petersburg, where he passed the rest of his life. His later scientific work bore on the variety of the races of man, but had no relation to embryology. All his writings give the impression of great depth and intellectual power.

Four important advances in embryology are associated with von Baer's name. They are:

(*a*) The discovery of the mammalian ovum.

(*b*) The proposition known as 'the germ-layer theory'.

(*c*) The law of corresponding stages in the development of embryos.

(*d*) The discovery of the notochord.

Of these the first appeared in his *De ovi mammalium et hominis genesi* ('On the origin of the mammalian and human egg', Leipzig, 1827). The other three were adumbrated in his great *Entwickelungsgdschichte der Thiere* ('Developmental history of animals,' 2 vols., Königsberg, 1828–37).

Harvey had seen nothing of the ova of mammals (p. 466). Apart from the large eggs of birds, reptiles, and amphibia, he knew of the ova of fishes. These are quite visible to the naked eye, as are those of insects and of the other larger invertebrates. It was naturally thought that the mammalian ovum must be of size comparable at least to that of fish. Now it happens that during the sexual cycle of mammals, there form in the ovaries certain peculiar bodies, easily visible to the naked eye. These were described by Regnier de Graaf (1641–73). The 'Graafian follicles', as they are called, were erroneously hailed as the mammalian ova.

The error was corrected by von Baer. He saw minute objects which he took to be ova in the tubes or oviducts leading from ovary to womb. He tried to detect similar structures in the ovary. They evidently could not be the 'Graafian follicles', which were far too large. But, with his very near-sighted eyes, examining one day the intact ovaries, he 'clearly discerned in each follicle a yellowish point floating freely in the fluid. Opening the follicle I was able to examine this point microscopically. It proved to be the body already known from the oviduct.'

Thus at last the reproductive processes of mammals and of man himself were brought into line with those of other animals.

§ 4. *The Germ-Layer Theory*

A doctrine associated with von Baer is known as the 'germ-layer theory'. The word 'theory' has several senses. Commonly it indicates an hypothesis which, if accepted, would explain a number of observed phenomena. The 'Origin of Species by Natural Selection' is a theory of this order. Its supporters hold that the variety of species is explained if the action of Natural Selection be accepted. The 'germ-layer theory' is no theory in this sense. Structures known as 'germ-layers' are visible in the embryo. They demonstrably give rise to various organs of the body. On these matters there is no room for doubt. The germ-layer theory is thus rather a description, in general terms, of events which have been observed in many organisms and can safely be assumed in others. It is a generalization and can only be questioned in the sense that we may doubt whether it is adequate. We might legitimately say, for instance, that it summarizes the facts too simply, or we might doubt whether the germ-layers are constant in their action or comparable from group to group.

Von Baer, in the matter of the germ-layers, was carrying farther the views of Wolff and Pander. He showed that, during the development of certain types of animal, the egg or ovum divides into layers of tissue. These layers give rise to the various organs of the body. Further, each layer gives rise to the same organs throughout various groups of animals. Von Baer distinguished four germ-layers, an inner and outer which were formed first and from which were budded off the two intermediate layers.

At a later date (1845) Robert Remak (1815–65), a pupil of Johannes Müller, showed that the two middle layers were best treated as one. Thus Remak distinguished an *ectoderm* (Greek, 'outside skin'), an *endoderm* ('inside skin'), and a *mesoderm* ('middle skin'). He confirmed and extended the teaching of von Baer that the skin and nervous system are formed from the ectoderm; that the notochord (pp. 476–77) and the lining membrane of the digestive canal and of the digestive organs are from the endoderm; and that the muscular and skeletal and excretory systems are from the mesoderm. This is a remarkably wide generalization—one of the widest in the whole domain of animal biology. It needs some qualification but has, on the whole, preserved its usefulness and held its ground to the present day.

§ 5. *The Biogenetic Law*

A contribution to embryological thought for which science owes much to von Baer was the view of corresponding stages in the development of the embryo. Von Baer applied the doctrine specially to the vertebrate embryo. There are traces of this attitude among earlier workers. Thus Harvey wrote that Nature 'by steps similar in all animals, goes through the forms of egg, worm, embryo, gradually acquiring perfection at each step' (1628). Again, John Hunter about 1790 had put down in his notes 'If we were to take a series of animals from the more imperfect to the perfect, we should probably find an imperfect animal corresponding with some stage of the most perfect'. Meckel (p. 469) expressed the view (1811) that the embryonic stages of higher animals actually resemble lower animals. Somewhat similar teaching was that of the French anatomist E. R. A. Serres (1787–1868) who included man in his list. Von Baer corrected this by pointing out that lower forms never resemble embryos of higher forms, but that embryos of higher and lower forms resemble each other more closely the farther we go back in their development.

In a famous passage which drew the attention of Darwin, von Baer wrote:[1]

'The embryos of mammalia, of birds, lizards, and snakes are, in their earliest states, exceedingly like one another, both as a whole and in the mode of development of their parts; indeed we can often distinguish such embryos only by their size. I have two little embryos in spirit, to which I have omitted to attach the names. I am now quite unable to say to what class they belong. They may be lizards or small birds, or very young mammals, so complete is the similarity in the mode of development of the head and trunk in these animals. The extremities are still absent, but, even if present, we should learn nothing from them in this early stage of development, for the feet of lizards and mammals, the wings and feet of birds, no less than the hands and feet of man, originate on the same fundamental plan.'

It is doubtful if this position of von Baer can be rigidly maintained. It is true that embryos of different vertebrate types are built on the same fundamental plan and bear a strong resemblance to each other. But it is equally true that, if sufficiently carefully studied, they can be distinguished from each other.

Nevertheless, von Baer's embryological doctrine is of great historic and practical importance. Later, when the doctine of Organic Evolution had become generally accepted, structural relationship came to be explained as due to relationship of descent. The process of development then came to be a test of structural relationship. Some who stressed this aspect ascribed to von Baer the view, epitomized later by Haeckel (pp. 487–91) that *Ontogeny* (individual development) *is an epitome of Phylogeny* (race development). To ascribe this 'Recapitulation Theory' to von Baer is, however, to read into his work something to which his ideas did not extend. Von Baer, though

[1] In the earlier editions of the *Origin of Species* Darwin unaccountably ascribes this passage to Agassiz! In his notes, however, he had the ascription correct.

he lived till 1876, did not accept the doctrine of organic evolution.

Like most German writers of his day, von Baer was saturated with *Naturphilosophie*. His views on the geometrical and numerical relations of the developing organs savour of Pythagoreanism. We need not follow these fancies. He has, however, profoundly influenced biology through what has become known as his *biogenetic law*. This consists, in effect, of four propositions:

(*a*) In development, general characters appear before special.

(*b*) From the more general characters are developed the less general and finally the special.

(*c*) In the course of development an animal of one species diverges continuously from one of another.

(*d*) A higher animal during development passes through stages which resemble *stages in development* of lower animals [Contra Meckel & Serres, p. 474].

One of the most remarkable of von Baer's contributions was in connexion with the *notochord*, or as he called it, *chorda dorsalis*. This cellular rod runs the length of the vertebrate embryo, between alimentary canal below and nervous system above. It cannot be seen in the adult, save in certain fish. Previous investigators had seen it in the embryo, but did not perceive its importance. Von Baer followed the formation and fate of the *chorda dorsalis*. He showed that it arises from the layer that produces also the lining of the alimentary canal, the endoderm (Fig. 165). He showed that at a later stage it is broken up by the formation of the vertebrae. It thus has no place, or at most a very minor one, in the anatomy of the higher adult vertebrate. Its demonstration in embryos, however, not only established a criterion for the vertebrate nature of an organism but had some extremely important implications for the relationship and classification of the vertebrates (pp. 481–4 and Figs. 165–7).

§ 6. *The Earlier Morphological Embryologists*

With the knowledge of the mammalian ovum and the notochord secure, and with such important generalizations as the germ-layer theory and the biogenetic law, embryology advanced rapidly. Development as the key to morphology became the watchword of biology even before evolutionary teaching provided its special explanation of this relationship. The immediate followers of von Baer were essentially morphologists.

Heinrich Rathke (1793–1860) succeeded both to the chair and to the tradition of von Baer at Königsberg. The general tendency of his numerous embryological observations was to destroy the view based on the 'ideal forms' to which the embryo was attaining as expressed by the Naturphilosophen. They took the adult as basis of their 'archetype'. Rathke forced back the archetypal conception to earlier developmental stages. Thus in 1829 he described gill-slits and gill-arches in embryos of birds and of mammals as corresponding to a fish-like stage. This exposition of Rathke linked up with observations of Malpighi and illustrated the biogenetic law of von Baer (Fig. 76, p. 152 and Fig. 162, p. 478).

In a classical treatise on the lampreys and their allies, Johannes Müller recognized these creatures as the most primitive of true vertebrates (1835–8). He set forth the relationships of their gill-arches, though his views were vitiated by his continued adhesion to the vertebral theory of the skull (pp. 222–3).

The transformation of the embryonic structures which in fish form parts of the gills into other organs of higher animals—Eustachian tubes, tonsils, thymus gland, hyoid, &c.—is now part of stock anatomical teaching. It took

its place there through the exposition of the leading morphological teacher of the later nineteenth century, Carl
Gegenbaur (1826–1903).

The germ-layer theory was also soon brought into the

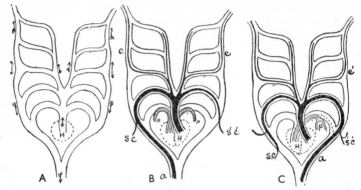

FIG. 162. Diagrams to illustrate fate of arteries supplying gill arches
(Rathke 1829).

A, Ideal diagram of primitive condition still encountered in some fish. From
the single chambered heart *H* a great vessel goes forth and divides into gill
arches right and left. These unite and distribute blood to the body.

In the subsequent diagrams the dark or shaded lines show the vessels which
persist. The others disappear.

B, Modification in bird. The heart, *H,* now has two ventricles. The great
vessel from it has divided into a pulmonary artery *p* (shaded) and aorta *a* (black).
The former is formed from the first, the latter from the second arch.

C, Modification in mammal.

In *B* and *C* the artery to the head, *c,* as well as the artery to the fore-limb, *sc,*
are formed, in part, by remains of the arches.

general field of discussion. Almost the first contribution
made to science by T. H. Huxley bore on this very
topic. He perceived that the jelly-fish and their allies
consist of two cellular layers separated by the jelly-like
substance. In 1849, writing *On the affinities of the
Medusae,* he made the happy suggestion that these layers
correspond to the ectoderm and endoderm (p. 473–4) of
the embryos of higher animals. Thus the lowly jelly-fish

are devoid of true mesoderm, and Huxley suggested that presence or absence of this layer can be made the primary basis for classifying multicellular animals according to both adult structure and embryonic development. The idea has won acceptance.

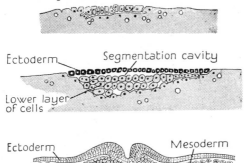

FIG. 163. Sections of three early stages in development of hen's egg. Contrast Fig. 164.

A, Section through germinal disk of egg about the middle of its stay in the uterus. The cells are shown segmenting off from the yolk. At the edges of the disk, are seen nuclei still embedded in yolk.

B, Section toward the end of segmentation. The ectoderm and endoderm are definitely separated and a segmentation cavity has made its appearance. This corresponds to the beginning of the gastrula stage, greatly deformed by the mass of yolk ectoderm represented as black. Compare Figs. 164 *E* and *F*, 167 *A*, and 168 *D*.

C, Transverse section of the embryo about the twentieth hour of incubation. The ectoderm and endoderm are clearly differentiated from each other. The neural groove has appeared. From the junction of ectoderm and endoderm the mesoderm is being given off.

The cellular view now began to have its influence on the study of development. The earlier embryologists worked without any clear conception of a community in microscopic structure of the tissues. Wolff and von Baer speak but vaguely of 'a cellulosity' of certain parts of animals. A fine exposition of the new cell doctrine in all its bearings was made by Huxley in 1853. It was immediately followed by advances in the conception of the development of the embryo. In 1861 Gegenbaur set forth the view that all eggs are simple cells. In the same year Kölliker published his fine general text-book of embryology on a cellular basis.[1] It is valuable for its lucid account of the earliest stages of division of the egg-cell in mammals—the process of *segmentation* as it is now called (Fig. 163)—and for its complete grasp and exposition of the germ-layer theory.

§ 7. *First Reaction of Evolutionary Theory on Embryology*

Embryological science at this stage came within the orbit of the great evolutionary movement. Darwin himself early perceived its bearings on his theory of descent. In his note-books of 1842 and 1844 he devotes much space to embryological discussion relying on von Baer.

In the *Origin of Species* (1859) Darwin exhibits a grasp of the biogenetic law. He was, however, more at home with living nature than with laboratory specimens. Characteristically he points out that

'A trace of the law of embryonic resemblance sometimes lasts till a late age: thus birds of the same genus, and of closely allied genera, often resemble each other in their immature plumage; as we see in the spotted feathers in the [young of the] thrush group. In the cat tribe, most of the species [when adult] are striped or spotted in lines; and stripes or spots can be plainly distinguished in the whelp of the lion [and the puma].' [*Bracketed clauses added from later edition.*]
'The points of structure, in which the embryos of widely different animals within the same class resemble each other, often have no direct relation to their conditions of existence. We cannot, for instance, suppose that in the embryos of the vertebrata, the peculiar loop-like courses of the arteries near the branchial slits (Fig. 162) are related to similar conditions in the young mammal which is nourished in the womb of its mother, in the egg of the bird which

---

[1] Earlier summaries were made by G. Valentin, 1835, Th. L. W. Bischoff, 1842, L. Agassiz, 1859, and H. Rathke, 1861.

is hatched in a nest, and in the spawn of a frog under water. No one will suppose that the stripes of the whelp of a lion, or the spots on the young blackbird, are of any use to these animals, or are related to the conditions to which they are exposed.'

The immediate disciples of Darwin also perceived the bearing of individual development on evolutionary theory. Several took up the subject with enthusiasm. Prominent among these was Fritz Müller (1821–97), an eccentric pupil of Johannes Müller. Fritz Müller went as a merchant to Brazil. There he studied the development of the Crustacea, and published some of his results in his little book, *For Darwin* (1864). This curious document is conversational in form but contains much original observation.

Fritz Müller convinced himself that the complex process of development of certain Crustacea is of the nature of an 'historical document'. The various and startling phases traversed by these forms during development closely resembled, he thought, those through which their ancestors had passed. Striking instances of these phases had long before been adduced by Vaughan Thompson (Fig. 170), who ascribed no evolutionary significance to them. Fritz Müller's views on ancestral recapitulation attracted Haeckel and excited in him an enthusiasm for embryology. They were followed by a series of magnificent studies from Haeckel's laboratory by Alexander Kowalewsky (1840–1901), whose papers on the development of Amphioxus (1866–77) and of Tunicates (1866–71) are among the most striking in embryological literature. They largely determined the general systematic position of the vertebrates (Figs. 164–7).

Amphioxus is a small lancet-shaped marine creature which lives half buried in sand and, at times, darts about freely in the water. In its adult state it has many features in common with the lowest vertebrates and with all vertebrate embryos. It has a persistent notochord. It has an unsegmented nervous system, the front of which is slightly enlarged but is hardly a brain.   The muscular system

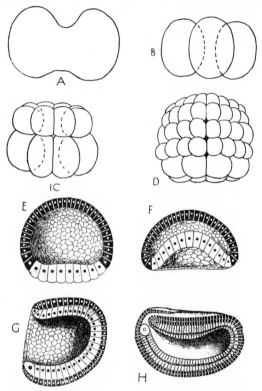

FIG. 164, *A–H*.   Early development of *Amphioxus*.

A–D, Segmentation of the egg.   *E–M*, Later stages shown in section.   The cells of the ectoderm are represented as black, with white nuclei, the cells of the endoderm and mesoderm as white, with black nuclei.   *E*, *F*, and *G*, Formation of gastrula (cf. p. 491).   *H*, Beginning of formation of neural canal (that is, the formation of central nervous system).   In *G* and *H* a large posterior cell is seen which is the beginning of the mesoderm.   For *J–M* see Fig. 165.

is segmented much as in a fish.   Like a fish, its gullet is pierced by a series of gill-slits (Fig. 166).

The development of Amphioxus, like its structure, presents many features that suggest those of the primitive vertebrate. Thus the notochord is pinched off from the endoderm; the mesoderm is formed in a series of

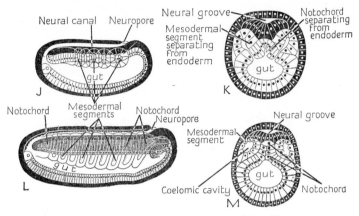

FIG. 165, *J–M*. Sections of Embryos of Amphioxus.

J, Embryo just after hatching. The neural canal is now complete and opens anteriorly by a 'neuropore'. Segments of mesoderm have formed at the side of the gut. *K*, Transverse section through middle of *J*. The notochord is shown separating from the endoderm. The mesodermal segments, each with a hollow cavity ('coelom'), are also separating off from the endoderm, a somewhat unusual developmental relation. In most coelomates the mesoderm arises from junction of endoderm and mesoderm and splits after separation. *L*, More advanced larva with nine mesodermal segments, the last of which is still connected with the point of junction of ectoderm and endoderm. Notochord well developed. *M*, Transverse section through the middle of *L*. The mesodermal segments are now quite separate and show each its own cavity.

segments that are budded off between endoderm and ectoderm; these mesodermal segments give rise to the segmented and muscular system. Even in a late larval stage of Amphioxus, the nervous system is hollow and opens anteriorly (neuropore); the cavities of the intestine and of the nervous system are continuous, and the liver is a mere diverticulum of the gut (Fig. 166).

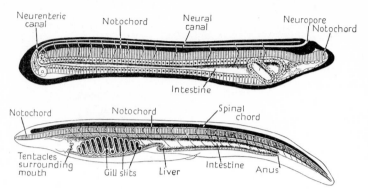

FIG. 166. *Above,* Late embryo of *Amphioxus* with 14 body segments indicated by angular lines. *Below,* Young *Amphioxus.* Body segments indicated by angular lines. Gill slits in process of formation. Kowalewsky (1867).

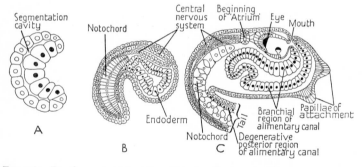

FIG. 167. Development of Ascidian, Kowalewsky (1871). The adult organism bears no trace of its chordate origin. The early stages strongly resemble similar stages of *Amphioxus.*

Compare *A* to Fig. 164 *F.*

Compare *B* to Fig. 165 *L* (omitting segmentation).

Compare *C* to Fig. 165 *L* and Fig. 166.

The structure marked 'Atrium' ultimately comes in contact with the 'branchial region' of the alimentary canal where gill slits form. The posterior region of the alimentary canal degenerates and disappears. In the early stages of the Ascidian, *B* and *C,* there is a well-defined and characteristic notochord. From Kowalewsky.

The further investigation of Amphioxus has been undertaken by many able observers, who have greatly refined the work of Kowalewsky. Notably we have learnt of the peculiar excretory system of Amphioxus, which resembles that of certain worms, and of the high degree to which the creature as a whole is adapted to its mode of life. But the attitude towards Amphioxus which Kowalewsky suggested has remained substantially unchanged (pp. 486–7). The developmental history of Amphioxus is still held to be much like that of a very primitive vertebrate, and that to which the development of less primitive vertebrates must be compared.

No less extraordinary were Kowalewsky's discoveries concerning the development of the Ascidians. These are curious fixed forms, of which the affinities were then quite uncertain. Most naturalists had placed them near the molluscs. Kowalewsky showed (Fig. 167) that they have a free-swimming larva, that this larva has a notochord and has many other features which unite it with the Chordata and separate it from the mollusca. The Ascidian larva has been compared to a tadpole, which it somewhat resembles in general structure and appearance. But, unlike a tadpole, its metamorphosis to the adult state is reached by regressive stages comparable in degree to those gone through by the barnacle (Fig. 170).

Since Kowalewsky's time much has been learned concerning the Tunicates, the group to which the Ascidians belong. Some forms have been distinguished which retain throughout life their larval organization together with their notochord and associated structures. Others have acquired a secondary free-swimming habit. Yet others have been shown to go through extraordinary methods of reproduction by branching, by fission, and by other exceptional methods. They are now united together under the term 'Urochorda', and for this group as a whole the observations and conclusions of Kowalew-sky have fairly held their ground.

These sensational discoveries made it necessary to erect a new group which should include Amphioxus, the Tunicates, and Vertebrates. This group is of the order of a Phylum, to which the name *Chordata* was given by Balfour in 1880.

There have been claims from time to time for the inclusion of other classes among the Chordates besides Vertebrates, Amphioxus, and Tunicates. Mention may specially be made of a worm-like creature *Balanoglossus*. In 1884 the English investigator, William Bateson (1861–1926), showed that it presents a notochord in its free part or proboscis only. This creature and its allies he therefore placed in a special Class 'Hemichorda'. It has also been claimed that the free-swimming ciliated larvae of the Hemichorda bear strong resemblances to the larvae of certain Echinoderms, thereby uniting the vertebrate stock with certain non-vertebrate forms.

Thus we may set forth the current modern classification of the Phylum Chordata. It is based on von Baer's biogenetic law, is derived from his discovery of the origin of the notochord, and involves the work of Kowalewsky and of Bateson.

PHYLUM: CHORDATA (Balfour, 1880)

Class I. Hemichorda (Bateson, 1884).	*Balanoglossus and allies.* Notochord of larva and adult in proboscis only. Larva ciliated, free swimming, and resembling that of Echinoderms.
Class II. Urochorda (Lankester, 1877).	*Tunicata and allies.* Notochord in tail only of larva and disappears in adult. Larva tadpole-like. Adult a degenerate form, fixed or secondarily may float or swim, or in some cases preserve a free life (Fig. 167).

Class III. Cephalochorda *Amphioxus and allies.* Notochord from
   (Lankester, 1877).   tip to tail, develops early and persists
       throughout life (Figs. 165–6). Larva
       free swimming and ciliated. Adult
       active and free swimming. Primitive
       kidney (Wolffian body) persists.

Class IV. Vertebrata.   Notochord runs the length of the body
       in embryo, but is broken up by
       vertebrae in adult. Limbs in all
       except lowest forms. Primitive kid-
       ney replaced by newer formations.
       Brain surrounded by cartilaginous
       or bony cranium.

§ 8. *The Systematic Evolutionary Embryologists*

Ernst Haeckel (1834–1919) studied with Johannes Müller, Kölliker, Virchow, and Gegenbaur. After an expedition to Sicily, he brought out his first important contribution. It was on the Radiolaria of the Mediterranean (1860). The Radiolaria are a group of Protozoa which have very beautiful and symmetrical skeletons. Haeckel was an admirable artist, whose imagination was ever prone to rule his brush. His beautifully illustrated memoir earned him the chair of Zoology at Jena, which he held till 1909.

It is difficult to estimate Haeckel's place in the history of biology. His faults are not hard to see. For a generation and more he purveyed to the semi-educated public a system of the crudest philosophy—if a mass of contradictions can be called by that name. He founded something that wore the habiliments of a religion, of which he was at once the high priest and the congregation. A large part of his insatiable energy was devoted to propaganda for the great liberal intellectual movement of his time, the essential nature of which he misunderstood. In science, his peculiar employment of hypothesis had close affinities with that of the scholasticism that he denounced and that vitiated alike his observations and his inferences. He confidently constructed genealogical trees of living things, which now give rise only to a smile. He habitually perverted scientific truth to make certain of his doctrines more easily assimilable. He invented a bizarre philo-sophico-scientific nomenclature, now happily forgotten. His graphic talents led him to portray the more beautiful forms of minute life in which he saw things less visible to eyes devoid of that fire that burned within his own.

The works of the German apostle of Darwinism now rest peacefully on the less accessible shelves of libraries. And yet they contain contributions which are still funda-mental to the fabric of scientific thought.

In the department of embryology, Haeckel's strong points are his great experience of diverse forms of life and his fertile and ingenious imagination. He seized on von Baer's law (pp. 475–6) and interpreted it along his own lines. Stimulated by the work of Fritz Müller, he transformed the biogenetic law into a series of doctrines to which von Baer would never have assented.

'Ontogeny, or the development of the organic individual, being the series of form-changes which each individual traverses, is immedi-ately conditioned by phylogeny, or the development of the organic stock to which it belongs.

'Ontogeny is the short and rapid recapitulation of phylogeny, con-ditioned by the physiological functions of heredity (reproduction) and adaptation (nutrition). The individual repeats during the rapid and short course of its development the most important of the form-changes which its ancestors traversed during the long and slow course of their palaeontological evolution.

'The complete and accurate repetition is obliterated and abbreviated by secondary contraction, as ontogeny strikes out for itself an ever straighter course; accordingly, the repetition is the more complete the longer the series of young stages successively passed through' (*Generelle Morphologie*, 1866).

All this is briefly stated in the phrase 'Ontogeny is an epitome of phylogeny'. If this rule be correct, there should be three biological parallels: (*a*) the history of the individual, (*b*) the history of the species to which the individual belongs, and (*c*) a natural classification or system of organic forms which include the species.

These doctrines were neither peculiar to Haeckel nor first suggested by him. In the very year of the publication of the *Origin of Species* the great Swiss naturalist, Louis Agassiz, published his *Essay on Classification* (1859). The book is, in large part, a summary of the embryological knowledge of its day. Agassiz, though firmly opposed to evolutionary interpretations, considers that 'Embryology furnishes the best measure of the true affinities existing between animals. It affords a test of homologies in contra-distinction to analogies.'

To this familiar conception of classificatory relation-ship, Haeckel, following Fritz Müller, added the idea of evolution. He developed the theme with more detail than discretion. But by the vigour, bulk, and controversial character of his writings, he turned the eyes of naturalists to the study of development. He failed to give full weight to the disturbing factor of the mechanical element in the process of development. He failed to explain the dropping out of many stages and the modifications of others—factors concerning many of which we are still in the dark. Naturalists since his time have concluded that we cannot, by embryological study alone, reach a complete or com-prehensive summary of the relationships of living forms. Yet there can be little doubt that Haeckel, more than any other man, initiated the modern movement which has, in effect, transformed comparative anatomy into comparative embryology. In this respect his work is to be contrasted with that of his more pedestrian teacher, Carl Gegenbaur the representative of comparative anatomy in its older and more classical sense.

Haeckel made valuable contributions to the knowledge of several important animal groups, notably the Radiolaria, the Sponges, and the Hydrozoa. His greatest work on the Radiolaria is his magnificent monograph of the 'Challenger' material. Certain of his embryological doctrines that are still of value arose from these investigations. Of special importance is his *gastraea theory*.

Kowalewsky, when investigating the development of *Amphioxus* and of the Tunicates, observes that these creatures develop by the segmentation of the egg into a hollow sphere. This sphere in due course invaginates into a cup, much as a hollow ball may be invaginated. The inner part of the resulting invaginated stage is the endoderm, the outer the ectoderm. The mesoderm is given off between the two. During subsequent development the invaginated sphere elongates. The opening narrows and forms the anus of the larva or of the adult. All these stages were noted by other observers in the early development of many organisms unrelated to the Chordates (Figs. 164, 167–8).

Huxley had suggested that jelly-fish have not three germ-layers but two (pp. 478–9). The matter was taken farther by Haeckel in a monograph of sponges (1872). He maintained that the invaginated form, the *gastrula*, as he termed it (diminutive of Greek *gaster*, 'stomach'), occurs as a stage in all multicellular animals. Looking always for lines of descent, he pointed out that in the sponges and jelly-fish the two-layered gastrula, in a state of greater or less complication, is the final body structure. Thus the passage of the higher animals through this stage represents the historical passage of their ancestors through a similar stage. The 'gastrula' or 'gastrea' theory thus illustrated Haeckel's version of the biogenetic law (Fig. 168).

Moreover, Haeckel pointed out that many multi-cellular organisms go through stages represented by B a single cell, C a spherical conglomeration of cells, D a hollow sphere, and E a *gastrula* (Fig. 168). He saw other stages and invented names for them all.

Haeckel further propounded the view that the definitive body cavity of many animals, *coelom*, as he called it

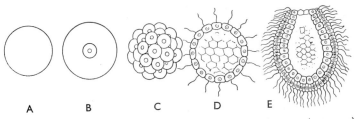

A   B   C   D   E

FIG. 168. Haeckel's five primary stages of individual development (ontogeny) corresponding to five types of independent organisms (1877).

A, Monerula, a hypothetical non-nucleated form that has, as we now know, no existence. *B, Cytula* or one-celled nucleated form. The term is no longer in use. *C, Morula,* a solid sphere of cells of equal size. *D, Blastula,* a hollow sphere of cells, ciliated in free living form. *E, Gastrula* formed by invagination of *D*.

(Greek, 'hollow'), has a common origin in hollow buds from the endoderm. The coelom is a cavity in the mesoderm of higher organisms. Other organisms, such as the two-layered sponges and jelly-fish, have no mesoderm and therefore no coelom. Yet others, such as the flatworms, possess a mesoderm which shows no tendency to form a coelom. This provides a basis for classification of many forms.

The 'coelom theory' was developed by others and especially by Haeckel's pupils, the brothers O. and R. Hertwig. Although neither the *gastrea theory* nor the *coelom theory* are now accepted as originally propounded, they still form the main basis of the classificatory arrangement of multicellular animals.

Despite his very early death, the best representative of systematic evolutionary embryology is the Cambridge investigator, Francis Maitland Balfour (1851–82). He developed the more profitable embryonic theories of Haeckel.

Balfour's first work was on the group of cartilaginous fishes, to which the sharks and rays belong. They have large eggs, loaded with a yolk, comparable in size to that of birds. Balfour demonstrated many important parallels between the development of these fish and that of Amphioxus. Some of the differences between them were, he showed, due to distortion by the yolk (1878).

Balfour's great *Comparative Embryology* (1880–1) was issued a few months before his death. It is a masterly summary of the subject, a mass of erudition and yet lucid and critical, covering the whole field of the development both of vertebrates and of invertebrates. It is generally acknowledged as the foundation of the modern study of embryology. Balfour's death took place within a few weeks of that of the aged Darwin.

The traditions of Balfour and the theories of Haeckel were alike developed and expounded by Edwin Ray Lankester (1847–1929). He was, after Huxley, the most prominent English biological teacher of the nineteenth century. Lankester's great collaborative treatise on morphology remained unfinished, but by an immense number of separate studies, he did much to give their modern form to the gastrea theory, the germ-layer theory, the coelom theory, and the recapitulation theory. His researches on the development of the Mollusca, Annelids, and Arthropoda are largely responsible for the generally accepted internal classification of these groups. He did a somewhat similar service for the Protozoa and the Acoelomate Metazoa.

In Germany the work of Balfour was carried on by O. Hertwig. His great embryological text-book appeared first in 1886. It is distinguished by clarity of exposition, by excellence of arrangement, and by its careful historical material. It has established a standard for embryological writing.

§ 9. *Digression on Metamorphosis*

In most beings the progress from embryo to adult is gradual. Some, however, exhibit a series of rapid and fundamental changes. Such is the case with the Ascidians (pp. 484–6). More familiar phenomena of the same order are presented by the insects. Thus, the caterpillar is transformed into a chrysalis and the chrysalis into a winged insect in what seems a sudden manner. Such events are termed 'metamorphoses' (Greek, 'complete change'). This was the title of the great poem by Ovid (43 B.C.– A.D. 17). It recounts legends of miraculous change of shape and opens:

In nova fert animus mutatas dicere formas corpora.
'My mind is bent to tell of bodies changed into new forms.'

Thus the term is peculiarly appropriate to these insect changes. Moufet (pp. 97–8) was the first in modern times to use the word in this sense, and it was popularized by Jan Goedart (1662, Fig. 166). Insect metamorphoses were studied by Swammerdam and others in the seventeenth century. He classified insects into four main groups based on the modes of metamorphoses.

Of eighteenth-century entomologists none was more noteworthy than the Frenchman, René Antoine Ferchault de Réaumur (1683–1757). After a long Jesuit training, he applied his varied talents to the whole range of the sciences. Among his biological activities we note his observations on regeneration in the Crustacea, on the locomotion of star-fish, on the electric apparatus of the torpedo, on marine phosphorescence, on the growth of Algae, on the digestion of birds, on spiders and silk, and on the nature of coral organisms. Réaumur's great *Contributions to the History of Insects* (6 volumes, 1737–48) is remarkable among such works for its physiological inter-

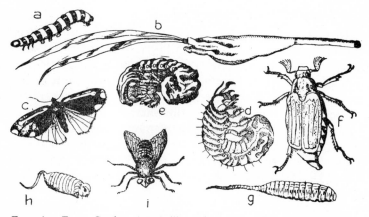

FIG. 169. From Goedart (1662) illustrating three different types of insect metamorphosis: *a* is the caterpillar, *b* the chrysalis of *c* the cinnabar moth, *Euchelia jacoboeae; d* and *e* are larvae of the cock-chafer, *Melolontha vulgaris; g* and *h* are the rat-tailed maggots of *i*, a hover-fly, a species of *Eristalis*.

est. In it he discusses the effect of heat on the development of insects and their larvae, and has much to say on the leaf-boring and gall-forming insects. Throughout he emphasizes development and especially metamorphosis.

Several contemporary naturalists worked on similar lines. The Dutch lawyer, Pieter Lyonet (1707–89), produced with incredible labour in 1740 his famous monograph of the goat moth caterpillar. As a refined anatomical examination it is still unapproached. The Parisian physician Étienne Geoffroy (1725–1810) described and figured beautifully the metamorphoses of a number of forms.

Thus by the beginning of the nineteenth century the conception of metamorphosis was quite familiar.

The subject was put on a new footing by the English army surgeon, John Vaughan Thompson (1779–1847), who was stationed in Cork in 1816. For twenty years he conducted investigations in marine zoology, which both

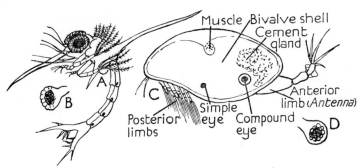

FIG. 170. Developing forms of Crustacea from Vaughan Thompson (1828–30).

A. 'Zoea' shown by Thompson to be the larval form of the shore crab *Carcinus moenas.*

B. Stalked compound eye of slightly older 'zoea'. This type of eye is characteristic of certain groups of higher Crustacea.

C. Free-swimming Crustacean, proved by Thompson to be the larval form of a fixed 'barnacle', an organism classed among the Cirripedia. The larva attaches itself by its anterior limb or 'antenna'. This becomes converted into the foot or 'peduncle' of the adult barnacle. The foot becomes attached to a fixed object by cement from the cement gland. The larval form possesses eyes both simple and compound. These disappear in the adult barnacle. (Compare Figs. 179 and 180.)

D. Compound eye of larval barnacle, which may be compared to *B.*

modified and extended the current conception of metamorphosis beyond the range of entomology. It has been said of him that 'no great naturalist has ever written so little and that so good'.

Thompson showed that the common shore crab, during its development, passes through a series of changes not less remarkable than those of insects. He extended his discoveries to other groups of Crustacea. Most revolutionary was his discovery of the nature of the Cirripedia or barnacles.

The Cirripedia had already been much studied. Cuvier had treated them as a class of his phylum *Mollusca* (p. 234). Cirripedes are immobile, without head, eyes, limbs, or obvious joints. They have a shell formed of a few calcareous plaques. In general appearance they resemble bivalve molluscs. Yet Thompson showed convincingly that they pass through a young or larval stage, in which they swim freely and are equipped with eyes and jointed limbs. These larvae closely resemble certain stages of the development of the shore crab and other Crustacea (Fig. 170). With that group they now became classed.

Since Thompson's time, metamorphoses no less remarkable have been traced in a variety of animal forms. The most extreme types are encountered among parasitic worms, many of which dwell at different periods of their lives in several different hosts.

Such a very extreme type of metamorphosis was investigated by Thompson himself. Shore crabs are sometimes found with a bag-like growth on the abdomen. This is the degenerated parasitic body, limbless, eyeless, motionless, devoid of mouth, alimentary canal, or sense organs, of what was once a free swimming creature. It is hermaphrodite, that is to say, it contains the sexual organs of both sexes—testis and ovary. It contains little else. Thompson saw its larval stage and interpreted it as a crustacean. It is now classed near to the Cirripedia.

These observations drew attention to the profound effects of parasitism on structure and life history. Numerous and varied investigations have since been made on the subject. The general laws of parasitism and the types of degeneration that parasitism involves are now well recognized. The phenomena occur throughout the animal and vegetable kingdoms and are found to be ever more complex as knowledge advances (pp. 323–5).

§ 10. *Developmental Mechanics*

There has always been a school that seeks to refer all biological phenomena to known physical laws. With such an end in view the process of development has long been the subject of experimental investigation.

An important pioneer in this work was the Swiss, Wilhelm His (1831–1904), who professed anatomy at Leipzig. No one before him described so diligently, lucidly, and exactly the minutiae of the development of the higher animals and especially of man. He was the chief agent in the introduction of the instrument for cutting serial sections known as the *microtome*.[1] Its use has become an indispensable method of biological research in general and of embryological research in particular (*Beschreibung eines Mikrotoms*, 1870). His very exact knowledge of the developmental history of man enabled him to attempt an explanation of the entire process on mechanical principles.

In a work on *Our bodily structure and the physiological problem of its formation* (1874), he compared the various layers and organs of the embryo to a series of more or less elastic tubes and plates. The local inequalities of growth on the one hand and differences in the consistency of the tissues on the other, might, he conceived, account for the formation of the various organs and structures.

If we were to accept such forces as in themselves adequate to explain development, we should find ourselves involved in an ancient fallacy; we should have to admit that the originals (primordia or rudiments) of the various organs are represented in the germ layers, at least as regards their local behaviour. The germ layers themselves

---

[1] An instrument for cutting sections, though not serial, had been invented by John Hill in 1770.

would be similarly represented in the parts of the embryo in a yet earlier stage, and these would be represented in the egg. We thus reach the old doctrine of preformation in another dress. Apart from theoretical objections to this conception, there is experimental evidence that renders it untenable. For a time, however, the evidence that was forthcoming, notably that produced by Wilhelm Roux (1850–1924), was in its favour.

Roux was a pupil of Haeckel. In a series of important writings collected as *The developmental struggle of the parts within the organism* (1881) he worked out in detail his view of a mechanical basis for the functional adaptation of the parts to each other. The investigation of function implied experiment. Thus in seeking the basis of 'developmental mechanics' (Entwickelungsmechanik), with which his name is specially associated, Roux passed naturally to experiment on the embryo.

An examination of the process of development along the lines suggested by Roux soon revealed the fact that development cannot be treated as a uniform process. It is divisible into two periods. The first is the true *embryonic period*. During it there are formed the structures that are *predetermined* (Roux's own term, 1905). It includes all that part of the development which takes place until the organs function. The second is the *period of functional development*. It is in this second period that the formation of the parts is brought about by their specific action. Thus structure, according to Roux, determines function, but function also determines structure. The statement is doubtless true, but as thus analysed it gets us no farther towards an explanation of either. A mystery is no less a mystery because it is limited in its action in space and time, nor even when re-expressed in terms of the unknown. Nevertheless, as the creator of the science of Developmental Mechanics and of its handmaid Experimental Embryology, Roux rendered enormous service.

Roux, like his contemporary, Weismann and like many other naturalists to this day, held that in the fertilized egg-cell there exists a very complex structure or machine. He held further that this structure or machine is divided or disintegrated into its constituent but still complex parts, structures or machines, by the process of division or segmentation of the egg-cell.

A famous experiment by Roux raised again in a new way the old antithesis between preformation and epigenesis. In a developing frog's egg, which had just segmented into two cells, Roux succeeded in destroying one without injuring the other (1888). The remaining cell developed as a half embryo.

This result would appear to be interpretable on the basis of preformation, one cell containing the germ of one half of the frog, the other of the other half. Nevertheless this view proved to be untenable.

O. Hertwig had shown the value of sea-urchin's eggs for the investigation of developmental problems. Hans Driesch (1867–1941), one of the most important architects of modern biological theory, then began to experiment with these sea-urchins' eggs in 1891. He succeeded in separating the two cells into which a sea-urchin's egg had segmented, and from each half egg he reared a complete larva, half the normal size (1900 *onward*). Further, he obtained complete embryos from yet later stages of segmentation—1/4th, 1/8th, and even 1/16th or 1/32nd the normal size. Moreover, it was shown subsequently that if in such a frog's egg as that on which Roux had experimented the dead cell be carefully removed, the remaining cell will develop as a whole embryo, though of course half the size of the normal. These results, contradicting the work of Roux, focussed the attention of naturalists on the experimental method in embryology.

§ 11. *Experimental Embryology*

Thus arose Experimental Embryology as a separate science. It arose late in the nineteenth century and its findings remained equivocal to the end of our period. We shall glance at a few of its earlier results.

The primary phenomenon of development is the segmentation of the egg-cell. This exhibits great variety in eggs of different types. The main factor is the presence of yolk. Some eggs, like that of the hen, are masses of yolk with the protoplasm of the egg at one pole (Fig. 163). The segmentation of such eggs begins at the protoplasmic pole, and the yolk is only gradually included in the embryo. Other eggs, as that of Amphioxus, have almost no yolk and segment almost uniformly (Fig. 164). Yet other eggs, as that of the frog, are intermediate. The segmentation in the frog's egg is more active in the region where there is less yolk than in the region where there is more yolk.

That the yolk itself causes these differences by its mechanical action was proved by O. Hertwig (1897). He centrifugalized frogs' eggs so that the yolk collected at one pole and the protoplasm at the other. The result was that the frog's egg segmented only at the protoplasmic pole. This view has been since confirmed in many other ways on many other animals.

A characteristic of animals, as distinct from plants, is their bilateral symmetry. This feature appears far down in scale. Most groups even of protozoa exhibit some trace of bilateral symmetry. In higher metazoa, bilateral symmetry may become masked, as in the radial sea urchins and starfishes or in the spiral snails or hermit crabs. Yet all these begin embryonic life as symmetric forms. Bilateral symmetry is often less apparent internally, as in the intestines and abdominal parts of man, than externally. Nevertheless developmentally, bilateral symmetry is deep rooted among almost all animals. Some eggs, as those of insects, are visibly bilaterally symmetrical from the beginning. Eggs of many other animals can be shown to have hidden bilateratility. The plane of symmetry is, in some cases at least, determined by the point of entry of the spermatozoon, as was shown for the frog's egg by Roux (1903). Work on this theme was done by the Oxford embryologist J. W. Jenkinson (1871–1915), who demonstrated that in some forms the axis of symmetry is determined by external mechanical causes, such as light and gravity (1909).

Much of the confusion in the early days of experimental embryology arose, as we can now see, from the fact that eggs belong to one of two classes, according to their physiological behaviour during segmentation. Thus there are eggs, as those of the sea-urchins, which segment into cells each of which can produce a complete animal. (It is evident that this power can extend only to a certain degree of division.) Such are *regulative eggs* as distinguished from *mosaic eggs* in which the different regions of the embryo develop independently of one another—the loss of one piece spoiling the pattern. There are, however, all stages between completely regulative eggs and completely mosaic eggs. The loss of the regulative power takes place at different dates in different eggs and can be either retarded or hastened experimentally.

The factors that determine the development of an egg are of different classes. Certain of them, which we may call 'internal factors', are transmitted from the parents. These we shall consider elsewhere (chap. xv). The development of these factors depends, however, upon their environment. This environment is in part truly biological, that is to say, it is made up of the interaction of different parts of the embryo upon each other. A simple case of this was described by the American cytologist E. B. Wilson in the Mollusc *Dentalium* (1904, Fig. 171). It is, in fact, characteristic of living things that they carry with them the power of such mutual interaction of their parts.

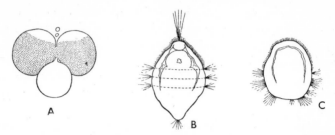

FIG. 171. *A*, Early stage of cleavage of *Dentalium*. The egg has segmented into two nucleated cells, from one of which depends a non-nucleated proto- plasmic lobe. *B*, Normal larva of *Dentalium*. *C*, Larva obtained after removal of non-nucleated lobe. Neither the apical organ nor the hind region is then developed (Wilson, 1904).

This we may call 'internal environment' (pp. 399–400).

Over and above internal factors and internal environ- ment there are external factors that have an influence on development. Among these are such forces as gravity, temperature, electrical and chemical actions, and osmotic pressure. All vary within wide limits during normal development. All have been experimentally varied within wider limits with significant results. The variation of each of these factors produces its own characteristic disturbances.

Extremely interesting results were early elicited in con- nexion with the power of healing or regeneration of lost parts, especially in embryonic and growing animals.

After the changes of the 'embryonic' period, there comes a time in the life of every being when the stimulus of function is necessary for the development of the part. This was emphasized by Roux (pp. 498–9). The first appearance of the main trunks of the blood-vessels is an event of the embryonic period. The formation of the elaborate connexions of their branches depends on the development of the parts that they supply. This has been proved experimentally for these vessels and for numerous other organs by many workers.

A particularly interesting series of events of the 'functional' period is provided by the nerves and the tissues that they supply. It is a remarkable fact that certain nerves are very faithful to certain structures. Despite changes in the form, position, and function of these structures in the course of evolution, their nerve-supply remains constant. Thus the normal nerve-supply of organs and notably of certain muscles is very complex, and the normal course of their nerves is correspondingly intricate. It might be thought, therefore, that there is some natural affinity between a particular organ and a particular nerve, so that each attracts the other and refuses co-operation with more conveniently placed neighbours.

This is in fact in some degree the case. Thus the limb of a newt grafted on to another becomes innervated after the normal fashion. It is, however, also true that some organs can, if suitably ingrafted, be persuaded to form quite abnormal connexions. Thus the eye of the amphibian *Amblystoma* can be removed and engrafted in another individual in such a way as to acquire its innervation from the nerve that normally supplies the tongue.

The whole subject of grafts and their innervation is in course of active investigation. Particularly interesting but very difficult are experiments of early grafts in the 'embryonic' as distinct from the functional stage. The very skilful experimenter Hans Spemann (1869–1941) has had important results on the newt. Before the formation of the gastrula, he removed a piece of the ectoderm destined for the formation of the nerve-cord. For this he substituted a piece of ordinary ectoderm (not destined to form nerve cord) from a newt of another species and colour. The difference in colour enabled the fate of the engrafted fragment to be traced. It was found to take its place in the nerve cord which was formed and completed in the normal way (1918). Thus the work of Spemann demonstrated that the fate of the ectoderm of the newt is not irrevocably determined until after the period of development at which this experiment was made.

The determination of the parts continues throughout development. It is a process which never reaches finality. Its last exhibition is the power of repair which ceases only with death. The process is indeed a reflex of life as a whole. What we are *in esse* we can no longer be *in posse*. The very fact of achievement means a limitation of possibilities of development. The parallel is more than an accidental one. Development, evolution, progress, ageing, bring gains that are problematic, but losses that are certain.

XIV

SEX

§ 1. *First Attempts to Analyse the Nature of Generation*

THE process of coming into being has always roused human curiosity. Aristotle devoted two of his finest treatises to the subject. After ages have not been wanting in theories as to the nature of the process, but it was many centuries before observation led to doctrines more comprehensive than his.

Without a microscope, it was inevitable that Aristotle should hold some organisms to be 'spontaneously generated'. Apart from this, he recognized that most animals reproduce their kind through the intervention of two sexes. It was evident that in reproduction an egg or something like it was frequently involved. In seeking analogy to the egg in those animals in which he had seen no egg, Aristotle compared an embryo in its membranes to an egg surrounded by its characteristic coverings (Figs. 63–4). The egg, he thought, was rendered capable of development by the sperm of the male. He believed that this sperm had in itself the power to form all parts of the body by its action on the 'egg'. The male element was thus a potentiality and not necessarily material. The male gave *form*, the female *substance*, just as the artist 'imparts shape and form to the material by means of the motion he sets up' (*De generatione animalium*, i, § 22).

Opposed to Aristotle were those that held the embryo to be formed of material elements derived from *both* parents. This view was especially acceptable to the philosophical sect known as 'Epicureans'. Their representative, Lucretius (*c.* 60 B.C.), assures us that 'the embryo is always formed of two seeds, and whichever parent that which is so formed more resembles, of that parent it hath greater share'. To this view Galen and his Arabian followers conformed.

As the Middle Ages and the Renaissance derived their biological ideas in part from Aristotle and in part from the Arabian Galenists, confusion ensued. So the matter remained until the time of Harvey, who adopted something in the nature of a compromise. He greatly stressed the role of the egg, thus yielding to the Galenists, but he followed Aristotle in regarding the female as impregnated 'by a kind of contagion which the male communicates, almost as the lodestone does to iron'. He held, with Aristotle, that for impregnation nothing material need pass from male to female. Fertilization was an act of a mysterious essence, the *aura seminalis*.

In the later seventeenth century the erroneous doctrine of 'preformation' unfortunately came to hold the field (pp. 467–8). It was now taught that a complete being lay in the egg. Only the suitable stimulus was needed to cause it to unfold.

But complication again arose. If the being in the egg be complete, it must contain ovaries and ova. And if such secondary ova are present, they also must contain complete beings, and so on in a process of *emboîtement*, like a series of Chinese boxes. Eve, it was held, had within her ovary the forms of all the men and women that were to be. One writer estimated them at the very moderate figure of twenty-seven million. This remarkable theory long held the field. The philosophers and men of science who professed it were named 'ovists'. Their views had certain evident theological implications with which we are not here concerned.

A further complication arose. It was found that the semen contained *animalcules*, as they were called, *spermatozoa*, as we name them today. These were depicted by

Leeuwenhoek in 1679 (Fig. 172 *a–c*). It was now claimed that the spermatozoon, and not the ovum, contained the preformed organism. This was a justification of the dignity of the male! It was also an interpretation—on the basis of a misunderstanding—of Aristotle's view that the 'male contributes *form* and the female *substance*'. Some claimed to descry a human form in the spermatozoon itself. Thus arose the school of the 'spermatists'. These held that the testis of Adam must have contained all mankind, as the

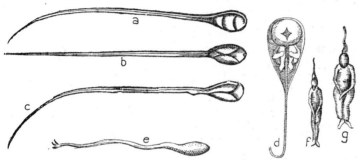

FIG. 172. Spermatozoa as seen in the seventeenth century. *a, b, c,* Leeuwenhoek from the dog (1679). *d,* Hartsoeker from man showing the 'homunculus' (1694). *e, f, g,* Francois Plantades (Delenpatius) from man. *e,* intact, *f* and *g,* broken to show the 'homunculi' (1699).

ovists had held for the ovary of Eve! The literature of these extraordinary preformationist doctrines is considerable (Fig. 172 *d–g*).

During the eighteenth century Wolff gave a new turn to embryology by attacking the doctrine of preformation and presenting epigenesis in its stead (pp. 469–71). Many, Haller among them, remained unconvinced. The knowledge of the essential nature of sexual generation was not greatly forwarded by the new discovery, for, though spermatozoa were demonstrated in a variety of animals, the ovum of mammalia had not been seen, or rather had been seen wrongly. The cellular nature of the ovum was

still unsuspected. The spermatozoa were misunderstood
and the most bizarre forms sometimes ascribed to them.
Buffon and Needham, misled by their faith in spontaneous generation (chap. xii), compared spermatozoa to the
superficially similar organisms yielded by infusions. They
thought that somehow these 'filaments' represented the
primordia of higher forms of life (p. 296).

§ 2. *Early Writers on Pollination*

With this confusion in regard to animal forms it is not
surprising that the reproductive processes of plants were
little understood. The general sexual character of flowers
was foreshadowed by Millington and Grew (p. 161),
opposed by Malpighi, and supported by Ray. None of
these contributed any facts of great importance to the
subject. In 1691, the Roman Jesuit Filippo Buonanni
(1655–1725) published a polemical treatise attacking
Redi's destructive criticism of spontaneous generation
(p. 440). Buonanni figures the styles of several plants with
pollen grains actually adherent to them. He found insects
sometimes in association with these parts of plants, but
he advanced no further (Fig. 173).

The conception of sex in plants was first lucidly set
forth a few years later in a small tract by a professor at
Tübingen, Rudolph Jacob Camerarius (1665–1721).
The work is balanced, based firmly on experiment, and
frank in its presentation of difficulties. Camerarius first
made observations on such plants as Dog's Mercury in
which the flowers are of different sexes. He says:

'In plants in which the male flowers are separated from the female
on the same plant, I have learnt by two examples the bad effect
produced by removing the anthers. When I removed the male
flowers before the anthers had expanded, and also prevented the
growth of the younger male flowers, but preserved the ovaries,

I never obtained perfect seeds, but only empty vessels, which finally fell exhausted and dessicated.'

He goes on to say that:

'In plants no production of seeds takes place unless the anthers have prepared the young plant in the seed. It is thus justifiable to regard the anthers as male, while the ovary with its style represents the female part.'

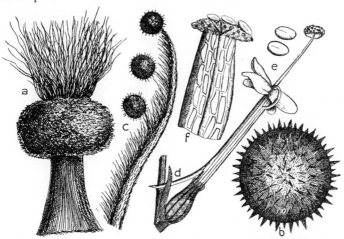

FIG. 173. Process of pollination from Filippo Buonanni (1691). The degree of magnification varies greatly in the different figures.

a, Sexual parts of the flower of a mallow, *Althaea hirsuta* Linn.; *calyx* and *corolla* being removed. There are numerous *styles*, or separate female organs, shown as hair-like lines above. They are surrounded by a large cushion, formed of a mass of *stamens*, or male structures.

b, Pollen grain of mallow very highly magnified.

c, Highly magnified style of mallow on which are three pollen grains.

d, Flower of a valerian, *Centranthus ruber*. The five petals forming the corolla are joined at their bases into a spurred tube. Near its attachment this tube is surrounded by the calyx. Projecting from between the free ends of the petals are the sexual parts which consist here of but one style and one stamen. On the end of the stamen is an *anther* bearing pollen grains, shown as black dots.

e, Pollen grains of valerian, further enlarged.

f, Style of valerian with three pollen grains in place on its expanded end or *stigma*. Between the cells, of which the style is composed, a tube, shown as a black line, leads to the ovary.

Camerarius judiciously added that his theory did not apply to non-flowering plants.

The sexual character of flowers was widely canvassed in the early eighteenth century. It was formally accepted by Linnaeus and incorporated into his system, though in a mechanical way. Linnaeus was not interested in physiology, and he used the sexual parts merely as a convenient basis for his system of classification.

During the eighteenth century, several botanists succeeded in using the pollen of one species to fertilize the flowers of another. The products had some of the characteristics of both species. The word *hybrid* was introduced to describe these cross-breeds. The Latin *hybrida* means the offspring of a domestic sow and a wild boar, that is a cross between varieties, not species. Flowers that are hybrids, in the strict sense, that is crosses between species, occur occasionally in nature.[1] The observation of these and their formation under cultivation provided additional evidence of the sexual character of parts of flowers.

Joseph Gottlieb Koelreuter (1733–1806), director of the grand-ducal botanic garden at Carlsruhe, was interested in hybrids. His book on the sex of plants (1761) acknowledges a debt to Camerarius. Köelreuter rightly considered that a very important agent in the fertilization of flowers is the wind, that some flowers fertilize themselves, but that

'in flowers in which pollination is not by immediate contact, *insects* are the usual agents, and consequently they alone bring about fertilization. It is probable that they render this important service if not to the majority of plants, at least to many, for all the flowers which we discuss have in them something agreeable to insects'.

Koelreuter pointed out that there are certain plants with flowers of different sexes in which pollination by wind is

---

[1] The word 'hybrid' has come to be used much more loosely (see chap. xv.)

almost impossible. Here he thinks insects must be the agents. Such is the case with the mistletoe. He drew attention to the distribution of its seeds, later investigated by Darwin. The mistletoe is parasitic, growing only on certain species of tree. There must be some means by which the seeds find their way from one host to another. Koelreuter showed that the carriers are birds. Thus the mistletoe depends on two groups of animals, insects and birds, for its existence. The ancients believed the seed of mistletoe to be conveyed by the excrement of birds. It is in fact conveyed by birds wiping their beaks on the bark.

Koelreuter's lead was taken up by several German workers, but especially by Christian Conrad Sprengel (1750–1816), Rector of Spandau. He became so devoted to botany that he neglected his duties. He was removed from his parish and he removed himself to Berlin, where he led a solitary life and was regarded as a crank. His book on sex in plants is a work of genius. Its title may be translated *The Newly Revealed Mystery of Nature in the Structure and Fertilization of Flowers* (1793). He was so depressed by its reception that he abandoned botany for philology. There is no evidence that philology was the gainer.

'In 1787', says Sprengel, 'I observed the lower part of the petals of *Geranium sylvaticum* to be provided with slender rough hairs on the inside and on both edges. Convinced that the Wise Framer of Nature has not produced a single hair in vain, I pondered on their purpose. Suppose the five drops of juice secreted by the five glands are for food of certain insects, it is likely that there are means of protecting this juice from rain. The flower is upright, but drops falling into it cannot reach the juice, being stopped by the hairs. An insect is not so hindered.

'I examined other flowers and found that several had structures for a like end. I now saw that flowers which contain this juice are contrived so that it may be accessible to insects but not to rain;

and that it is for the insects that these flowers secrete the juice, which is so secured that they may enjoy it pure and unspoilt.' [*Much abbreviated.*]

The flowers of Forget-me-not suggested that the spots on the corolla are placed in relation to the honey glands, and he concluded that

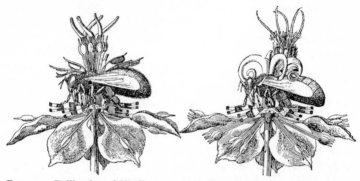

Fig. 174. Pollination of *Nigella arvensis* as represented by Sprengel (1793). In the centre is the multiple pistil in a single whorl. Around these are several whorls of stamens. Around these again is a circle of very small tubular petals filled with honey. These are surrounded by a circle of large coloured sepals. When the flower first opens the filaments of the outermost whorl bend so that their anthers touch the visiting bee. These stamens, being exhausted of pollen, come to lie upon the sepals. Next day their place is taken by the next whorl, and so on. When all the whorls of anthers are thus exhausted the segments of the pistil come into play. Bending down to touch the back of the visiting bee they are duly pollinated.

'the corolla has a particular colour in particular spots, to indicate the juice to the insects, that when in search of food they may see the flowers from afar and know them for receptacles of juice'. [*Abbreviated.*]

Sprengel afterwards found that the stigmas of a species of Iris cannot be fertilized save by insects, and

'that many, perhaps all flowers which have this juice, are fertilized by the insects which feed on it. The whole structure of such flowers can be explained if we consider these points; first, that flowers were intended to be fertilized by the agency of insects; second, that insects in seeking the juice were intended to sweep off the dust from the anthers with their hairy bodies and convey it to the stigma, which is provided either with short delicate hairs or with a sticky substance to hold the pollen.'

He paid special attention to those plants in which the two sexes, while occurring on one blossom, yet mature at different periods. This process, *dichogamy*, Sprengel first observed in the Rose-bay. He found 'the individual flowers fertilized by the humble bee, but not by their own pollen, for the older flowers are fertilized by the pollen which the insects carry to them from the younger flowers'. Later he found exactly the opposite arrangement to prevail in a *Euphorbia*. There the pollen is brought by insects from the older flowers only. Sprengel's observations on dichogamy are specially striking in the case of Love-in-a-mist (*Nigella*, Fig. 174).

Sprengel concluded that the whole structure of nectar-bearing flowers is directed to fertilization by insects and can be fully explained by it. He found the chief proof of this in dichogamy, the details of which he traced skilfully.

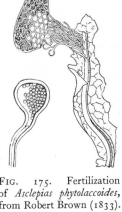

FIG. 175. Fertilization of *Asclepias phytolaccoides*, from Robert Brown (1833). *Right*, Pollen mass which has germinated. The pollen grains have sent forth their tubes, some of which are entering and descending the style which has been laid open. *Left*, Enlarged detail of pollen grain.

Sprengel noted that there are flowers which imprison and even destroy the insects that serve them. He also observed that flowers devoid of attractions for insects are fertilized by the wind or other mechanical means. These wind-fertilized flowers always produce, he saw, great quantities of very light pollen. He believed that his principles explain all the characters of flowers—position, size, colour, smell, form, season of flowering and the like.

At about the turn of the eighteenth century, several workers were occupying themselves with experiments in fertilization of flowers. T. A. Knight (p. 376) wrote his *Experiments on the fecundation of vegetables* (1799) which drew far more attention than the incomparably greater treatise of Sprengel. Working primarily on peas, Knight observed

'the variety of methods which nature has taken to disperse the [pollen], even of those plants in which it has placed the male and female plants within the same empalement. It is often scattered by an elastic exertion of the filaments which support it, on the first opening of the blossom; and its excessive lightness renders it capable of being carried to a great distance by the wind. Its position within the blossom is generally well adapted to place it on the bodies of insects, and the vellous coat of the numerous family of bees is not less well calculated to carry it. The [pollen] is often so placed that it can never reach the summit of the [pistil] unless by adventitious means; and many trials have convinced me that it has no action on any other part.'

Knight formulated the view, which long prevailed but must now be abandoned, that in no species of plant can interbreeding between male and female elements in the same flower continue to be productive through many generations.

§ 3. *Nineteenth-century Study of Pollination*

The sexual character of flowers being determined, it remained to analyse the sexual process. At the beginning of the nineteenth century the microscope was still very defective. Several physicists were at work upon it. Among them was the Italian, Giovanni Battista Amici

(1784–1860), who had already done good work in astronomy and was a professor of mathematics. This versatile and gifted man succeeded in introducing many

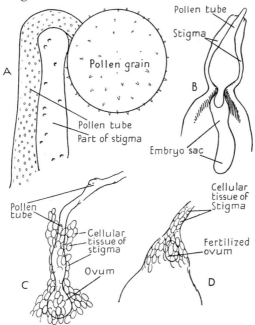

FIG. 176. Early figures of pollination.

A, Amici's drawing of formation of pollen tube in Purslane, 1823. *B, C, D,*
Schleiden's figures of fertilization in the pumpkin (1845). Schleiden regarded the
ovum as the male, the pollen as the female element. We have lettered the figures,
however, according to modern views.

improvements in the microscope, some of which, such as
the immersion lens, are still in daily use. No less skilled
was J. J. Lister (1786–1869) father of Lord Lister.
After much labour, both experimental and mathematical,
he succeeded in combining a pair of compound achromatic
lenses with perfect security against spherical error (1830).

Helped by Robert Brown (p. 350) he made many other improvements in the microscope. Of his observations the most important was perhaps that demonstrating the red blood corpuscles of the form of a biconcave lens with rounded edges (1834).

With one of his improved microscopes, Amici had seen streaming movements of protoplasm in the Alga *Chara*. In 1823 he was examining the hairs on the stigma of Purslane. He then saw a tube given off by the pollen grain, and the granular contents perform streaming movements like those in *Chara* (Fig. 176 *A*). In 1830 he even followed the pollen tubes into the ovary and he observed one to find its way into the ovule.

These observations were repeated a short time after by Robert Brown (Fig. 175) and Schleiden (Fig. 176). Unfortunately, the Scot, Brown, was too cautious, and the German, Schleiden, too bemused by theory to make the right inference from their observations. The work of both on the pollen tubes was overshadowed by further investigations by the brilliant Amici.

In 1846 Amici placed the whole matter on a firm foundation by his observation of Orchids, which are peculiarly suitable for the investigation of pollination. He demonstrated that an egg cell is present in the 'embryo sac' of the ovule before the pollen tube reaches it; that the egg cell is stimulated to further development by something brought by the pollen tube; and that this further development results in the formation of the embryo. In the meantime a firm foundation had been given to the doctrine of the nature of 'cells' (chap. ix). The examination of the sexual process was thus facilitated, and further light on the sexual process in plants was thrown by von Mohl (p. 340) and by Hofmeister (pp. 525–30).

Sprengel's great work remained generally neglected during the early nineteenth century. In 1837, Charles

Darwin became persuaded that intercrossing from flower to flower plays an important part in keeping the forms of species constant. In 1841, on the advice of Robert Brown,

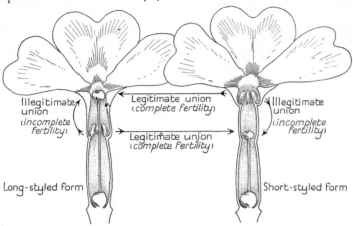

FIG. 177. The two sexual types of cowslip, *Primula veris* Linn., in vertical section. Slightly modified from Darwin (1876). The two types were observed by the South African naturalist C. H. Persoon (1770–1836) as early as 1794. Their interpretation in relation to the visits of insects and the associated process of fertilization was provided by Darwin 80 years later. The style leads down to the ovary, around which are honey-glands. The proboscis of the insect is inserted within the flower to reach this honey. If the insect visit a long-styled form, its proboscis will be dusted near the middle with pollen from the anthers. This pollen will then be in position to adhere to the stigma of a short-styled form, but not to the stigma of a long-styled form. On the other hand, if a short-styled form be visited, the proboscis will be dusted near its base with pollen. This pollen will then be in position to adhere to the stigma of a long-styled form, but not to the stigma of a short-styled form.

he read Sprengel's book. It impressed him deeply, and much of his later work on flowers arose out of it.

Darwin produced his *Fertilization of Orchids* in 1862. The book is the first in which the pollination process is set forth for a whole group of plants. Like so much of Darwin's work, it opened out a new field of research for others. Fourteen years later, Darwin published his *Effects of Cross*

H.B.—18*

and Self-Fertilization (1876), which is the complement of
its predecessor. The earlier book had shown how perfect
are the means for ensuring cross-fertilization. The later
discusses why cross-fertilization is important. Darwin had
been experimenting from 1851 on the lines suggested by
Knight (p. 514).

Some extremely interesting discoveries were made by
Darwin in certain plants which exhibit variations in the
distribution of the sexual parts. Of such species the com-

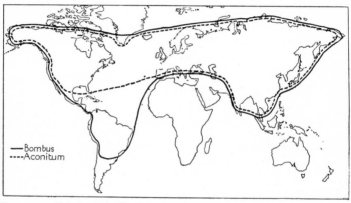

FIG. 178.   Relative Distribution of Aconite and bumble bee, from Knuth
(1898).

mon cowslip is a very good example. Cowslip plants
accord to one of two definite types. Their flowers have
either a long style and short stamens or a short style and
long stamens (Fig. 177). Such species are spoken of as
heterostyled. There are heterostyled species such as the
Purple Loosestrife which exhibit three types of individuals.

Darwin showed that in the cowslip a long-styled individual crossed by another long-styled individual is relatively infertile. To obtain complete fertility the crossing
of the two forms is needed. This is brought about by the

adaptation of the different types of flower to the form of the visiting insect, so that the pollination of the style of one type with the pollen of the other is encouraged. Thus the structure of the flower is fully adapted to that of the insect (1876).

Such work created much interest, and several naturalists threw themselves into an intensive study of the phenomena of pollination. Among these were the brothers Fritz and Hermann Müller (1829–83). The latter had produced in 1873 his classical treatise *Fertilization of Flowers by Insects and their Reciprocal Adaptation*. This was largely used by Darwin in his botanical work. Müller expressed in condensed form the doctrine toward which Knight and Darwin had been striving, by his enunciation of the law: 'whenever offspring resulting from crossing comes into serious conflict with offspring resulting from self-fertilization, the former is victorious. Only when there is no struggle for existence does self-fertilization prove satisfactory for many generations.'

Investigations along such lines as those of the Müllers were given encyclopaedic shape by Paul Knuth (1854–1900) in his *Handbook of Flower Pollination* (1898). Not only are the actual relationships of insects to many plants there demonstrated, but a geographical correspondence between many species of plants and insects is tabulated (Fig. 178).

§ 4. *Sexual Dimorphism*

In 1851 Charles Darwin published his first independent scientific treatise, *A Monograph of the Cirripedia*. Vaughan Thompson had already made important observations on these marine creatures (pp. 495–6). Darwin now discovered among the cirripedes or barnacles some remarkable sexual relations. The majority of the individuals of this extraordinary group have both male and female organs,

that is, they are 'hermaphrodite'. The Cirripedia discharge their male sexual elements into the sea and thus
fertilize one another. This process of cross-fertilization
is possible to them since they are fixed like plants, and
like many plants, live in groups. The currents of the sea
act for them as the wind acts for many plants. But while

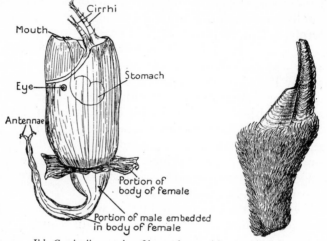

FIG. 179. *Ibla Cumingii*, a species of barnacle parasitic on other barnacles. The
male is parasitic on the female. *Right*, female × 1. *Left*, male × 40. From
Darwin.

most species of Cirripedia are cross-fertilizing, some
species are capable of self-fertilization.

The most remarkable sexual phenomena presented by
the Cirripedia have, however, yet to be considered. Dar-

Description of FIG. 180.

A, Diagram of hermaphrodite Cirripede. *B,* Hermaphrodite *Scalpellum
peronii* × 2 after Darwin. *C,* Complemental male of *S. peronii* × 20; general
appearance similar to hermaphrodite form. *D,* Complemental male of allied
species *S. vulgare* × 30 after Darwin; structure very different from hermaphrodite form. *E,* Female of *Alcippe lampas* × 10 after Darwin. *F,* Male of
lampas × 50 after Darwin.

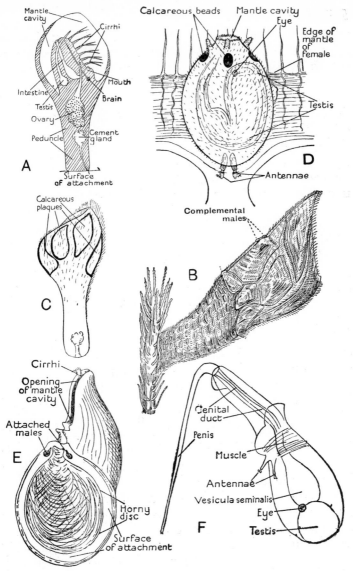

FIG. 180. Sexual Dimorphism in Cirripedia. (*See opposite page*.)

win showed that in some species there are numbers of minute and degenerate males attached to the hermaphrodite forms and living a parasitic life upon them. These he called 'complemental males', since they are not essential to the completion of the sexual process. Such complemental forms possess, in some cases, certain of the external organs of the free-swimming larvae, but their internal organization exhibits little beyond the sexual glands. In yet other species, the major form shows an ovary only and is, therefore, female and not hermaphrodite, but she has parasitic upon her a number of males which, though degenerate dwarfs, are not complemental. Lastly, it should be added that one form of barnacle described by Darwin is parasitic within the body of another species of barnacle. The female of this parasitic species exhibits some, but not all, of the degenerative changes which have been described in the complemental and dwarf males of other species, in addition to the degenerative changes common to all Cirripedia (Figs. 179 and 180).

The investigations of Darwin drew attention to the extremer manifestations of sexual differentiation which are encountered here and there throughout the animal kingdom. Such 'sexual dimorphism' is often very marked in those organisms which exhibit more or less advanced stages of parasitism. It is also known, however, in other forms as, for instance, in certain oceanic fishes (Fig. 181), in rotifers, and in spiders. It is usually the male that is dependent on the female. Sexual dimorphism manifests itself even among protozoa.

There is a type of minor sexual dimorphism that was much stressed by Darwin. Among many species of birds and mammals are forms of ornament or armament, such as the clumsy tail of the male peacock or the unwieldy antlers of the stag, the beauty and size of which may give the possessor advantage over rivals for favour of the females. The special development of these organs thus leads to more numerous progeny. This process Darwin called 'sexual selection'. Evidently it can act only within the species and may be a handicap as against other species. In introducing the subject in the *Origin of Species* (1859), Darwin draws attention to what we should now call 'sex-

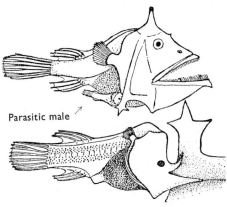

Parasitic male

FIG. 181. Sexual dimorphism in Oceanic Angler-fish *Edriolychnus Schmidtii*. Above, female with parasitic male attached. Below, parasitic male which lives attached upside down to the female. The creature inhabits middle depths of the Western Atlantic, and was first discovered in 1926.

linked characters'. 'Peculiarities', he says, 'often appear under domestication in one sex and become hereditarily attached to that sex.'

§ 5. *Alternation of Generations*

A curious aspect of animal life came into view toward the middle of the nineteenth century. The Franco-German poet Louis Adelaide de Chamisso (1781–1838), author of the famous story *Peter Schlemihl*, 'the man who sold his shadow', applied himself with effect, like his contemporary Goethe, to biological research. In a work

On certain animals of the Linnaean class Vermes (1819), he described, for the first time, the very peculiar life cycle of certain Tunicates, introducing the expression 'alternation of generations'. In 1842 the Copenhagen zoologist Japetus Steenstrup (1813–97) brought out in Danish his work *On the Alternation of Generations, or the Propagation and Development of Animals through Alternate Generations.* He showed that

'certain animals, notably jelly-fish and certain parasitic worms, habitually produce offspring, which never resemble their parent, but which, on the other hand, themselves bring forth progeny, which return in form and nature to the grandparents or more distant ancestors. Thus the maternal animal does not meet with its resemblance in its own brood, but in its descendants of the second, third or fourth generation.'

Steenstrup wrongly believed that this peculiar process 'always takes place with the intervention of a determinate number of generations'.

Steenstrup's observations drew much attention. Further evidence of such alternation of generations was rapidly accumulated from a number of different groups in the animal kingdom. It was seen to be frequently related to an alternation of methods of reproduction. A well-known instance is in the common Aphides of roses. These creatures are normally female. They produce their young without the intervention of any males, that is to say, they are parthenogenetic (Greek, 'pertaining to virgin birth'). To this Leeuwenhoek had long ago called attention (p. 168). It was now found that at times true males appear among the Aphides, and the young are then produced by means of a sexual process. Thus there is an alternation of parthenogenetic and sexual generations. Alternation of generations has since been shown to exist in a number of other animal forms. In the plant world the process has proved to have a particular importance in some forms and may be applicable in all (Fig. 190).

During the early nineteenth century there were many random observations and suggestions directed to the elucidation of the sexual process of the non-flowering plants. Carl Nägeli (1817–91) made some advance. He examined the tiny leaf-like bodies or *prothalli* that are liable to form in moist earth in the neighbourhood of ferns. These had been generally regarded as 'cotyledons', though no clear picture had been formed of the manner of reproduction of these plants. On the prothalli Nägeli observed free swimming spiral objects (1844–6). These we now know to be the male elements or *spermatozoids* of the fern (Fig. 183). They could not at the time be thus interpreted, for the female elements were unknown until Hofmeister revealed them (Fig. 184).

The scientific career of Wilhelm Hofmeister (1824–77) is unique. Much of the most important biological research in England has been done by amateurs. In Germany the amateur has taken a much less important place. Yet Hofmeister, entirely without academic training, not only won for himself the leading position among German botanists, but also came to hold an important University chair. The consensus of expert opinion places him among the very greatest of modern botanists. He was notably distinguished for the refinement and beauty of his technical methods, for which, it might be thought, training is especially needed.

Hofmeister left school at fifteen to assist in a shop, but used every spare minute for study. He read Schleiden's *Outlines of Scientific Botany* (1842), which was based on the cellular view of the plant. Inspired by this he started at nineteen, entirely self-taught, on a series of studies which placed him at one bound in the front rank of biologists.

His first important publication, *On the Embryology of*

Flowering Plants, appeared in 1849 when he was only twenty-four. It brought him immediate notice. At that time, despite the work of Amici (pp. 515–6), the essential nature of the process of fertilization was still little understood. Schleiden himself had put forth the perverse notion

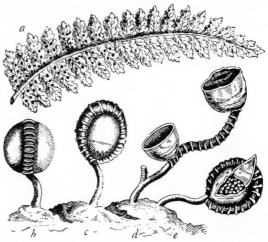

Fig. 182. Spore-bearing apparatus of *Polypodium* from Swammerdam. *a* is part of the frond of natural size. The black spots mark the collections of spore-cases, each covered in with its special roof or 'indusium'. Below, can be seen a highly magnified series of spore-cases or 'sporangia'. Each sporangium is surrounded by a ring-like structure seen on end in *b* and in its full course in *c*. The splitting or 'dehiscing' of this ring-like structure causes the dispersal of the spores. The condition after dehiscence is shown, somewhat inaccurately, in *d*, The sporangium at *e* is torn open and the spores within can be seen.

that the pollen tube was the female element, formed the embryo, and was fertilized by the ovule which was male (Fig· 176)! Hofmeister corrected this. His investigation of the process of fertilization in the flowering plants contains but one important error. He held the ancient view of fertilization by a seminal 'aura'. Thus he still conceived fertilization as a result of 'diffusion', thinking 'the direct flow of the contents of the pollen tube into the germinal vesicle to be simply impossible'.

The most important of Hofmeister's works marked an epoch in the history of botany. Its title may be translated '*Comparative researches on the germination, development, and fructification of the higher non-flowering plants (Cryptogams)*' (1851). Up to Hofmeister's time, non-flowering plants were treated in relation to the known conditions of the flowering plants. It was assumed that flowers and seeds

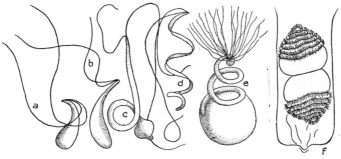

FIG. 183. A series of male cells or 'spermatozoids' exhibiting essential similarity throughout a series of lower and higher plant forms. *a* and *b* from club moss, *c* from moss, *d* from liverwort. All these have two flagella and the last is of spiral form. *e*, from a fern, is spiral and has numerous short flagella gathered at one end. *f*, from a Cycad, is top-shaped with cilia arranged along a spiral, and is double.

are concealed in non-flowering plants, which were therefore called *Cryptogams* (Greek, 'hidden marriage'), a term still current. Hofmeister, examining the development of these plants, revealed a regular and definite alternation of sexual and asexual generations in the mosses, ferns, horsetails, and liverworts. Throughout these groups, hitherto held to be diverse, he demonstrated an essential unity of developmental plan. In them all he showed that a sexless generation, which propagates itself by means of spores, alternates with another generation, which exhibits the phenomenon of sexual union of motile spermatozoids with ova contained in characteristically shaped receptacles known as *Archegonia* (Figs. 183–5).

Hofmeister was able to give a consecutive account of the life-history of the fern. He traced the spore through its development into the little green *prothallus* or true sexual generation. He saw how the prothallus develops spermatozoids and archegonia; and how the ovum in the latter, under the influence of the spermatozoid, gives rise to the more conspicuous and asexual form that we know as a 'fern'. He went on to show that the mode of production of the embryo in the pines and their allies is, in certain senses, intermediate between that of flowering plants and that of ferns and other higher cryptogams.

Hofmeister extended his investigations into the ovule of flowering plants. After tracing the elaborate process of its development, he made the suggestion that its earlier cell-divisions correspond closely to the 'prothallus' of ferns, which in flowering plants never leaves the asexual generation, is inconspicuous and reduced to a few cells (1851–62, Fig. 190).

It had already been recognized that the conifers have affinities with the flowering plants. The old distinction between the flowering and flowerless plants had now to be abandoned. What this has meant for botany may be gauged by comparing the classification of A. P. de Candolle (p. 198) with a modern classification of plants (see below). The latter is as much based upon reproduction as is the classification of animals foreshadowed by Aristotle.

Hofmeister spent little space in discussing the wider implications of his valuable contributions to science. The formal recognition of the alternating character of the generations in all plants was thus left to others. The conception was given wide circulation by the influential textbook of Sachs, especially from its fourth edition (1874) onward. It was adopted and developed by Strasburger.

On the basis of alternation of generations, we may, with Strasburger, set forth the great groups of plants in a series thus:

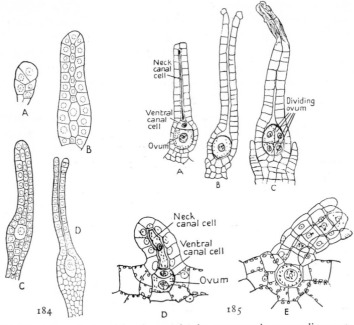

To illustrate similarity of female reproductive apparatus in mosses, liverworts, and ferns.

FIG. 184. *A–D*, successive stages in formation of archegonium of a moss. From Hofmeister, 1851.

FIG. 185. *A–C*, Archegonium of a liverwort. *A*, young form. *B*, mature with neck open. *C*, fertilized ovum dividing. *D–E*, archegonium of a fern. *D*, young form. *E*, mature. From Strasburger, 1869–71.

(a) *Thallophyta*. Relations between asexual and sexual generations in most cases irregular. Sexual reproduction by motile spermatozoids. Ovum in archegonium. Water the medium of fertilization.

(b) *Mosses*. Asexual generation alternates regularly with sexual. Sexual the more conspicuous. Asexual generation dependent on sexual. Sexual reproduction by motile spermatozoids. Ovum in archegonium. Water the medium of fertilization.

(c) *Ferns*. Asexual generation alternates with sexual ('prothallus'). Asexual more conspicuous and may increase vegetatively to an enormous degree. Sexual generation has independent life. Sexual reproduction by motile spermatozoids. Ovum in archegonium. Water the medium of fertilization.

(d) *Conifers*. Asexual generation enormously over-shadows sexual. Sexual generation entirely dependent on asexual. Sexual reproduction by non-motile generative cell. Ovum in archegonium. Air the medium of fertilization.

(e) *Flowering Plants*. Asexual generation completely conceals sexual. Sexual generation may be long deferred when reproduction becomes continuously vegetative. Sexual reproduction by non-motile generative cell. Ovum not in archegonium. Air—sometimes with insect aid —the medium of fertilization.

This striking series of sexual relationships is susceptible of exposition on an evolutionary basis. Such has been specially the work of F. O. Bower (1855–1948) from 1890 onwards. He supposed that plants with archegonia originated from algae, as is borne out by their mode of reproduction. Gradually spreading to drier and yet drier regions—invading the land from the sea—they developed a method of reproduction that depended less and less upon water as the carrier of their sexual elements. Thus the asexual generation has gradually increased in importance, finally, in the higher plants, completely concealing the sexual. Bower's views of the origin of the land flora have proved generally acceptable.

§ 6. *Early Observations on Cellular Phenomena of Sexual Union*

Despite the emphasis laid on spermatozoa in controversial writings, their essential role in fertilization was still unproven when the nineteenth century dawned. Early in the century, filtered semen was shown to be infertile, and the semen of infertile males was shown to be devoid of spermatozoa. Thus upon the spermatozoa was fastened

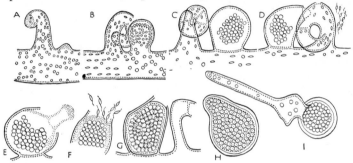

FIG. 186. Sexual process of Alga *Vaucheria sessilis*, from Pringsheim (1855), who indicates no nuclear elements. *A–D* are stages in development of the organs of sex. In *D* the spermatozoids escape and the protoplasm of the female cell is exposed at one point. In *E* the female cell sends forth a process of the nature of a 'polar body'. In *F* they are seen clustering round the opening in the cell wall of the ovum. In *G* fertilization is complete and a new cell membrane has been formed around the ovum. *H* and *I* are early stages in germination.

the power of awakening the development of the ovum. In 1838 Schwann interpreted the ovum as a cell, a conception more fully expounded later by Gegenbaur (p. 342). In 1841 Kölliker at last succeeded in tracing the development of spermatozoa. He showed that they, too, are of cellular origin and nature and dissipated much ancient myth.

At this time good progress in the knowledge of fertilization was being made from the botanical side. The general situation in 1855 was thus expressed by the German amateur Nathaniel Pringsheim (1823–94):

'The existence of sexuality in plants is now admitted. In flowering plants, the necessity of conjunction of pollen tube and ovule for the production of the embryo can no longer be denied. The sexual organs of the higher flowerless plants are also known. But with regard to the manner in which the organs participate *materially* in the act of impregnation, and even as regards the necessity for their co-operation, there are but vague surmises.'

This state of affairs was ended by Pringsheim's own observations on a fresh water alga *Vaucheria* (1855). This plant grows in filaments. He showed that, under certain circumstances, there separate off adjacent processes which develop respectively as female and male organ. The female organ contains a single protoplasmic, cell-like mass, walled off from the filament. The protoplasm in the male organ or cell divides into numerous ciliated spermatozoids. The wall of the female cell becomes thin at one point and finally opens. Pringsheim saw the spermatozoids cluster around this opening and apparently enter the female cell (Fig. 186).

This, then, was the true cellular expression of the process of fertilization. Immediately on fertilization the female cell formed around itself a delicate membrane, impermeable to the further intrusions of spermatozoids. Later, a definite cell wall was formed around the fertilized cell. The fertilized cell was then set free and Pringsheim was able to watch it germinating to form a new *Vaucheria* plant. Pringsheim's observations were seconded by those of Cohn. The cellular aspect of sexual conjugation then became known for a considerable number of cryptogams.

Pringsheim and his followers were moreover perfectly aware of the parallel character of the phenomena in the animal and vegetable kingdoms. 'Sex', he said, 'is a universal property of all organisms manifesting a wonderful analogy in the most highly organized animals, as well as in the simplest cellular plants' (1855).

In the meantime, advances in the knowledge of the sexual process in animals had been made from several quarters. Thus, in a series of communications, Newport had related the planes of segmentation of the egg of Amphibia to the point of entry of the spermatozoon (1850–4). Johannes Müller had noted that the spermatozoon in animals may enter the ovum through a special pore. This point of entry, the so-called 'micropyle' (Greek, 'small gate') Leuckart compared, in the case of insects, to a similar structure in the ovule of plants (1855).

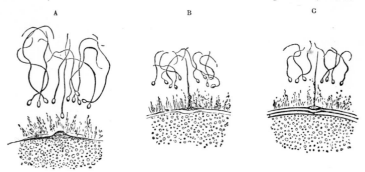

FIG. 187. Successive stages of fertilization in *Asterias glacialis* as depicted by Fol (1879).

§ 7. *Nuclear Phenomena of Sex*

Advances in technique in the 'seventies greatly improved the scope of microscopic observations. Immersion lenses of high power came into general use. New staining methods facilitated the examination of nucleus and cell contents. A searching examination of the phenomena of sexual reproduction was now undertaken. The cellular phenomena gradually became reduced to their nuclear expression.

That the result of the entry of the spermatozoon would be nuclear fusion was foreseen by Oscar Hertwig

(1875). In 1879 Hermann Fol of Geneva saw for the first time a spermatozoon penetrate an ovum, thus capping the observation of Pringsheim in plants (Fig. 187).

The extrusion of the polar cells had been seen as long ago as 1824, and in 1877 Hertwig showed that they are formed by division of the nucleus of the ovum. Intra-nuclear research on the germ-cells was initiated by Walter Flemming in 1879, when he described the splitting of the chromosomes (p. 346).

Very important additions to the knowledge of the re-productive process were made in 1887 by the researches of the Belgian cytologist, Edouard van Beneden (1845–1910) of Liége. He demonstrated that the number of chromosomes was the same for each cell in a given body, and that the number was probably characteristic for each species. He further showed that the number of chromo-somes was reduced during maturation and restored during the sexual process. For this purpose he used the intestinal round-worm of the horse, *Ascaris megalocephala*. It has the advantage for purposes of research, of possessing chromo-somes which are large and few—four in the 'diploid', two in the 'haploid' stage.

Cytological investigation during the current century has focussed attention on the behaviour of the nucleus during cell-division. A so-called *diploid* (Greek, 'double') cell generation with 2 *n* chromosomes, after reproducing itself for a while by a process of mitosis, at length gives rise to a *haploid* (Greek, 'single') cell generation, equipped with *n* chromosomes. The two types are separated by a process of *meiosis* (Greek, 'reduction'). During meiosis the number of chromosomes is halved. Pairs of certain haploid cells—the sexual cells—presently unite to form diploid cells. The process of division by mitosis is then again initiated.

In the same year (1887) as the publication of van Beneden's epoch-making work, Theodor Boveri (1862–1918), a pupil of O. Hertwig, was investigating the centrosomes and their surrounding asters. He showed their relation to the processes of division, treating them as the dynamic system of the cell. In 1890 Hertwig showed that the nuclear and cellular phenomena displayed in the formation of ovum and spermatozoon, though superficially widely different are basically alike. In 1892 Boveri drew up the general diagram of spermatogenesis and oogenesis that is still current (Fig. 188).

In 1894 Strasburger made a searching examination of

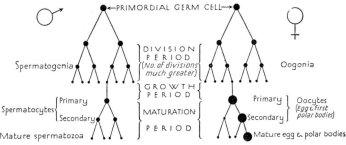

FIG. 188. Diagram of origin of spermatozoa and ova.

spore formation in higher non-flowering plants (ferns, &c.). He observed a process of reduction in the number of chromosomes similar to that distinguished by van Beneden in the maturation of the egg of *Ascaris megalocephala*. He inferred that, in the plants with which he was dealing, the chromosome number in the 'sporophyte' must be half that of the 'gametophyte'.

It is very interesting to observe that Weismann had approached this position, on theoretical grounds, in 1887. 'There must be', he said, 'another kind [of cell division] in which the primary equatorial loops are not split longitudinally, but are separated without division into two groups.' Transverse division does not, in fact, take place, but 'reduction', *i.e.* separation without division, is essential.

In 1895 Karl Rabl (1853–1917) made an important suggestion based on theoretical grounds. He expressed the view that the chromosomes retain their individuality even when the nucleus is in the 'resting' stage, at which time the chromatin usually presents the appearance of a network. This persistence of the chromosomes is presupposed by the modern doctrine of the nature of heredity (chap. xv).

A certain sequence seems universal in organisms exhibiting the phenomena of sex. The details of its expression fall into three distinct groups:

(a) *Animals, meiosis being gametic or terminal.*

In ordinary division by mitosis chromosomes split along their length (Fig. 145, p. 347). In the process which leads up to the formation of reproductive cells in animals there are two successive characteristic divisions (Fig. 189):

(i) In the former of these characteristic divisions the paternal member of each of the paired chromosomes separates from its maternal partner. They move toward opposite poles. This cell division, in which maternal and paternal chromosomes pass into their respective daughter cells, is called *reduction division*, since the number of chromosomes is thereby reduced to one-half.

(ii) The latter characteristic division is like an ordinary process of mitosis in that daughter halves of each chromosome move to opposite poles and go to form the nuclei of separate cells. These separate cells develop as the true sexual cells. The characters of these cells, however, differ fundamentally in the two sexes.

In the male the four resulting haploid cells are identical in form and size. They are smaller than the ovum and contain little protoplasm. They become spermatozoa which, typically, consist of a head in which the nucleus is packed, a neck which contains the centrosome, and a vibratile protoplasmic tail by which the spermatozoon is propelled (Fig. 189).

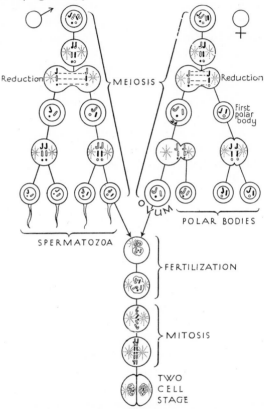

Fig. 189. Diagram illustrating the fate of the chromosomes in animals during meiosis and subsequent fertilization.

In the female the four haploid cells differ from one another. One, the ovum, is always relatively large. The three smaller—the polar cells—play no further important part (Fig. 189).

Spermatozoa are formed in far greater numbers than ova. At the time of fertilization they beset the ovum. One enters it.

The ovum now surrounds itself with a membrane, excluding other male elements. The sperm nucleus that has gained admittance enlarges. It approaches the egg nucleus. The boundary walls of both nuclei fade. The chromosomes become apparent. As they do so a mitotic spindle appears around them. Thus the original diploid number, 2 *n*, of chromosomes is restored. Each chromosome now splits lengthwise, preparatory to the first division of the egg. Divisions are henceforth by mitosis. They produce diploid cells until the time comes again for the formation of reproductive cells (Fig. 189).

(b) *Plants, meiosis being sporic or intermediate.*

There are certain algae in which haploid (*gametophyte*) and diploid (*sporophyte*) cell generations form separate plants of comparable size and appearance. Higher forms, however, present a series on which there is a progressive recession of the gametophyte until it becomes vestigial.

In flowering plants the gametophyte has become minute. In the female it is represented by the products of the embryo sac (*megaspore*). In the male it has shrunk yet farther, and is represented only by the pollen grain (*microspore*) which produces a tube containing three nuclei of which two are the generative nuclei. If these vestigial structures were to disappear, the life cycle of the higher plants would resemble that of animals and be devoid of an independent haploid generation. The suggestion was made by Strasburger that animals had in fact dropped their alternate generations, and that the polar cells (Fig. 190) represented the last remains of an asexual generation (1894).

The development of the sporophyte is characterized by meiosis leading to spore formation. These spores, like gametes of animals, are formed in quartets. There is also some parallel to the formation of polar cells, for the

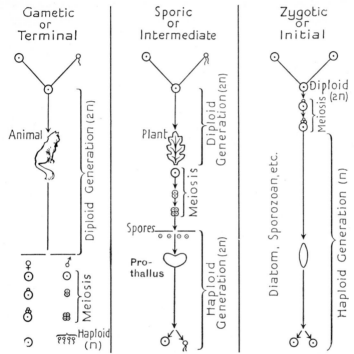

Fig. 190. Diagram of the three types of alternation of cell generations.

mother-cell in plants typically undergoes the meiotic division to form a series of which only one becomes sexually functional.

(c) *A small group of lowly organisms in which meiosis is zygotic or initial.*

In animals a diploid organized being intervenes between one conjugation of haploid cells and the next. In plants there intervene characteristically, as in ferns, two independent beings, one diploid, the *sporophyte*, and one haploid, the *gametophyte*. There is a small group of organisms in which meiosis does not take place till *after* conjugation. The products of the zygote in this group are diploid, but immediately become haploid. The organism itself is haploid, and becomes diploid only for a brief interval. The group contains a few lowly animals and plants—*Spirogyra*, some diatoms, certain protozoa (the Sporozoa). The organisms that exhibit this sequence do not form a concrete group (Fig. 190).

Great historical importance attaches to the knowledge of the sexual process in Protozoa. Weismann set forth the view that Protozoa are essentially 'immortal', being without somatic (non-germinal or body) elements (1881, p. 531). This view was opposed by Emil Maupas (1844–1916), who traced the actual history of a series of generations of protozoa (1888–9). He found in *Paramecium* that after generations of division, the stock becomes enfeebled. Ultimately the individuals die, unless they meet partners for conjugation. When two such enfeebled individuals meet, their bodies are approximated and they interchange nuclear elements. Next, division takes place. The stock has regained its vigour. No differentiation of sexes is evident in the conjugatory process, though such differentiation can be traced in other protozoa.

Maupas regarded conjugation as essential to the continued life of protozoa. Subsequent research has shown the matter to be far less simple. The American protozoologist, G. N. Calkins (1869–1943), postponed indefinitely the process of enfeeblement by adding suitable substances to the medium in which the organisms live (1904). *Paramecium* has since been traced through thousands of generations without conjugation.

§ 8. *Factors Determining Sex and Sex Characters*

The development of a fertilized egg has been found to proceed along a line that is determinate according to species. But species are in most cases made up of male and female individuals. What determines which of these two directions development shall take?

From an early time there have been theories on this point. The first observations that led to productive discussion, however, were made by the German, H. Henking. In the division that leads to the formation of spermatozoa of an insect he detected (1891) 'a peculiar chromatin element'. The other chromosomes divide and separate in anaphase, but this remains undivided and passes to one pole. Thus half of the spermatozoa have this element and the others are without it. Henking labelled it X. He was unaware of its nature.

In 1901 one American cytologist T. H. Montgomery (1873–1912), and in 1902 another, W. S. Sutton (1876–1916), working on grasshoppers, found the body chromosomes to be individually recognizable from cell to cell by their sizes and shapes. Sutton noted their tendency at meiosis to arrange themselves in pairs—in his species eleven pairs and one unpaired. Since then investigation of the body cells of a large number of animals has shown that in one sex there is usually a pair of chromosomes of which one member is not identical in size with the other. Of this pair the larger is called the X-chromosome, the smaller the Y-chromosome. In the other sex there is a corresponding pair of chromosomes, but both members are alike and are, moreover, both like the X-chromosome of the other sex. It has repeatedly been clearly shown that this chromosome pair is primarily responsible for determining the sex of the organism. In some animals there is no Y-chromosome: then one sex has a single X-chromosome, the other two. During the process of meiosis in gamete-formation (p. 539) the members of all pairs of somatic chromosomes are separated from each other, one of each pair going into a different gamete.

An animal, the sex-chromosome constitution of which is XX in its body cells, will therefore form only one sort of gamete and all the gametes will contain one X. The sex in which the somatic constitution is either XY or XO will form two types of gamete, one will contain one X-chromosome, while the other will contain either one Y-chromosome or else no sex chromosome at all. The sex which forms two types of gamete is known as the 'heterogametic' sex; that which forms only one type is the 'homogametic' sex.

In 1901 C. E. McClung (1870–1946), working on grasshoppers, described two types of spermatozoa of which one had one chromosome more than the other. In this animal, therefore, the male is the heterogametic sex. This is true in most groups of animals. The male is heterogametic, there are two types of sperm but only one type of egg, and the sex of the offspring is determined by the type of sperm that fertilises the egg. An X-sperm fertilising the egg will produce an XX, i.e. female zygote. A Y or O-sperm will produce an XY or XO, i.e. male zygote. But in some insects, as the butterflies and moths, in certain fishes and amphibia, and in birds, the female is heterogametic, there being two types of egg and one of sperm.

Although the sex-chromosomes are chiefly concerned in sex-determination, part of them also controls certain body characteristics unconnected with sex. These latter are known as 'sex-linked characters' Most bisexual animals possess a chromosomal sex-determining mechanism of one of the two types described above.

When an organism is said to be of a certain sex three things are implied. First, that it produces one type of gamete, either eggs or sperms. Second, that it has a certain characteristic structure of its sexual organs, necessary for the production of the gametes concerned and for the ensurance of their union with those of the opposite sex. And third, that it has certain characteristics (as plumage-colour in birds or horns in deer or voice in man) always associated with the formation of one type of gamete. These last are obviously less closely related to the sexual processes than are the structure and character of the sex-organs. They are known as 'secondary sexual characters'.

The differentiation of sex organs, such as ducts and external genitalia, and of the secondary sex characters, is not directly under the control of the sex-chromosomes in the cells which compose them: it depends on substances secreted by the developing gonad. These substances are the *sex-hormones*, which induce formation of the organs of the appropriate sex from the indifferent rudiments in the embryo. Thus in vertebrates interference with the normal hormone-production, by injury or disease, will upset sex determination in spite of the normality of the sex mechanism. In insects, on the other hand, hormones are less important in the development of sex characters. An abnormality of somatic cell division during development may produce cells with the chromosomal constitution of the cells of the opposite sex. These cells may then form a mosaic area showing the characters of one sex on the body of an insect that elsewhere shows those of the other. This shows a more direct effect of chromosomes on sex characters than one mediated through the gonad and its hormones. The action of hormones, however, is beyond our scope since these 'chemical messengers' were hardly recognised till 1903.

§ 9. *Parthenogenesis*

It was pointed out in the previous section that the males

of certain animals arise parthenogenetically, i.e. from unfertilised eggs. In such cases no male parent takes part in the process, and the offspring are haploid. Complete parthenogenesis is also found among certain animals. In some animals sexuality has completely disappeared, only diploid females exist, and these arise from unfertilised eggs. Thus all Rotifers of the group Bdelloidea and some insects such as the 'walking sticks' show complete parthenogenesis. A third type is 'cyclical parthenogenesis': in which generations of diploid females, parthenogenetically produced, alternate with ordinary bisexual generations. Cyclical parthenogenesis occurs in aphids. Here diploid sexually produced females give rise during the summer to an indefinite number of parthenogenetic diploid generations. As the season wanes individuals appear which produce parthenogenetic diploid females of two types. One of these types produces sexual females only, the other sexual males only. The ova and sperms from these diploid sexual individuals undergo normal meiosis, and subsequently unite to form diploid fertilised eggs which rest through the winter and yield parthenogenetic females in the new year. Similar phenomena are found in other insects and in some Rotifers.

The existence of parthenogenetic phenomena has been recognized since the time of Bonnet (*Traité d'insectolologie*, 1745). But in the later nineteenth century interest passed from the evident external phenomena to the cell, and especially to the nucleus. It was observed as early as 1869 that the parthenogenetic eggs of Aphids formed only one polar body (Fig. 189), while eggs normally fertilised formed two. Weismann confirmed these observations, and regarded the formation of only one polar body as characteristic of all types of parthenogenesis (1886). This fits well with our modern knowledge that the ommission of meiosis will keep the eggs diploid since only one cell division will be concerned in gametogenesis. If eggs are to be fertilised meiosis will take place, the cells will become haploid, and two polar bodies will be formed. The diploid number of chromosomes is restored by union with the sperm.

Weismann's observations and conclusions stand as true for most cases of parthenogenesis. In them meiosis is normally omitted and the eggs are diploid. There are cases of parthenogenesis, however, in which the eggs do undergo meiosis, and give off two polar bodies. In these cases the diploid chromosome number is restored either by fusion of one polar body with the egg, as observed by O. Hertwig in a starfish (1890), or by a fusion in pairs of the cells of the developing embryo. The organism is thus exercising in these cases a sort of self-fertilisation.

Summarising the above facts we see that there are three main types of parthenogenesis:

(*a*) Haploid parthenogenesis. Male and female individuals occur; the former are haploid and formed from parthenogenetic eggs, the latter are diploid and formed from fertilised eggs.

(*b*) Diploid parthenogenesis—complete. Only females are known.

(*c*) Diploid parthenogenesis—cyclical. Generations of parthenogenetic females alternate with bisexual generations. Diploid parthenogenesis may be either meiotic or ameiotic. In the former case meiosis occurs, but the chromosome-number is restored either by fusion of the egg with a polar body or between the dividing cells of the zygote. In ameiotic parthenogenesis meiosis does not occur, and only one polar body is formed.

Animals in which complete parthenogenesis is found often show polyploidy. That is, that whereas the normal diploid organism contains two of each kind of chromosome, in polyploids each chromosome is represented three, four or more times, so that the total chromosome number is a multiple of the normal. This condition is found in Crustacea in the parthenogenetic *Artemia*, the brine shrimp, in the Ostracod *Cypris*, and in some insects. Polypolidy is common in plants; it can evidently occur in animals only when a chromosomal sex-determining mechanism has been abandoned.

The phenomena of sex, despite much that is common to them all, show considerable variety in the different groups of living organisms. It is not, therefore, surprising that no single hypothesis has yet been put forward that adequately covers all the phenomena of biparental reproduction. A widely discussed theory of the genetic control of sexuality is that of Goldschmidt, developed from his work on the Gypsy moth, *Lymantria dispar;* it has been supported by the result of the work by Bridges on intersexes in *Drosophila*. All cells of both sexes are regarded as containing sets of factors giving the potentiality for both male and female development. These factors are controlled by 'realizers'—factors present in different amounts in the different sexes. But it is not to be supposed that a single hypothesis will adequately cover all phenomena of the genetic determination of sex in living organisms. It is clear that sex is a vital process in the distribution of variation in living organisms (chap. xv) and thus a prime factor in evolution. Nevertheless many organisms, such as bacteria and viruses, get on perfectly well without sex and show inheritable variation. These simple forms of life have survived in the struggle for existence, and if actual numbers be the test, they are the most successful of all organisms. It seems clear however that sex has evolved as a necessary adjunct to the specialised method of the transmission of hereditary characters as adapted by the higher organisms.

FIG. 191. Gregor Mendel (1822–84).

XV

MECHANISM OF HEREDITY

1. *Earlier Conceptions of Heredity*

WE have glanced at some of the factors which have from time to time been thought to have determined the formation of species (chap. viii). Now it is evident that there must be some inward means for conveying factors that are the outward signs of evolution. This means the early followers of Darwin vaguely considered as 'inheritance', the process by which, in the course of reproduction, the offspring come to resemble the parents.

It needs to be emphasized that not only the word *inheritance* and its fellow *heredity* but also the ideas first implied by them are of legal origin. The legal conceptions of *inheritance* have no resemblance to the biological facts of *heredity*. Nevertheless in the course of history, the legal conception has influenced the biological. Strange to say, biology has still not quite shaken itself free from this extraordinary misalliance. The power of words, terminology, over scientific ideas is very great—far greater than most men of science are willing to admit.

Despite the manifest fact that offspring resemble their parents, it has always been recognized that they are never exactly like their parents. Moreover, brothers and sisters are not exactly like one another, though they are usually more like one another than they are like other folk. Apart from the influence of environment—of the scars that our personal history leaves upon us—we are made up of qualities that we inherit from our ancestors and particularly from our parents, together with qualities that are peculiarly our own and in regard to which we differ or vary from all other beings. *Heredity* and *variation* are the two elements that determine the stream of life. So much was recognized by Darwin and his immediate followers.

A great part of their work consisted in a search for examples of variation within the limits of species. Moreover, the systematists among them occupied themselves largely in describing the varieties—local and other—of the species with which they had to deal. Many of the leading exponents of evolution sought within the tissues for evidence of some apparatus to explain the mechanism alike of inheritance and variation. For centuries too, stockbreeders and farmers have been occupied in producing and preserving strains of animals and plants each of which had its own peculiar excellences. Such men were not given to literary expression and were moreover often unwilling to disclose the nature of their processes. However it is evident that they were very successful in producing specialized stocks, such as the racehorse and the carthorse, the bulldog and the greyhound, which like the merino sheep, the maize, the wheat and the beet of commerce were the results of careful selection long before scientific men began to discuss the mechanism of heredity.

In 1863 Herbert Spencer expounded his theory of 'physiological units'. These units were reminiscent, he believed, of the atoms of Lucretius. Spencer conceived that every species is endowed with its own type of physiological unit, each unit being capable, under certain circumstances, of reproducing the whole organism. He regarded the units as larger and more complex than the supposed protoplasmic molecule.

In 1868 Darwin, in his *Variation in Animals and Plants under Domestication*, set forth his theory of 'Pangenesis'. It is perhaps the great naturalist's least satisfactory effort. The doctrine assumes that, through a large part of the life of the organism, particles representative of the various organs pass from them to the reproductive elements and convey their own nature to ovum and spermatozoon. Thus, Darwin thought, characters acquired by the parent might be passed on to the offspring.

On this theory three observations may be made. First, it is described in all essentials by an ancient writer of about 400 B.C. whose work comes down to us under the name of Hippocrates (pp. 1 ff). The theory is repeated by Aristotle about 350 B.C. Several writers of the eighteenth century, Buffon among them, anticipated this theory. Second, no mechanism has been demonstrated in the animal body by which the action of pangenesis could be explained. Third, the facts of experimental embryology are against the existence of any such mechanism. A fertilized egg is a single cell which, by division, forms the body of the offspring. It has, and may convey to its cell descendants, potentiality to become a complete organism. But, among multicellular organisms, this capacity of the cells decreases as division advances, though the rate and manner of decrease vary with the organism and with its conditions. The rate, manner, and causes of cellular differentiation are among the prime problems of the science of experimental embryology (chap. xiii).

A new theory destined to have more influence than pangenesis was the theory of the 'germ-plasm' enunciated by Weismann (pp. 553–5). In the meantime attention was diverted from the essence of the problem to statistical study of the phenomena.

§ 2. *Galton* (1822–1911) *and the Statistical Study of the Phenomena of Heredity*

The importance of Francis Galton, a grandson of Erasmus Darwin and a cousin of Charles Darwin, is chiefly due to his appreciation of the value of statistical treatment of biological material.

That method had been introduced into biology by the very able Belgian astronomer Lambert Quetelet (1796–1874). He made statistical investigations on the development of the physical and intellectual qualities of man and on the 'average man' both physically and intellectually considered (1835–46). He followed this in 1848 by his treatise, *On the Social System and the Laws which govern it*. In this work Quetelet shows how the relation of the numbers representing the individual qualities of many men to the numbers referring to the same qualities in the average man can be shown to conform to the principles of the theory of probabilities. This conception, elaborated and further analysed, has formed the basis of all subsequent researches in vital statistics.

Francis Galton had a long and varied scientific career. Thus he made important contributions to meteorology, introducing the word and the idea of 'anticyclone' (1863). From 1865 onward, however, he devoted himself to investigating the laws of heredity. He demonstrated that many attributes, often regarded as purely qualitative, may be given exact numerical expression and may thus be treated statistically.

Among Galton's successes were his demonstration of the individuality and permanent character of the ridges on the balls of the fingers (1892–5), his introduction of composite photographs (1879), his division of mental images as 'visual' and 'auditory' (1879–81), his work on the hereditary nature of ability (1869–89), and his conception that the human race is susceptible of indefinite improvement by breeding, the study of which he named *Eugenics* (1883). His leading contribution to the science of statistics is his demonstration, with Karl Pearson (1857–1936), that the degree of relationship between any two attributes in any set of individuals is expressible as a numerical factor, the 'correlation'. In his *Inquiries into human faculty and its development* (1883) and in his *Natural Inheritance* (1889) and subsequent works he applied this mathematical method, notably to the inheritance of human stature.

Galton made an attempt to reach an exact expression for the proportionate contribution, by each generation of ancestors, to the make-up of the individual. He worked especially on the colour of a race of dogs of known pedigree. On this investigation he based his 'Law of Ancestral Inheritance'. According to this law, to the total heritage of the individual the parents together, on the average, contribute $\frac{1}{2}$, the grandparents together $\frac{1}{4}$, the great-grandparents $\frac{1}{8}$, and so on, the summation of the series $\frac{1}{2}+\frac{1}{4}+\frac{1}{8}+\frac{1}{16}$ . . . to infinity being 1. The law can be applied, in any event, only to certain characters and within narrow bounds. Even as thus limited it has needed modification. Such significance as it may have, however, has been entirely changed by the rediscovery of the work of Mendel (p. 563 ff).

Galton sometimes allowed his faith in the possibilities of measurement to mislead him. Thus he drew up a statistical table of the incidence of mutual ill-temper in married couples. He made a map of the distribution of beauty in Great Britain, and he attempted to estimate statistically the efficacy of prayer. His determination to treat biological material only in the mass prevented him from examining cases of individual discontinuous variation. Thus, though he worked on the heredity of eye colour—one of the classical illustrations of Mendelian inheritance—and experimented with peas—on which Mendel's own work was done—and even contrasted, at times, 'blended inheritance' and 'sports', yet he and his followers entirely missed the Mendelian phenomena. The general failure to appreciate Mendel's work for thirty-five years is one of the most extraordinary events of nineteenth century science.

The work of Galton was continued by Karl Pearson (1857–1936). He and his pupils have made many contributions to the exact and statistical expression of biological phenomena. The special method of study founded by Pearson and Galton has become known as *Biometrics* (Greek, 'life measurement').

§3. *Weismann* (1834–1914) *and the Germ-Plasm*

The elaboration of the conception of protoplasm was contemporary with the diffusion of the doctrine of Organic Evolution. These conceptions led naturally to the idea of the historic continuity of living substance. But the protoplasm of the germ-cells is clearly in a different position from that of the other cells of the body. It conveys the hereditary elements, whatever those may be. Thus stress came to be laid specially on the protoplasm of ovum and spermatozoon. The other cells of the body ultimately die. But ovum and sperm go on and carry the hereditary factors from generation to generation in a continuous stream. This view was specially developed by Weismann in connexion with the problems of heredity and variation. It is well to remember, however, that it needed neither Darwin nor Weismann nor any theory of protoplasm, nor the science of cytology to make it fairly evident that 'we live again in our children' in a physical as well as a spiritual sense.

August Weismann of Freiburg was a pupil of Leuckart (p. 323). In middle life he began to suffer with eye trouble which made biological observation impossible to him. This explains some of the weaknesses and, strange to say, also some of the strong points of his work.

Weismann developed with great theoretic skill the conception of the continuity of substance from parent to offspring. He gave to the substance the name *germ-plasm*. By its means he endeavoured to explain the phenomena of heredity.   His views have exercised much influence on the development of biological thought.

For Weismann, the offspring resembles the parent simply because it is derived from the same substance, i.e. the germ-plasm, the essential germinal elements of the germ-cells.   The body with its other cells he looked on as a vehicle for the conveyance of the germ-plasm.   The body-cells do not convey any of their own substance, but they nourish, protect, and carry the germ-cells, and ultimately pass them to the succeeding generation.   These germ-cells in turn have been passed on from a preceding generation, and so on.

Fig. 192.   Diagram to illustrate Weismann's conception of the continuity of the germ-plasm.   Only the germ cells (black) carrying the germ-plasm are continued from generation to generation.   The body cells (white) are destined for death.

The picture that Weismann raised is of a continuous stream of life.   On its banks are set escapement channels from which great sheets or ponds of water are, at times, filled.   It is for the sake of these sheets or ponds that the conduit has been made, but nothing that happens in them alters the stream.   And, while the stream goes on, they evaporate and disappear.   The stream is the germ-plasm, the escapements are the developmental processes, the watery expansions are the individuals.   Samuel Butler (1835–1902) put it well when he said 'a hen is only an egg's way of producing another egg' (Fig. 192).

These seemingly simple germ-cells transmit an inconceivably complex inheritance.   The germ substance has passed through the countless generations that have been. It contains, wrapped in its mysterious complexity, all the possibilities of the generations that are to be. Some such conclusion can hardly be evaded. It raises a picture of mystery and wonder that is very different from that atmosphere of simple, comprehensible, measurable factors with which men of science habitually deal, or think they deal. The picture is not very far from those mysteries associated with religion.

The question arose whether the germ-plasm is homogeneous, or whether some parts of it, rather than or more than others, carry the hereditary qualities? The question could only be answered by the examination of the structure and behaviour of the germ-cells and notably of their nuclei. This task occupied many observers during the

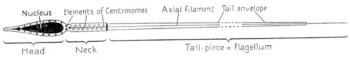

Nucleus   Elements of Centrosomes   Axial filament   Tail envelope

Head   Neck   Tail-piece = Flagellum

FIG. 193. Diagram of spermatozoon.

later nineteenth century and has been continued with great activity to our own time.

It soon became clear that the nuclei of the generative cells carry at least most of the hereditary elements. This must be so, since the spermatozoon, which conveys inheritance as much as the ovum, consists of little but nucleus (Fig. 193). Moreover, it was found that part of the protoplasm of the egg could be removed without affecting the general course of development.

Weismann's stress on the importance of the germ-plasm led him to regard it as the sole transmitter of the heritage of the past. Could it be altered in the present? In other words, could characters impressed on it from without be passed on to the offspring?

The matter seemed susceptible of experimental proof. The experiments applied in Weismann's day were of the nature of mutilations of the adult parent. All available evidence shows that characters thus modified are *not* inherited. Experiments so far confirmed Weismann's theoretical view that the germ-plasm cannot be so modified from without as to communicate the modification to subsequent generations. The range of variation from the parents, like the resemblance to the parents, must then, according to Weismann, be transmitted by the germ-plasm and by the germ-plasm only.

Weismann elaborated a theory of passage, through the germ-plasm, of particulate elements representative of the body as a whole. His view was based upon evidence derived from the study of variation. These particulate hereditary elements, or *ids* as he called them, he came to identify with small parts of the chromosomes or, in some cases, with the chromosomes themselves. Each id, he believed, contains all the elements essential for development as an individual, though the ids differ in representing different ancestral types of individual. Variation he regarded as due to different combinations of ids, the result of union of ovum and spermatozoon. The number of ids would double at each such union if it were not for meiosis, the existence of which was predicted by Weismann (Fig. 189, p. 537). This was a very great intellectual achievement. The services of Weismann to biological thought may be thus summarized:

(*a*) He challenged Lamarckianism (pp. 297–300) and thus stimulated the experimental study of variation.
(*b*) He introduced the valuable conception of the isolation and continuity of the germ-plasm (p. 554).
(*c*) He fastened on the chromosomes the responsibility for conveying the hereditary elements (p. 547).
(*d*) He predicted the phenomena of meiosis (pp. 535–6).
(*e*) His theory as to the mechanism of inheritance has helped to shape the current doctrine.

Following on Weismann, biologists became sharply divided into two schools, one accepting, the other rejecting, belief in the inheritance of acquired characters. More recent work has avoided the method of mutilation. It has investigated rather changes in the organism induced by varying the environment in such ways as may be expected in nature. Whether such induced changes are inherited or not is still a matter of controversy. The interest of the question has, however, faded somewhat. The problem shifted with new knowledge of the phenomena of Mendelism and mutations. The theory of the gene seeks to cover these phenomena. The question is whether the distribution of genes can or cannot be altered in the gametes or zygotes. The answer of 1958 is that it can.

§ 4. *Discontinuous Variation*, 1859–1900

Darwin's views gained immediate acceptance on the publication of the *Origin* (1859). The essential element in his teaching was that within a species there are innumerable small variations. These variations he held are 'accidental', of a 'chance' character. The words 'accidental' and 'chance' merely express ignorance of immediate causation. It is remarkable that the Darwinians themselves did not seek what lay at the back of the words. This failure accounts for the divorce between the philosophy and the science of the period. The philosophers saw that Darwinian 'variations' explain nothing until the elements 'chance' and 'accident' in them can be given some meaning. Biologists did not see this. In our own time, and following on Mendel's work, the occurrence of some variations, at least, has been reduced to exact numerical expression. Further, the physical factors that these mathematical rules express have become the main quarry of biological pursuit. The biologist is thus seeking the immediate causes of variation. He is hard upon the trail, and we hear far less of that *deus ex machina*, the Darwinian 'chance'.

Darwinian theory demanded that species should produce endless variations, some, at least, favourable to the needs of the organism. It was an essential corollary that species could not be exactly delimited. They would tend always to be giving off varieties which would gradually pass into new species. Thus the work of the systematist was merely formal, at least when regarded *sub specie aeternitatis*. The systematist, according to the Darwinian view, is not really distinguishing species from each other, but only species as they exist at his particular point of time. Since new species must be ever in process of formation in nature, we ought to be able to trace the process.

In 1859 Darwin and his followers confidently hoped that this would be possible. That was a hundred years ago. No clear case has, however, been reported, despite all the search and research that has been made. During all these hundred years, there has accumulated a body of irresistible historical evidence of a geological character that species do actually evolve. Yet we are still far from any knowledge of how the process takes place. It is only within the last generation that hopeful lines of work on this topic have come into view. We can hardly discuss them here.

There is a certain extraordinary phenomenon in the history of Darwinism. That body of teaching should have influenced, above all others, those occupied in examining and classifying species. Almost all important systematists of the later nineteenth century were professed Darwinians. And yet Darwinian doctrine influenced their work surprisingly little.

One example will suffice. J. D. Hooker died in 1911 at the age of 94. He was one of the ablest and most voluminous botanical systematists that has ever lived. He was on terms of intimate friendship with Darwin for forty-two years. He saw the drafts of the *Origin of Species* long before the book was published. From the very first he declared his adhesion to Darwinian doctrine. His most important work appeared in 1860 and contains an 'Introductory essay' expressing his Darwinian views. The first systematic work from his pen appeared in 1840 and the last in 1911—a stretch of seventy-one years. Yet in all this mass, Hooker proceeded on his systematizing way almost as though Darwin had never arisen. It would seem as though such a man should have said to himself 'This systematizing is of no use. Darwin has shown us a better way. We must concentrate on one or two species and see how they vary and pass into new species. It is now evident that species have no objective existence apart from the immediate present.' But neither Hooker nor any of the great systematists did anything of the sort. They acted as though, in their inmost minds, they were convinced that somehow or other species were real existences with real boundaries. The march of knowledge has, on the whole, supported them. How species have become realities is another matter and a matter on which we must, for the present, say simply *ignoramus*. There is very good reason to believe that we need not add *ignorabimus*.

Kölliker and Nägeli in criticism of Darwin (pp. 310–11) had seized upon the question of variation, and expressed the view that species change suddenly or discontinuously. In 1870 Darwin's own lieutenant, Huxley, declared his belief in discontinuous variation. 'We greatly suspect', he wrote in 1870, 'that [Nature] does make considerable jumps in the way of variation now and then, and that these *saltations* give rise to some of the gaps which appear to exist in the series of known forms.' Galton expressed himself guardedly in somewhat the same way (1889).

During the decade preceding 1900, examples of discontinuous variation or 'saltations' were collected by the Cambridge biologist, William Bateson (1861–1926). The results were set forth in his great book *Materials for the study of variation treated with especial regard to discontinuity in the origin of species* (1894). It is interesting to follow Bateson's procedure in trying to find a solution of the kind of discontinuity present in serial homology (p. 222, &c.) and kindred morphological series. From his vague comparisons of the segmentation of animals as exhibited in their back bones, &c., with ripple marks and other physical phenomena, it is evident that, despite his attacks on the methods of the old ideational morphology, he was still under the influence of that school. There was much of the *Naturphilosoph* in him. From this he was freed by the rediscovery of Mendel's paper (1900), with its experimental proof of the discontinuous inheritance of variation. The clear-cut experiments of Mendel are the very antipodes of the ideational morphology of Goethe and Oken.

§ 5. *De Vries* (1848–1935) *and the Doctrine of Mutations*

A new period in the investigation of the phenomena of heredity was opened by Hugo De Vries, a pupil of Julius Sachs (1832–97) and professor of botany at Amsterdam. He developed in his *Intracellulare Pangenesis* (1889) a special theory of heredity related to that of Darwin. De Vries emphasized that in individuals, hereditary characters may vary independently of one another and, on the other hand, may combine with each other. His *pangens*, introduced to explain phenomena of this order, were hypothetical material entities representing these individual variations. Of them he said:

'Every hereditary character has its special kind of pangen. In the nucleus every kind of pangen of the given individual is represented, the rest of the protoplasm in each cell containing only those that are to be active in that cell.'

The pangens of De Vries could multiply. Thus he believed that the form and activity of the tissue cell are determined by those pangens which leave the nucleus. The pangen community which remain within the nucleus retain there the characters of the individual. 'All living protoplasms consist of such pangens derived at different times from the nucleus, together with their descendants.'

De Vries saw, however, that despite the independent variation in individual characters, species in nature remain extraordinarily constant. They do not exhibit those slight but frequent and cumulative variations which the Darwinian theory demands for the formation of new species. In nature, series of individuals leading from one species to another are conspicuous by their absence. It occurred to De Vries therefore to seek illustrations of Kolliker's view that species change suddenly or discontinuously (p. 310).

In a waste meadow he found (1886) a colony of plants that clearly presented discontinuous variation. The American evening primrose, *Oenothera lamarckiana*, was growing there freely as a garden escape. Some of the individuals differed very markedly from the normal type. De Vries permitted one of these variants, *O. gigas*, as well as *O. lamarckiana*, to multiply by self-fertilization, under close observation. From each of these two, there arose in the next few generations many new, true breeding and definite types. The species seemed to be disintegrating into a number of forms so different from the old as to justify their description as 'species'. This was a very different thing from the gradual production of domestic 'varieties' on which so much stress had been laid by Darwin.

After many years of study, De Vries expounded his views in his *Mutationslehre* (1901–3). According to his 'mutation theory':

'The properties of the organism are made up of units sharply distinguishable from one another. These units are bound up in groups and, in related species, the same units and groups of units recur. Transitions, such as are seen in the outer forms of animals and plants, no more exist between the units than between the molecules of the chemist.

'Species are not continuously connected but arise through sudden changes or steps. Each new unit added to those already present forms a step and separates the type as a species independent from the species from which it arises. The new species is a sudden appearance. It arises without visible preparation and without transitions.'

While De Vries was thus investigating 'mutations', Bateson was also reaching the point of view that many variations are produced by sudden jumps and not by the finely graded minute differences that Darwin postulated. Thus he, as well as De Vries, was prepared to receive Mendel's doctrine of discontinuous variation on its rediscovery.

During the decade following publication of the theory of De Vries, the Dane, W. L. Johannsen (1857–1929), put to experimental test the Darwinian assumption that the characters of a race could be changed by selection in a given direction. He showed that the beans produced from a pure line give a definite range of weights distributed according to the normal curve of error. With self-fertilizing beans this continued through generation after generation, however the beans were selected from each generation of the line. The heaviest beans and the lightest produced offspring that ranged in the same way between the same weights. In one case he observed that the range shifted spontaneously and the change was preserved in the descendants. This change was a true *mutation*, susceptible of numerical and statistical expression. This was the first observation of the kind in a whole population.

§ 6. *The Work of Mendel* (1822–84) *and its Rediscovery* (1900)

Gregor Johann Mendel was of peasant stock. He entered a convent at Brünn in Moravia in 1843. Later he went to the University of Vienna. There he studied the physical sciences. He returned in 1851 to his convent, where he passed the rest of his life, first as a teacher and later as Abbot. The experiments by which Mendel became posthumously famous were performed in the monastery garden. They were begun about 1857 and terminated in 1868. He published his results in two communications to a local journal in 1866 and 1869. They were unnoticed by the scientific world. Darwin did not know of their existence. Nägeli, who corresponded with Mendel, ignored them. They were discovered in 1900 and his appearance in the world of science is therefore posthumous and effectively of that year.

Examining pea plants, Mendel found two sharply marked races, the tall and the short. Normally pea flowers fertilize themselves, but Mendel experimentally fertilized flowers of tall plants with pollen of short. The offspring, *the first filial generation*, were all tall plants. Tallness prevailed over shortness. Tallness is *dominant* as we now say, and shortness *recessive*.

Mendel next let the flowers of this 'first filial generation' be fertilized with their own pollen. In the following or 'second filial generation', shortness reappeared in some of the offspring. The opposed characters of tallness and shortness were, however, distributed not at random but in a definite, constant, and simple ratio. This ratio gave an average of three dominant talls to one recessive short.

Moreover, Mendel showed that if the recessives of the second filial generation be self-fertilized, they invariably breed true. If, however, the dominants of the same generation be self-fertilized, they breed true only in a certain fixed proportion of cases. The proportion is thus expressed: One of every three talls breeds true, and the remaining two behave as the first filial generation in giving three talls to one short. In fact, so far as these characters of tallness and shortness are concerned, there are only three kinds of peas. These are:

(*a*) Shorts which breed true.

(*b*) Talls which breed true.

(*c*) Talls which give a fixed proportion of talls and shorts.

These results may be expressed in a genealogical tree:

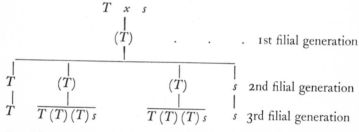

In this tree it is not necessary to exhibit both parents except those of the parental generation, since all other generations are self-fertilized. Thus in each generation, except the first, the male and female elements carry the same inheritance. (*T*) is here used as a sign for a tall giving a fixed proportion of talls and shorts, *T* for a tall that breeds true, *s* for short that breeds true.

Mendel showed that the same constant numerical distribution of opposed characters persists throughout successive generations bred from a (*T*). He elicited an identical numerical distribution for a number of other opposed characters, among them being round and wrinkled form and green and yellow colour of the peas themselves. Opposed characters that behave thus are now known as 'Mendelian',

Now came the question of the interpretation of these phenomena. Mendel concluded that they could be explained on the following four suppositions.

(*a*) Each of these pea plants carries within itself two height-influencing factors. These two are either one for tallness together with one for shortness, or again two for tallness, or again two for shortness.

(*b*) Of the two factors carried by each parent, one and one only passes to each offspring. Thus the offspring derives one of its two factors from each of its two parents.

(*c*) The chances of the combination of either factor from one parent with either factor from the other parent are equal.

(*d*) In the case of dominant factors, whether either one or two be carried, the individual will manifest the dominant character. In the case of recessive factors, the individual will bear the recessive character only if two be carried.

These suppositions, generalized for a number of other characters, are known collectively as *Mendel's First Law*. They have since been found to hold good for a very large number of characters and in many organisms.

While developing these suppositions, Mendel made further experiments of mating hybrids (such as that which we have expressed as (T)) with pure lines (such as T or s) and hybrids with hybrids. The results aided him to formulate his suppositions and helped him to confirm them. These results may be expressed graphically in a simple manner. Let T represent a tallness-carrying dominant factor and the small letter s a shortness-carrying recessive factor. Then every possible type of parent is covered by the formulae TT, Ts, sT, or ss. Of these four types Ts and sT are indistinguishable and are, in fact, the forms represented above by (T).

Let us now consider the offspring from all kinds of

mating of these three types. Let each square represent one offspring which carries two height factors, one from each parent. Only four offspring are represented in each case, but for higher numbers the average results are statistically as indicated.

(*a*) Result expected and found when *TT* is mated with *ss*:

	T	T
s	*Ts*	*Ts*
s	*Ts*	*Ts*

i.e. 4 *Ts*

(*b*) Result expected and found when *Ts* is mated with *Ts*:

	T	s
T	*TT*	*sT*
s	*Ts*	*ss*

i.e. 1 *TT*, 2 *Ts*, 1 *ss*

(*c*) Result expected and found when *TT* is mated with *Ts*:

	T	T
T	*TT*	*TT*
s	*Ts*	*Ts*

i.e. 2 *TT*, 2 *Ts*

(*d*) Result expected and found when *Ts* is mated with *ss*:

	T	s
s	*Ts*	*ss*
s	*Ts*	*ss*

i.e. 2 *Ts*, 2 *ss*

The simple suppositions of Mendel are capable of exhibition for other pairs of opposed characters which do not interfere with each other. Thus tallness and shortness, wrinkledness and smoothness, greenness and yellowness, each behaves in a Mendelian fashion without regard to the other pairs.

Mendel also made experiments with two and more pairs of opposed characters. Thus he crossed peas whose seeds were yellow Y and round R with peas whose seeds were green g and wrinkled w, Y being dominant to g and R to w. The offspring were all yellow and round.

The numerical results in the second filial generation may be thus expressed:

When the first filial generation, usually indicated as F_1, was self-fertilized, the result in the progeny was as follows:

Yellow and Round   9
Yellow and $w$rinkled   3
green and Round   3
green and $w$rinkled   1

The meaning of this distribution can be grasped from the accompanying diagram (Fig. 193). It will be seen from the diagram that either one of the colour factors may meet equally often with either one of the shape factors, the manifestation in the offspring being, however, controlled by the operation of *dominance* and *recessiveness*. Comparable but more complex results can be obtained from three or more pairs of opposed characters.

The statement that a pair of opposed characters will behave independently of another pair and yield results on a numerical basis which illustrate this independence is known as *Mendel's Second Law*. It has needed considerable modification in the light of subsequent knowledge notably in connexion with 'linked characters'. Nevertheless, since Mendel's time an enormous number of simple Mendelian characters in many other organisms, both animals and plants, have come to light.

That the papers of Mendel remained unnoticed at the time of their publication is a very remarkable fact. He was a man of good personal repute. Though not widely known among men of science, he was yet well trained at a good University. He published, it is true, only in a local journal, but his article found its way into the bibliographies. He corresponded, moreover, with the distinguished botanist Nägeli, who was himself interested in the problem of heredity.

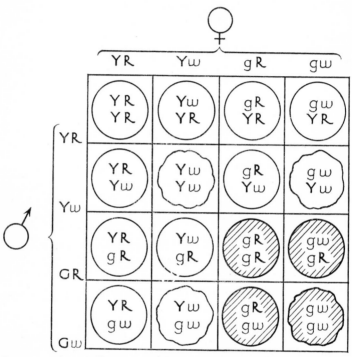

FIG. 194. Diagram illustrating the sixteen combinations that result when the four kinds of egg cells are fertilized by the four kinds of pollen grains (see pp. 565-7).

The explanation lies perhaps in the mental orientation of the naturalists of the time. They were looking always for *series* of variations which led gradually from one species to another. Their experimentation was always between varieties that differed from each other in a large number of characters. These differences concealed such simple numerical relationships as may have existed.

Mendel succeeded because he simplified his problem so that at first only a single difference was under observation. Darwin had dismissed large discontinuous changes as unlikely to be of service to the species in relation to its environment. The search was thus turned in another direction. The statistical method of Galton and Pearson, involving as it did the use of large numbers of individuals, had the effect of concealing the rare mutations and thus further removing discontinuous variation from the scientific purview.

The paper of Mendel was discovered in 1900 by De Vries and others who were working on discontinuous variation. Their discovery was not a mere literary find. It was the result of a deliberate search of the literature for confirmatory evidence for sequences of events that had already been observed. It is a moot point how long these would have had to await the interpretation which Mendel furnished, had not the search led to the recovery of his paper before the new results had become intelligible.

EPILOGUE

TOWARDS the end of the nineteenth century, several types of study were converging on the problem of heredity.

The cellular phenomena of meiosis, of union of gametes, of conjugation of male and female nuclei in the zygote, of subsequent division by mitosis, were being actively investigated. The general behaviour of the chromosomes during these processes was becoming familiar.

The old Darwinian formula of 'formation of species by imperceptibly minute continuous favourable variation' was becoming generally discredited. On the whole, naturalists had turned away from the Lamarckian point of view, and disbelieved in the inheritance of acquired characters. The view of variation by sudden jumps, saltations, or mutations of spontaneous origin was in favour in some quarters. The rediscovery of the work of Mendel greatly encouraged this point of view. The degree to which bodily characters could be analysed on a Mendelian basis was still unrecognized.

As regards the investigation of the mechanism of inheritance, much was still hoped from the statistical method. The only theory that held the field was, however, that of Weismann. Though many were interested in this theory, Weismann's insistence on the non-inheritance of acquired characters introduced a difficulty as regards the origin of new species. Weismann himself weakened on this point, though not consciously. Nevertheless, changes in the germ-plasm induced from without were seen to be inconsistent with the logical development of his views. It is true that natural selection was capable of acting on such variations as arose within the germ-plasm. In view, however, of the vast number of variations within every species demanded by Darwinian theory, the throwing of the responsibility for them all on the germ-plasm raised a picture of such prodigious complexity in its original constitution that naturalists hesitated to accept this solution.

At the dawn of the twentieth century these various lines of thought were brought into relation with certain others.

The general similarity of the chromosomes to each other had been assumed by investigators such as Flemming and Strasburger, who worked out the details of mitosis and meiosis. Weismann, on the other hand regarded the chromosomes as each containing slightly different germ-plasm corresponding to individual variations in the species.

Experiments and observations had been made by Fol, O. Hertwig, Boveri, and others on ova fertilized by more than one sperm and therefore containing abnormal numbers of chromosomes. The results were abnormal or monstrous embryos. Nevertheless it transpired, in various ways, that the mere *number* of the chromosomes does not in itself determine the character of the organism.

Between the years 1887 and 1902 Boveri worked out the doctrine of the individuality or genetic continuity of the chromosomes. As a result of a large and varied series of observations, he concluded that 'for every chromosome that enters into a nucleus there persists in the resting stage some kind of unit, which determines that from this nucleus come forth again exactly the same number of chromosomes that entered it, showing the same size-relations as before, and often also the same grouping'.

By 1901 T. H. Montgomery (1873–1912) had concluded that in the zygote the members of each pair of chromosomes are respectively of maternal and paternal origin. W. S. Sutton (1876–1916) now produced evidence that in the early stages of meiosis, preceding the disjunction into haploid cells, there is a *synapsis* or temporary union of corresponding maternal and paternal chromosomes (1902). After synapsis, the members of a pair of chromosomes pass to opposite poles of the spindle and thus always into different cells.

At this time little was known of Mendelian inheritance beyond what had been recovered in Mendel's own papers. Sutton, however, suggested that the synaptic mates contained the physical factors that correspond to the Mendelian alternative characters and added that 'in reduction division . . . any chromosome-pair may lie with maternal or paternal chromatid toward either pole. . . . Hence a large number of different combinations are possible in the mature gametes of an individual'.

Sutton suggested that the formula of inheritance of *characters* described by Mendel, and referred by writers of the new century to hypothetical *factors* in the gametes, could be applied without alteration to *chromosomes*.

Sutton also suggested some further relations of chromosomes to Mendelian characters.

(*a*) Some chromosomes are related to a number of different pairs of alternative Mendelian characters.

(*b*) All alternative Mendelian characters represented in any one chromosome must be inherited together.

(*c*) The same chromosome may carry some dominant and some recessive Mendelian factors.

Thus the Mendelian method of inheritance of opposed characters, the behaviour of the chromosomes, and the doctrine of the continuity of the germ-plasm were linked together. The subsequent development of the knowledge of the phenomena of heredity has confirmed this relationship in divers ways. But the further investigation of the inheritance of bodily characters and later investigation of chromosome behaviour in zygote and gamete are beyond the scope of this volume.

The emergence from obscurity of Mendel's great papers, with their clear-cut and lucky experiments, redirected the course of evolutionary thought. Almost contemporary (1903) was the demonstration of chemical messengers, 'hormones', which affected that analytical study of the working of the living body known as physiology. Along with this came the revelation of the integrative hierarchy of the nervous system (*c.* 1902) and the astonishing uncovering of some of the secret places of the mind (*c.* 1903). Contemporary, too, were the first inklings of the existence and nature of viruses which necessarily led to reconsideration of the nature of life (from 1898). In the same lustrum came the first demonstration of particles smaller than atoms (1897), and the not unassociated conceptions of relativity, of the uncertainty principle, and of the quantum theory. These and many other new things have necessarily led to a reconsideration not only of physical and biological principles but of the nature of objective reality itself.

Thus though time knows no periods—for these are creatures of historians' minds—it must still be decades before the open frontier between the science of the nineteenth and that of the twentieth century can be adequately surveyed, for it involves as great a change in the climate of thought as can perhaps anywhere be found in man's brief half million years on earth.

INDEX

Hooker, Sir J. D. (1817–1911), 142, 250, 254, 255–6, 258, 274, 302, 309, 558–9
Hooker, Sir W. J. (1785–1865), 142
Huerto, Garcia del (1490–1570), 133
Humboldt, von (1769–1859), 244, 274, 334
Hunter, John (1728–93), 146, 204, 210, **212–15**, 237, 333, 373–4, 375, 474
Hutton, James (1726–97), 246
Huxley, Julian S. (*b.* 1887), 312
Huxley, T. H. (1825—95), 224–6, 253, 309, 311, 312, 313, 315, 385, 418, 449, 478–9, 490, 492, 559
Huygens, Christiaan (1629–95), 147

INGENHOUSZ, Jan (1730–99), 374–5, 377, 378, 380

JENKINSON, J. W. (1871–1915), 501
Jennings, H. S. (*b.* 1868), 427
Johannsen, W. L. (1857–1929), 562
John of Trevisa (1326–1412), 39
Joule, James Prescott (1819–89), 401
Jung, Joachim (1587–1657), 47, 176, **181–3**, 184, 186, 192, 219, 293
Jussieu, Antoine de, senior (1686–1728), 186, 196
Jussieu, Antoine Laurent de (1748–1836), 196, 198
Jussieu, Bernard de (1699–1777), 196

KANT, Immanuel (1724–1804), **215–23**, 349
Kent, W. Saville, 332
Kepler, Johannes (1571–1630), 129, 135, 147, 172
Knight, Thomas Andrew (1759–1838), **376–7**, 514, 518, 519
Knuth, Paul (1854–1900), 518, 519
Koch, Robert (1843–1910), 173, **452–5**, 457
Koelreuter, Joseph Gottlieb (1733–1806), 510–11
Kölliker, Albrecht (1817–1905), 143, 310, 311, **343–5**, 404, 418, 480, 487, 531, 559, 561
Kowalewsky, Alexander (1840–1901), 481, 484–5, 486, 490
Kronecker, Hugo (1839–1914), 406
Kühne, Willy (1837–1900), 350, 390

LACAZE-DUTHIERS, H. de (1821–1901), 30

Lamarck, Jean Baptiste de Monet (1744–1829), 192, 199, 200, 227, 231, 234, 235, 247, 294, **296–301**, 306, 310, 414, 570
Lankester, Edwin Ray (1847–1929), 314, 486–7, 492
Laveran, Alphonset (1845–1922), 324
Lavoisier, Antoine (1743–94), 365, 374, 375, 377
Lawrence, Sir William (1783–1867), 298
Layard, Sir Austen Henry (1817–94), 52
Leeuwenhoek, Antony van (1632–1723), 141, 150, 151, **166–9**, 208, 329, 330, 379, 404, 405, 439, 468, 507, 524
Legallois, Julien (1770–1814), 420
Leibniz, Baron Gottfried Wilhelm von (1646–1716), 211, 295, 366
Leidy, Joseph (1823–91), 315
Leonardo da Vinci (1452–1519), 80, **82–4**, 98, 101, 116, 224, 226
Leuckart, Rudolph (1823–98), 323, 533, 553
Leydig, Franz von (1821–1905), 343, 417
Liebig, Justus von (1802–73), **377–9**, 382, 384, 385, 386, 393, 406, 409, 410, 415, 445   f
Linnaeus, Carolus (1707–78), 92, 176, 181, 182, 183, 186, **187–95**, 198, 199, 200, 231, 235, 240, 287, 293, 294, 295, 298, 305, 373, 510
Lippi, Fra Lippo (1406–69), 80
Lister, J. J. (1786–1869), 515
Lister, Joseph Lister, Baron (1827–1912), 515
Lister, Martin (1638–1712), 245
l'Obel, Matthias de (1538–1616), 175–6
Locke, John (1632–1704), 125–6
Loeb, Jacques (1859–1924), 377, 427
Loeffler, Friedrich (1852–1915), 462
Longinus (3rd cent. A.D.), 86
Lorenzo dei Medici (the Magnificent) (1449–92), 132
Lucretius (*c.* 98–55 B.C.), 438, 505, 549
Ludwig, Karl (1816–95), **395–7**, 406, 409
Lyell, Sir Charles (1797–1875), 247, 249, 255, 302, 308, 309, 316, 319
Lyonet, Pieter (1707–89), 494

MCCLUNG, C. E. (1870–1946), 542

Printed in Great Britain by
Hazell Watson & Viney Ltd, Aylesbury and Slough
for Abelard-Schuman Ltd
38 Russell Square, London, W.C. 1 and
404 Fourth Avenue, New York 16